T0257582

IET TELECOMMUNICATIONS SERIES 65

Advances in Body-Centric Wireless Communication

Other volumes in this series:

Advances in Body-Centric Wireless Communication

Applications and state-of-the-art

Edited by Qammer H. Abbasi, Masood Ur Rehman, Khalid Qaraqe and Akram Alomainy

The Institution of Engineering and Technology

Published by The Institution of Engineering and Technology, London, United Kingdom

The Institution of Engineering and Technology is registered as a Charity in England & Wales (no. 211014) and Scotland (no. SC038698).

The Institution of Engineering and Technology
Michael Faraday House
Six Hills Way, Stevenage
Herts, SG1 2AY, United Kingdom

www.theiet.org

British Library Cataloguing in Publication Data
A catalogue record for this product is available from the British Library

ISBN 978-1-84919-989-6 (hardback)
ISBN 978-1-84919-990-2 (PDF)

Typeset in India by MPS Limited

Contents

Acknowledgements

This publication was made possible by NPRP Grant #7-125-2-061 from the Qatar National Research Fund (a member of Qatar Foundation). The statements made herein are solely the responsibility of the authors.

Chapter 1

Introduction

Qammer H. Abbasi*, Masood Ur Rahman**,
Khalid Qaraqe† and Akram Alomainy††

Body-centric wireless networks (BCWNs) have recently gained substantial recognition and interest in both academic and industrial communities due to its direct and beneficial impact both economically and socially in various application domains. Such networks refer to a number of nodes/units scattered across the human body and the surrounding areas to provide communication on the body surface, to access points and wireless devices in the near vicinity and also to provide hieratical networking structure from implants to the main communication hub [1].

BCWN, in essence, is a combination of wireless body area networks (WBANs), wireless sensor networks (WSNs) and wireless personal area networks (WPANs) considering all their associated concepts and requirements. BCWN has got numerous number of applications in our every day life, including healthcare, entertainment, space exploration, military and so forth [2]. The topic of BCWN can be divided into three main domains based on wireless sensor nodes placement, i.e., communication between the nodes that are on the body surface; communication from the body-surface to nearby base station; and at least one node may be implanted within the body. These three domains have been called on-body, off-body and in-body, respectively, as shown in Figure 1.1.

The major drawback with current body-centric communication systems is the wired or limited wireless communication that is not suitable for some user and the restrictions on the data rate (like video streaming and heavy data communication, where we need to transfer large amount of data). Many other connection methods like communication by currents on the body and use of smart textile are proposed in the literature [2]. Communications using the minute body current suffers from low capacity, whereas smart textile method needs special garments and is less reliable; however, it is fair to say the latter is gaining more momentum specifically with the advances in smart materials and fabrications as it will be seen later in this book.

*Department of Electrical and Computer Engineering, Texas A & M University at Qatar, Qatar and School of Electronic Engineering and Computer Science, Queen Mary University of London, UK
**Departement of Electronics Engineering, University of Bedfordshire, UK
†Department of Electrical and Computer Engineering, Texas A & M University at Qatar, Qatar
††School of Electronic Engineering and Computer Science, Queen Mary University of London, UK

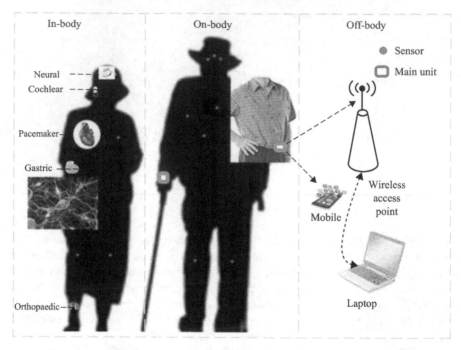

Figure 1.1　Envisioned body area network

BCWN provides the way forward for future smart, intelligent and effective communication technology. This is mainly owed to the less power requirements, re-configurability and unobtrusiveness to the user considering all requirements are met [3]. In order to make these networks optimal and less vulnerable, many challenges including scalability (in terms of power consumption, number of devices and data rates), interference mitigation, quality of service (QOS) and ultra-low power protocols and algorithms need to be considered. The radio channel in BCWN exhibits highly scattered paths and antenna near field effects due to body proximity conditions [4]. Radio transceiver systems used in BCWNs have to be low profile and light weight, while operating with low power for longer lifetime. The systems should also be designed with minimal restriction for the user, so that they can be used during regular day-to-day activities without inhibition. They should be easily integrated with the human body, or as a part of the clothing. Many currently existing short-range wireless technologies provide communication medium and cable replacement technologies for different transmission types. To design a suitable efficient radio interface for the wireless body-centric network, the understanding and integration of existing standards are required in order to bring to light the main areas in which new techniques are required to meet the harsh and demanding communication environment. In this book and subsequent chapters, the fundamentals behind BCWNs, theoretical limitations and explorations, experimental investigations from MHz to THz and also the interdisciplinary nature of the topic collaborating with material scientists, physicists, biological and chemical experts, human interaction and machine learning will be explored further and the current state-of-the-art in addition to the future directions will be discussed and detailed.

1.1 Frequency band allocation for body area communication

Wireless communications systems can operate in the unlicensed portions of the spectrum. However, the allocation of unlicensed frequencies is not the same in every country. Important frequency bands for BCWN are reported in Table 1.1 and they are:

- **Medical Implanted Communication System (MICS)**: In 1998, the International Telecommunication Radio sector (ITU-R) allocated the bandwidth 402–405 MHz for medical implants [5]. MICS devices can use up to 300 kHz of bandwidth at a time to accommodate future higher data rate communications.
- **Industrial, Scientific and Medical (ISM)**: ISM bands were originally preserved internationally for non-commercial use of radio frequency. However, nowadays it is used for many commercial standards because government approval is not required. This bandwidth is allocated by the ITU-R [6], and every country uses this band differently due to different regional regulation as shown in Table 1.1.
- **Wireless, Medical Telemetry Services (WMTS)**: Due to electromagnetic interference from licensed radio users such as emergency medical technicians or police, the Federal Communication Commission (FCC) has dedicated a portion of radio spectrum, 608–614 MHz, 1395–1400 MHz and 1427–1432 MHz for wireless telemetry devices in USA [7] for remote monitoring of patient's health; however, such frequency bands are not available in Europe. WMTS is approved for any biomedical emission appropriate for communications, except voice and video.

Table 1.1   Unlicensed frequencies available for personal area networks (Reproduced from Reference 11)

Name	Band (MHz)	Max Tx power (dBm EIRP)	Regions
MICS	402.0–405.0	−16	Worldwide
ISM	433.1–434.8	+7.85 E	Europe
ISM	868.0–868.8	+11.85	Europe
ISM	902.8–928.0	+36 w/spreading	Not in Europe
ISM	2400.0–2483.5	+36 w/spreading	Worldwide
ISM	5725.0–5875.0	+36 w/spreading	Worldwide
WMTS	608.0–614.0	+10.8	USA only
WMTS	1395.0–1400.0	+22.2	USA only
WMTS	1427.0–1432.0	+22.2	USA only
UWB	3100.0–10,600.0	−41.3	USA, etc.
UWB	3100.0–10,600.0	−41.3 (low duty cycle)	EU
MM WAVE	57–64 GHz	+82	USA
MM WAVE	57–64 GHz	+55	Europe
THZ	0.1–10 THz	+20	Worldwide

- **Ultra-WideBand (UWB)**: It is a communication system, whose spectral occupation is greater than 20%, or higher than 500 MHz. Initially, it was available only in USA and Singapore but on August 13, 2007, Office of Communications (OFCOM) approved the use of ultra-wideband wireless technology without a license for use in the UK.
- **Millimetre Wave (MMW)**: The FCC regulation 15.255, for devices operating in the 60 GHz band, initially specified Equivalent Isotropically Radiated Power (EIRP) up to a maximum average power level of +40 dBmi. In August 2013, a ruling by the FCC extended the EIRP for outdoor use between fixed points to as much as +82 dBm and later on same for indoor. In Europe, the European Telecommunication Standards Institute (ETSI) adopts recommendations for operation of devices in the 57–64 GHz band and calls for a maximum EIRP power level of +55 dBmi, In 2010, the UK OFCOMOFCOM approved the unlicensed use of the 57–64 GHz spectrum, although the spectrum allocation follows the FCC standard (maximum EIRP of +55 dBm) [8].
- **Terahertz (THz)**: The band above 275 GHz is the main part of terahertz band. Terahertz waves, also known as submillimetre radiation, usually refer to the frequency band between 0.1 THz and 10 THz with the corresponding wavelength of 0.03–3 mm [9]. In this direction, the IEEE 802.15 WPAN Study Group 100 Gbit/s Wireless (SG100G), formerly known as the IEEE 802.15 WPAN Terahertz Interest Group (IGThz), has been recently established. The ultimate goal of the SC100G is to work towards the first standard for THz band (0.1–10 THz) communication able to support multi-Gbps and Tbps links [10]. However, currently 20 dBm power is being used.

1.2 Book organization

The book is divided into two different parts: Part I (Chapters 2–7) deals with the state-of-the-art and recent advances in this area, whereas Part II deals with the applications of body-centric wireless communication (Chapters 8–12) and finally concluding remarks and the future of body-centric wireless communication is presented in Chapter 13. The details about the book chapters are given below:

Chapter 2 presents diversity and cooperative communications, and particularly cooperative diversity, for body-centric communications in WBANs.

Chapter 3 discusses about the various experimental investigations undertaken to thoroughly understand the UWB on-/off-body radio propagation channels for both static and dynamic scenarios.

Chapter 4 deals with characterization of body-centric wireless communication channels by applying sparse non-parametric model in addition to compressive sensing technique.

Chapter 5 presents a review of the state-of-the-art, recent advances and remaining challenges in the field of antenna/human body interactions in the 60 GHz band, with a particular emphasis on the near-field interactions that may occur in emerging body-centric millimetre wave applications.

Chapter 6 provides an overview of ingestible capsule wireless telemetry with the main focus on specific challenges and difficulties associated to the design of ingestible gastrointestinal capsule antenna systems.

Chapter 7 describes the state-of-the-art of *in vivo* channel characterization and several research challenges by considering various communication methods, operational frequencies and antenna designs are discussed. Furthermore, a numerical and experimental characterization of *in vivo* wireless communication channel is presented.

Chapter 8 discusses the use of antenna diversity and MIMO for on-body channels to support reliable and high data rate communication in addition to use of diversity for cancelling the co-channel interference.

Chapter 9 discusses the performance of on-body GPS antennas in real working scenarios discussing a recently developed statistical model and considering different body postures and antenna positions on the body.

Chapter 10 deals with the textile substrate integrated waveguide technology for the next generation wearable microwave systems. A state-of-the-art in this domain is presented, followed by various proposed design for next generation wearable systems.

Chapter 11 provides details about 3D localization by applying compact and cost-effective wearable antennas placed at different locations on the body by considering both numerical and experimental studies.

Chapter 12 presents a review of nano-scale communication for body area networks followed by a thorough simulation and experimental studies for nano-scale communication at terahertz frequencies. In addition, future directions in this topic are also presented.

Chapter 13 The road ahead for body-centric wireless communication and networks demand of WBANs is ever increasing. This chapter discusses different challenges faced by the WBANs that hurdle their expansion. It also identifies future venues and research trends for the application of WBAN systems.

References

[1] P. S. Hall, Y. Hao, and K. Ito, "Guest editorial for the special issue on antennas and propagation on body-centric wireless communications," *IEEE Transactions on Antenna and Propagation*, vol. 57, no. 4, pp. 834–836, 2009.

[2] P. S. Hall and Y. Hao, *Antennas and Propagation for Body-Centric Wireless Communications*, 2nd edition. Artech House, UK, 2012.

[3] B. Allen, M. Dohler, E. Okon, W. Q. Malik, A. K. Brown, and D. Edwards, *UWB Antenna and Propagation for Communications, Radar and Imaging*. John Wiley and Sons, USA, 2007.

[4] M. G. Benedetto, T. Kaiser, A. Molisch, I. Oppermann, and D. Porcino, *UWB Communication Systems: A Comprehensive Overview*. Hindawi Publishing Corporation, Cairo, Egypt, 2006.

[5]   "Federal communications commission (FCC), code of federal regulations (CFR), title 47 part 95, MCIS band plan," *URL: www.fcc.gov*, March 2003.

[6]   "International telecommunications union-radiocommunications (ITU-R), radio regulations, section 5.138 and 5.150," *URL: www.itu.int/home*.

[7]   "Federal communications commission (FCC), code of federal regulations (CFR), title 47, part 95, WMTS band plan," *URL: www.fcc.gov*, January 2003.

[8]   H. Willebrand, "Advantages of the 60 GHz frequency band and new 60 GHz backhaul radios," White paper, Light Pointe wireless, July 2015.

[9]   International telecommunications union-radio communications (ITU-R), "Technology trends of active services in the frequency range 275–3000 GHz," Report ITU-R SM.2352-0, June 2015.

[10]   OFCOM, "Managing spectrum above 275 GHz," December 2008.

[11]   Q. H. Abbasi, "Radio channel characterisation and system level modeling for ultra wideband body-centric wireless communications," PhD thesis, Queen Mary, University of London, 2012.

Chapter 2
Diversity and cooperative communications in body area networks

David B. Smith* and Mehran Abolhasan†

Abstract

In this chapter, we investigate diversity and cooperative communications, and particularly cooperative diversity, for body-centric communications in wireless body area networks (BANs). Cooperative diversity for BANs is vitally important for required communications reliability, as well as increasing network and sensor lifetime by potentially reducing energy consumption, as will be shown here. We describe what is meant by cooperative communications and cooperative diversity, including a brief survey of the state-of-the-art. Description and analysis of the benefits of cooperative diversity in BANs is mainly with respect to the physical layer, but there is also some brief discussion of the MAC layer and network layer. In terms of cooperative receive diversity, feasible in IEEE 802.15.6 Standard compliant BAN, several cooperative receive combining techniques are described, which are all beneficial over single-link communications in terms of first- and second-order statistics. A simple, practical, technique of switch-and-examine combining shows good performance in terms of important metrics, and this can be further enhanced when combined with a simple "sample-and-hold" transmit power control, which can help reduce energy consumption for sensor radios.

2.1 Introduction

Wireless body area networks (BANs) represent the forefront of personal area networks in body-centric communications, with communications networks of sensors and/or actuators placed in or on the human body. In this context in BANs, this communication can occur to other devices on the human body "on-body", from human body-to-body or off the body "off body" to another location, and from in the body [1]. In this chapter, in the context of cooperative diversity in body area networks the focus will

---

*Data61(NICTA) CSIRO, Australian Technology Park, NSW Eveleigh 2015, Australia and Australian National University (ANU), Canberra, ACT 0200, Australia
†University of Technology, Sydney (UTS), NSW 2007, Australia

be on-body communications, as this is expected to be most prevalent. In terms of cooperative communications, off-body nodes can be important in providing useful diversity, so these will also be addressed where appropriate.

One key main application in the development and applications of body area networks is in health-care [2, 3]. In health-care communications, reliability is vitally important, as this is potentially life-saving, and with this BANs have the potential to greatly improve the standard of health-care, providing very early intervention, and helping avoid medical emergency. Another further advantage of the implementation of BANs is from the aspect of potential home or community health-care rather than far more expensive hospital, or nursing-home care.

It has been shown that BAN communications can be effected by significant path losses for any given single-link of BAN communications [4, 5], furthermore the channel may be significantly attenuated for any given link, such that it is in outage for a duration longer than an acceptable delay for successfully transmitting vital communications packets [6]. Hence single-link, star topology communications, where there is only sensor-to-hub link communications, will often not provide the required reliability for BANs, particularly as have been prescribed for the IEEE 802.15.6 BAN standard [7, 8]. Thus cooperative diversity and cooperative communications can be vitally important for body area networks, where there can be one or more relay-links providing diversity gain along with the direct single-link communications. Furthermore, cooperative communications can be very beneficial to reducing circuit power consumption, and increasing network life-time, as it can enable sensor devices to transmit at lower power and reduce their circuit power consumption. The fact that this cooperative diversity is distributed spatially enables it to be more easily to be implemented.

When considering single-link communications, it may sometimes be beneficial to employ co-located diversity at the hub as a means of improving reliability and obtaining greater throughput from sensor communications – but due to sizes of typical sensor radios, and limits on power consumption at the sensor, co-located diversity will often not be practical at the sensor side. In some cases, co-located diversity may be practical at hubs and for on-body devices with less-limits on size and battery power consumption.

Maintaining reliability in different communications scenarios is very important for body area networks, in terms of enabling excellent users mobility, as well as enabling best performance in scenarios particularly common to BANs, such as those used for sleep monitoring. The best method to test such reliability is by employing extensive experimental data for everyday "mixed" activity body-channel measurement data [9, 10], as well as for particular BAN scenarios such as sleep monitoring [11]. Such data can be found in the open-access data found in Reference 10, which has extensive measurements, including those that can be used to test cooperative communications where multiple wireless nodes can act as transmitters and receivers. The measurements for the person sleeping channel, where sleep monitoring was evaluated for many sleeping subjects are also particularly useful for ascertaining the value of cooperative communications.

As further background on BANs good reviews of radio propagation and channel modelling can be found in References 1, 12 and 13 – hence the focus of this chapter will not be channel modelling, but rather the focus here will be system design in cooperative communications. However, in the context of system design, description of channel statistics is important, so the impact on radio communications schemes will be described with respect to typical channel statistics in BANs. This is in terms of measures such as outage probabilities, and second-order statistics such as fading durations and also switching rates (i.e., the rate of switching between different diversity branches).

In the majority of work here decode-and-forward communications is considered, potentially the simplest strategy for half-duplex channels [14]. The typical consideration therein is to (i) Source sends packet to destination; (ii) If the relay succeeds in decoding this packet in the same time, it sends a copy of it in a further slot. Note that the relay has to be listening; (iii) The sink decodes a packet with the two received copies.

Also, as stated in Reference 15, minimising energy consumption is critical in BANs, particularly for health-care, so relays should be used as efficiently as possible. Such a pursuit raises many interesting questions, for example: How does the hub decide that a node needs a relay (as opposed to attempting retransmissions)? Which node should act as a relay? Throughout this chapter, we will seek to answer these questions posed in Reference 15.

2.2 Cooperative on-body communications – illustrations

It now serves to illustrate what is meant by on-body, in-body and off-body communications. This is best visualised as in Figure 2.1, where we show a non-cooperative star topology. This follows from the illustration in Reference 13.

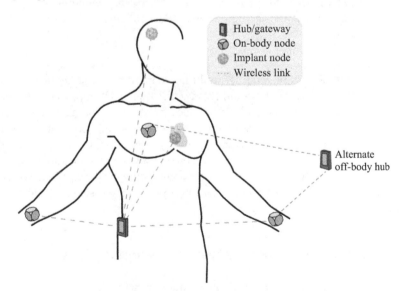

Figure 2.1   BAN on a male subject, illustrating gateway (hub), sensors and in-body, on-body and off-body links, following from Reference 13

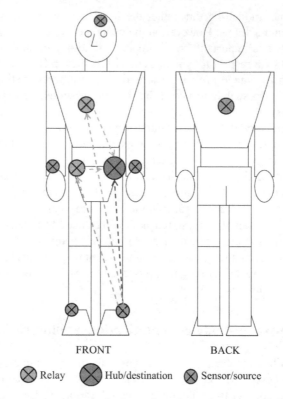

FRONT   BACK

⊗ Relay   ⊗ Hub/destination   ⊗ Sensor/source

*Figure 2.2   General set-up for uplink combined diverse and cooperative
communication of Tx (sensor)/relay/hub positions for on-body
measurements*

Next we present an illustration of multi-hop communications for sensors communicating with relays on right hip and chest, with a hub on left hip, where the uplink
is shown in Figure 2.2. One particular case concerning transmissions from the sensor
on the left ankle is illustrated. Importantly, it should be noted here that cooperative
communications for the uplink is significantly more important for the uplink than
the downlink – as sensor devices are significantly more energy constrained than hub
devices. Often hubs will be able to transmit at a maximum power of 0 dBm, where as
often it may be desirable to transmit at far lower powers from sensor devices, often
−10 dBm or less.

2.3   General overview of cooperative communications

In cooperative communications techniques, multiple network nodes' resources are
shared making use of wireless broadcast, to create effective diversity gain at the
receiver [16]. One or many nodes are used as relays between the source and destination

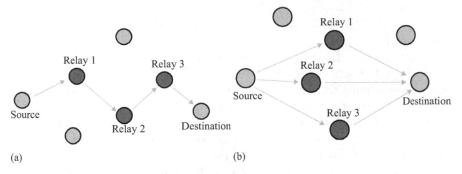

Figure 2.3   Sequential and parallel relaying topologies [19]. (a) Sequential relaying and (b) parallel relaying

to provide effective receive-diversity gain towards the receiver [17], this can be implemented in multiple-input multiple-output (MIMO) cooperatively, or single-input multiple-output or multiple-input single-output (MISO), as described in Reference 18, where there are three cases: (i) MIMO – Source (S) communicates with relay (R) and destination (D) simultaneously during the first time slot. In the second time slot, relay and source simultaneously communicate with destination. This protocol realises maximum degrees of broadcasting and receive collision. (ii) SIMO – In this protocol terminal S communicates with terminals R and D simultaneously over the first time slot. In the second time slot, only R communicates with D. This protocol realises maximum degree of broadcasting and exhibits no receive collision. (iii) MISO – The third protocol is identical to the first case apart from the fact that terminal D chooses not to receive the direct S→D link in the first time slot. This protocol does not implement broadcasting but realises receive collision.

As shown in Figure 2.3 from Reference 19, sequential, parallel relaying or a mixed of them can be used to form the multi-hop cooperative communications schemes. In terms of 802.15 task group 6 compliance, parallel relaying is typical, as three-or-more hops is not allowed in the 802.15.6 BAN standard [8] and may be overly complex to coordinate in BAN.[1]

The advantages of cooperative communications have been widely acknowledged in different types of network systems. In Reference 20, two intra-cell coordinated multi-point (CoMP) schemes are investigated for LTE-advanced systems, where relay nodes act as transmission points. Significant advantage is shown in capacity expansion. A potential drawback is that transmission from relays could possibly cause congestion to the centre-user in a cellular network. In Reference 21 a solution is provided by allowing a transmitter to be used as a relay at the same time as transmitting its own data with superposition coding in the downlink. In wireless ad-hoc networks, References 22–24 have shown that incorporating cooperative communications not

---

[1]This is also due to throughput degradation in half-duplex multi-hop links, with bandwidth halved at each hop.

only enhances the connectivity, but the network coverage is also extended significantly, both of which have a great impact on quality-of-service (QoS) of a wireless sensor network [25].

In Reference 26, it is argued that the use of an optimal relay in cooperative communication has equivalent performance to the use of all potential relays, which motivates the use of opportunistic relaying for WBANs. Opportunistic relaying reduces the complexity by adopting the concept that only the single relay with the best network path towards the destination forwards per hop [27]. It could also potentially avoid unnecessary interference, which is caused by all relays forwarding to the destination, to other operating nodes in the same vicinity [28]. In References 29–31, opportunistic relaying is proposed in a simple cooperative protocol where the best relay is selected and utilised among all the participating relay candidates.

2.4   State-of-the-art in BAN literature

Here, we present a survey of the state-of-the-art in the literature for diversity and cooperative communications for BANs with a particular emphasis on cooperative diversity, as it is anticipated that this will be the most common form of diversity in BAN communications. However, before moving to cooperative communications, we will provide a brief insight into co-located diversity, from the aspect of antenna system design.

2.4.1   Co-located spatial diversity in BANs

One such application for co-located diversity is presented in Reference 32 particularly with respect to antenna design and system measurement. This work highlights the problem of co-located diversity in BAN, with problems of on-body array footprint making unsuitable for everyday wearing and attachment to a typical sensor. A wearable integrated antenna (WIA) array and printed-F antenna array are presented, with co-located spatial diversity, hub or transmitter side, with 0.25 wavelengths (λ) separation between antennas, giving a somewhat cumbersome overall array dimension of 6×3 cm.

In Reference 32, the antenna arrays were tested for various receive positions, with two extremes, an anechoic chamber with no multipath and a reverberation chamber with excessive multipath. Equal gain combining, maximum-ratio combining (MRC) and selection combining (SC) were tested, maximum-ratio combining and selection combining will be described later, simply to state that MRC is a more optimal combiner and SC provides significantly more simplicity. Importantly SC only experienced ≈1 dB performance degradation from MRC. There were considerable gains for the WIA, using a mean diversity power level measure – and as might be expected there were considerably more performance gains with excessive multipath of the reverberation chamber, where there would typically be more channel gains over diverse paths, than the no-multipath case of the anechoic chamber.

2.4.2   Cooperative diversity

Cooperative diversity in wireless body area networks has been analysed for both ultra-wideband (UWB) (with typical bandwidths in BAN of 500 MHz) and narrowband (with typical bandwidths of 1 MHz). Due to the lower carrier frequencies and lower path losses, narrowband communications is more appropriate for applications in health-care, whereas UWB may be more appropriate for consumer entertainment applications where larger data rates may be desired and there are less stringent QoS requirements.

It is reasserted here that decode-and-forward communications protocol is most appropriate for BANs, but there have been some variants of decode-and-forward communications, and even amplify-and-forward communications (a potential problem in terms of energy efficiency for BANs).

One of the first works in cooperative communications for BAN appeared in Reference 33. In Reference 33, the authors discuss the use of virtual MIMO and postulate that with opportunistic relaying high capacity can be achieved at the cost of increased complexity and synchronisation requirements. The multi-hop approach consists of organising the network to route information. Unsurprisingly, but importantly, it is found that the S→R→D link alone is worse than the S→D link, because two successive transmissions are needed on the S→R→D a channel having the same conditions, but good gains combining both paths. Two-branch cooperative diversity is described, based on modelling with Rician fading and not on experimental conditions.

In Reference 34, the body-centric multipath channel is characterised, and diversity analysis is facilitated in a UWB cooperative BAN (CoBAN). The typical mode of operation is described, where in the first stage, the source transmits to the destination and the relays and in the next stage, the relays transmit to the destination. Measurement-based analysis is presented, in which the spatial diversity provided by the relays is dependent on the channel properties derived from the measurements around a human trunk. An intrinsic metric is then used to quantify the cooperative diversity of such networks, which is defined as the number of independent paths that can be averaged over to detect symbols. A form of detect-forward communications is presented, where there is relaying upon successful detection of a symbol. Of course there may be some further diversity benefits from the multiple channel taps available in UWB. Very large diversity gains are found in Reference 34 when the relay is cooperating with the source. In the presented scenario, little diversity can be achieved for link S→D without the help of R due to the severe path loss between S and D. Another important observation is that in general the relay should be placed in the middle of S and D to achieve higher diversity gains. One potential problem in a practical implementation in BAN communications is the large carrier frequency considering a 5 GHz bandwidth required for this UWB scenario.

In Reference 35, it is proposed that within a possible range of the channel quality in WBAN, cooperative communication is more energy efficient than direct transmission only when the path loss between the transmission pair (i.e., source and relay, and relay and destination) is higher than a threshold, which has been found for other wireless networks. Furthermore, for a practical WBAN, cooperative communication in most cases is still effective in reducing energy consumption, which is demonstrated in Reference 35. In Reference 36, an experimental campaign is presented, with theoretical analysis of packet error rate, transmit power and shows experimentally the valuable combination of S→R→D and S→D paths. One possible problem in the practicality of the analysis in Reference 36 is that the relays are not centrally located as they might be expected to be in a practical BAN, rather they are placed at the ear, thigh and shoulder. Prior to the release of the IEEE BAN standard specification, the Bluetooth LE PHY specification [37] was employed, which has higher power consumption than IEEE 802.15.6, partly due to its link layer specification. It is shown that using one two-hop path is less effective than only the direct link, but there are some good long-term advantages of combining multiple paths, by not choosing between different paths.

In Reference 38, a prediction-based dynamic relay transmission (PDRT) scheme that makes use of on-body channels correlation is proposed. It is claimed that in the PDRT scheme, that "when to relay" and "who to relay" are decided in an optimal way based on the last known channel states. Only links with bad quality are relayed. It is claimed that neither an extra signalling procedure nor dedicated channel sensing period is introduced by the proposed scheme as the relay allocation is fully controlled by the coordinator based on the last obtained channel states and is realised through beacon broadcasting. Consequently, the energy consumption and system complexity are further reduced over typical cooperative communications schemes.

The majority of cooperative communications analysis in BAN has been with respect to on-body communications, but in Reference 39 it is described with respect to in-body communications with UWB. With the aid of a relay node on the body surface, cooperative transmission can achieve a significant improvement on energy efficiency compared with direct transmission over a range of relay locations under various scenarios. The large existing path loss may still be a problem however in a practical implementation of the method described in Reference 39.

In Reference 40, a possible mode of operation for BAN cooperative communications is described accordingly at both the physical (PHY) and link (MAC) layers. Nodes select an adequate cooperator, when a node directly receives a hello packet from another node, the node measures the received-signal-strength-indicator (RSSI) of the packet and re-broadcasts the packet adding the measured RSSI in it only once. When a node receives a re-broadcast hello packet, the node never re-broadcasts it. In Reference 40, each node attempts only first packet transmission. Namely, when a node transmits a packet to a BAN coordinator, if it receives an acknowledgement (ACK) packet from the BAN coordinator, it recognises successful transmission of the packet. On the other hand, even if it does not receive an ACK packet, it does not re-transmit the same packet by itself. The re-transmission is made only by a cooperator selected by the node. Choice of BAN coordinator is also allowed, but this is not varied in experimental implementation as is appropriate for typical BANs where there is normally only a single hub/coordinator. Hence the hub is fixed for each of three subjects, two with hub in right hip front pocket and one with hub on finger. In the measurements in Reference 40, eight cooperating nodes are possible.

In Reference 41, an emergency services application for BAN is presented with respect to spatially separated receivers and SIMO operation, with instantaneous statistics combining for paths to four on-body locations. MRC has 8.7 dB gain, and SC gives 5.4 dB gain. A printed-F antenna is used, with dimensions not given. The raw signal envelopes for all receiver branches were combined in postprocessing to form virtual multiple-branch diversity receivers for each of the two experiments in Reference 41.

In Reference 42, incremental relaying is presented, which is somewhat similar to opportunistic relaying, where the stated aim is to save channel resources by ensuring that the relaying process adapts to the channel conditions. It relies on the broadcast nature of the wireless channel and exploits short feedback messages from the destination (in the form of an acknowledgement (ACK) or negative acknowledgement (NACK) packet indicating success or failure of the direct transmission). If the signal-to-noise ratio (SNR) of the source-to-destination (S→D) link is sufficiently high, the ACK feedback from the destination indicates that direct transmission over the S→D link is successful, and hence, relaying is not required. If the SNR over the S→D link is not sufficiently high for successful direct transmission, the NACK feedback from the destination indicates that the relay must decode and forward the data it received from the source in the previous phase. Such a protocol can efficiently use channel resources, as compared with some conventional cooperation schemes, because the relay will forward the signal to the destination only when it is necessary, similarly to opportunistic relaying. The results in Reference 42 show that an interesting threshold behaviour exists that separates regions where direct transmission is better from regions where this form of cooperation will be useful in terms of energy efficiency. For example, in the case of in-body communications, below a threshold of 16 cm, the overhead of cooperation out-weighs its gains and direct transmission is more energy efficient. For on-body LOS communication, this threshold distance is equal to 135 cm for the corresponding set of channel parameters – although such distance-based specification may be doubtful, as in the general cases for BANs, particularly on-body BAN, path losses are not distant dependent [43]. Above the distance threshold, cooperation gains can be achieved. Furthermore, the results in Reference 42 also show that choosing the best relay location for cooperation plays an important role in determining the overall energy efficiency. From the results of Reference 42, it appears that the gains of incremental relaying are less than those for opportunistic relaying, although to-date there is no work that has compared the two.

In Reference 44, amplify-and-forward is described for impulse-radio (IR)-UWB. According to theoretical BER analysis using Nakagami-m fading using direct and indirect links with one-relay gives 23 dB gain, two-relays 27 dB gain. Amplify-and-forward is described rather than the typical decode-and-forward for Reference 44. The amplification factor is not clear in Reference 44, and it is not obvious whether a constant power constraint is met.

In Reference 45, a network layer approach is presented with opportunistic routing for two-hops, with analysis in terms of IR-UWB. In the opportunistic routing approach in Reference 45, the sensor node examines if the sink is in line of sight by sending an RTS and receiving acknowledge from the sink. Then it decides to send the packet through the relay or send it through the sink. The sink is only considered at the hand, with relay at the waist and sensor on the chest. Only either a dual-hop link or a single-hop link is used at any time. According to the simulation settings therein, for one particular scenario with IR-UWB, BER is maintained compared to multi-hop, while energy consumption is reduced. In Reference 46, the broadcast nature of the wireless transmission is used, and a simple timer-based opportunistic routing for BANs is proposed. As soon as a packet is received at a relay, a timer that considers the quality of the second-hop channel is triggered. The relay with a timer that expires first will forward its packet to the hub, and the other relays do not send when hearing its transmission in Reference 46. In Reference 46, the next hop information uses broadcast.

In Reference 15, it is importantly noted that the IEEE 802.15.6 standard also allows a single relay (i.e., dual-hop) to be used in cases where the typical single-hop star topology cannot maintain the required levels of reliability. In contrast to other work, the use of relays in Reference 15 is particularly motivated for the case when the BAN subject is not moving, such as when they are sleeping, as some links may be attenuated below the receiver sensitivity for tens of minutes at a time. There is a cost, however, to the energy consumption of a relay node as it must remain awake, listening for packets to relay to the hub from the sensor that cannot reach the hub directly.

Sleeping subjects wearing on-body sensors and being monitored by an off-body device that is acting as the BAN hub have been considered in Reference 11. This would be a common health-care scenario where a monitoring device is placed beside the bedhead. In Reference 11, when there is channel attenuation of more than 90 dB resulting in outage, there is a node that could act as a successful relay 85% of the time the sensor node (Tx) to hub link (Rx) is in outage. If in Reference 11, instead, the hub is placed on the subjects left hip, there is a viable relay 80% of the time that the direct link is in outage considering an attenuation threshold of 90 dB. Measurements show that long outages occur for about 15% of the time that subjects are sleeping, and hence relays can play a significant role in improving BAN reliability for sleep monitoring [11].

2.5   Experimental method, gaining data for studies of cooperative communications

In the proceeding four sections, after this section, four case studies of the use of cooperative communications are presented that motivate such implementation. These are evaluated using open-access channel gain data with "everyday" mixed activity over long time periods (hours), where the effect of combined dual-hop communications could be accounted for. As summary of this method:

An experiment was set up to measure on-body BAN communication links while test subjects performed everyday activities over two hour periods. The subjects wore small body-mounted "channel sounder" radios that operated as both transmitters (Tx) and receivers (Rx), and the activities were mainly office work, some driving in a car, some walking and some activity at home. The measurements were made at 2360 MHz, one of the carrier frequencies in the draft IEEE 802.15.6 BAN standard [8].

Table 2.1   *Tx/Rx radio locations, x indicates a channel measurement.*
Lh – left hip, Rh – right hip, C – chest, Hd – head, Rw – right
wrist, Lw – left wrist, Lar – upper left arm, La – left ankle,
Ra – right ankle, B – back

	Tx-Hd	Tx-Rw	Tx-Lw	Tx-Lar	Tx-La
Rx-Lh	x	x	x	x	x
Rx-Rh	x	x	x	x	x
Rx-C	x	x	x	x	x
	Tx-Ra	Tx-B	Tx-C	Tx-Lh	Tx-Rh
Rx-Lh	x	x	x		x
Rx-Rh	x	x	x	x	
Rx-C	x	x		x	x

For each experiment, ten channel sounder radios [47] were placed on the bodies of three male and two female test subjects, with heights ranging from 1.65 m to 1.9 m. Three of the ten radios (representing relays/hubs, the remainder representing sensors) operated as both *Tx* and *Rx*, each one broadcasting in turn at 1 mW (0 dBm) in a round-robin fashion, with transmissions spaced 5 ms apart (hence, each individual transmitter would transmit every 15 ms). All of the radios, *Tx* and non-*Tx*, continuously listened for packets as a *Rx* and logged the RSSI value whenever successfully detecting a packet.

Reciprocity of the BAN channel means that any *Tx/Rx* link will exhibit the same channel gain, regardless of which device in the link actually transmitted the packet. We also note that the 15 ms round-robin transmission period is far less than the BAN channel coherence time, even for highly dynamic BAN channels [48], which means that we can effectively treat all communications within one such period as being concurrent. Hence, these measurements allow us to simulate the operation of a dual-hop sensor-to-relay-to-hub link. We explain this by example: consider a broadcast by the left hip in the experiment that is received by channel sounders on the left ankle and the chest; concurrency and reciprocity allow us to treat this as a *Tx* that originates from a sensor on the right wrist, is received by the relay on the right hip and relayed on to the hub on the chest. This idea is easily extended to a second relay as the *Rx* on the left hip would also overhear the *Tx* from the right wrist and relay that packet on to the hub on the chest. Table 2.1 gives all the *Tx/Rx* locations.

2.6   Coded GFSK on-body communications with cooperative diversity

Here the use of relays is described according to the method of Section 2.5 with either coherent selection combining and maximum-ratio combining for instance, with no change required to the transmission strategy, such that a SIMO system is realised. Possible performance improvements by using one or multiple -relays in BAN with a typical transmission strategy [8] following from Reference 49. One such strategy is to use [31,19] BCH-coded Gaussian frequency-shift keying (GFSK) modulation at 2.4 GHz carrier frequency, with one-bit per channel use. This method was first described in Reference 49 and is illustrated again here.

2.6.1 System model for coded GFSK CoBANs

With the use of the channel gain data including relayed data, a block fading channel model, which is justified based on the channel stability, is assumed for testing [31,19] BCH-coded GFSK modulation (with a modulation index of 0.5, i.e., GMSK). Over the fading channel, a phase from a uniform random distribution in $[0, 2\pi]$ is assumed for the channel. According to the block-fading model incorporated, the $Tx - Rx$ channel gain, $\mathbf{h}_{TxRx}$, is constant over each period of a transmitted GFSK modulated BCH-codeword (which is 5×31 samples); and the next $Tx - Rx$ channel gain block magnitude is based on the subsequent RSSI measurement for that link. With relay cooperation, *decode-and-forward* strategy is used where one or two relays are incorporated. A sensor (S) sends packets to a hub (or destination) (D), possibly via one or more relays (R). The procedure is for the sensor to broadcast a packet in the first time frame to one or more relays and a hub. In the second time frame, the relay(s) decode the packet and forward it to the hub if it is decoded correctly; e.g., passes a CRC check. The relayed packet(s) and direct-link packet are then combined at the hub using coherent SC or coherent MRC on a packet-wise basis, or a codeword basis. The time frames are assumed to be significantly shorter than 15 ms, hence each of the sensor to relay and relay to hub links are simulated with channel gains measured from the same 15 ms round-robin transmission period. We also assume a block fading channel, where the $Tx - Rx$ channel gain, $\mathbf{h}_{TxRx}$ is constant over the length of the transmitted packet. Block fading over the duration of a packet is a reasonable assumption given the temporal stability of the BAN channel [48].

Additive White Gaussian Noise (AWGN) is injected into the system, with receive noise variance varied from -70 dBm to -110 dBm. Instantaneous SNR is used for combining. Maximum-ratio combining (MRC, based on channel gain and SNR) and selection-combining (SC) are used to cooperatively coherently combine the relay channel/s and direct-link, i.e., source. Two- and three-branch MRC and SC are tested (where one of the branches is the direct link, and the other one or two branches is/are the decode-and-forward relay link).

For MRC, the decision phase variables $y_{\phi MRC}$ are GFSK demodulated (as GFSK demodulation only requires phase); and BCH-decoded; where $y_{\phi MRC}$ is found for each sample as

$$y_{\phi MRC} = \varphi \left(\varphi \left(h_{sd}^* r_{sd} \right) \gamma_{sd} + \sum_{k=1}^{K} \gamma_{r_k d} \varphi \left(h_{r_k d}^* r_{r_k d} \right) \right), \tag{2.1}$$

where $K = 1$ or $K = 2$; the subscripts s, r_k and d, represent source, kth relay and destination respectively; h represents channel gain coefficients; r is the received signal; $\varphi(x)$ implies the phase of x, γ represents instantaneous SNR, and $(\cdot)^*$ implies the conjugate. For selection combining, the detected phase variables $y_{\phi SC}$ are GFSK demodulated and then BCH-decoded, where $y_{\phi SC}$ is found for each sample as

$$j = \arg \max \left[\gamma_{1,min}, \gamma_{2,min}, \gamma_{sd} \right],$$

$$y_{\phi SC} = \varphi(y_j), \text{ where } y = \left[h_{r_1 d}^* r_{1d}, h_{r_2 d}^* r_{2d}, h_{sd}^* r_{sd} \right],$$

$$\gamma_{k,min} = \min \left[\gamma_{r_k d}, \gamma_{sr_k} \right], \text{ and with one relay } \gamma_{2,min} = h_{r_2 d}^* r_{2d} = 0. \qquad (2.2)$$

2.6.2   *Performance analysis*

Performance analysis is provided for two different scenarios in Reference 49, based on Monte Carlo simulations using the measured channel gain data according to the experiment described in Section 2.5 as a basis. The average bit-error-probability (BEP) performance, versus the ratio of median channel gain to receive (*Rx*) noise variance, is determined with respect to all seven *Tx*-only locations, i.e., the source locations, on the body. In the first scenario, it is assumed that the gateway or hub, i.e., the destination, is placed on the chest and cooperative receive diversity performance is tested: using one relay, where the relay is on left hip or right hip; and for two relays, using both left hip and right hip as relays; as shown in Figure 2.4(a). The non-cooperative direct link, or single-link, performance is provided as reference. MRC implies maximum-ratio combining, and SC implies selection combining. In the second scenario, it is assumed that the gateway is at the left hip; and cooperation performance is determined with one relay, on the chest or right hip, and two relays being on both the chest and right hip; as shown in Figure 2.4(b). In all cases in Figure 2.4, MRC performance is only marginally better than that of SC. Further, in both scenarios it is found that:

> With the hub at the chest, there are significant performance gains in all cases of cooperative receive diversity in Figure 2.4(a) At a BEP of 10^{-3}, there is 4 dB gain of both SC and MRC over "Direct Link" using the relay at the left hip; there is 11 dB gain of SC and 13 dB gain of MRC over "Direct Link", with one relay at right hip; and 14 dB gain of SC and MRC with two relays at the left and right hips. We note that the use of two relays gives some marginal performance improvement over the better performing one-relay at right hip strategy.
> With the hub at the left hip, there are significant performance gains in all cases of cooperative receive diversity in Figure 2.4(b). At a BEP of 10^{-3}: there is 4 dB gain of SC and 5 dB gain of MRC using the relay at the right hip over "Direct Link"; there is 8 dB gain of SC and MRC, with one relay at chest, over "Direct Link"; and 12 dB gain of SC and 13 dB gain of MRC using relays at the right hip and chest. We note that the use of two relays gives 4–5 dB improvement over the better performing one-relay at chest strategy.

In general, there is no real performance gain of cooperative maximum-ratio combining over selection combining. Depending on the hub placement, there may be some advantage to using two relays over one relay, if the added complexity is permissible. Well-considered relay placement with respect to the hub, with receive combining, is important; as shown in Reference 49 with up to 9 dB performance difference between two different relay locations when using just one relay for the same hub location.

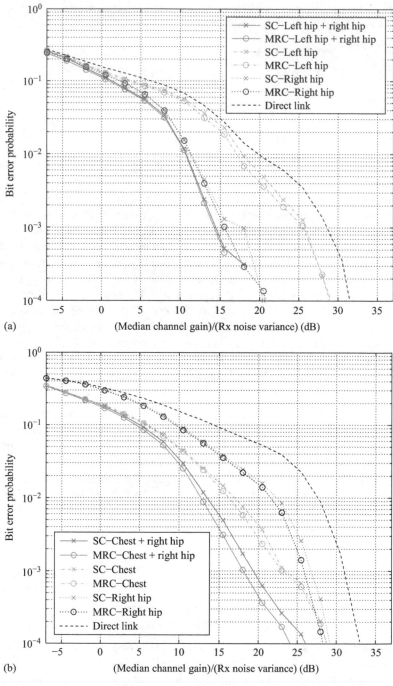

Figure 2.4   Bit-error-probability (BEP), MRC and SC, coded GFSK body area communications as in Reference 49. (a) Average BEP, cooperative receive diversity, hub-Rx at chest and (b) average BEP, cooperative receive diversity, hub-Rx at left hip

2.7 Outage analysis, cooperative selection combining and maximum-ratio combining

Here first- and second-order statistics of outages with cooperative receive diversity using experimental results of Section 2.5 are given, hence this is a more general presentation than that in the previous section, even though the two combining methods, selection and maximum-ratio, are the same. The outcomes in this section were first described in Reference 50 and are summarised here.

It is demonstrated that there is significant benefit for well-designed cooperative communications for the general BAN channel. Maximum-ratio combining and selection combining are both effective, but selection combining is generally sufficient. One relay improves performance over the direct link, but there are significant gains for using more than one relay. Well-considered relay placement is shown to be important. There is large variation in terms of outage probability for different modes of on-body cooperative receive diversity, whereas there is only some minor variation in average fade duration (AFD) for these different modes.

2.8 Implementation of cooperative selection and maximum-ratio combining

Once again decode-and-forward communications, as described in the previous section, is used. Here, for the sensor-to-relay link, it is assumed that a packet is decoded correctly if $|h_{SR}| > h_T$, where $|h_{SR}|$ denotes the constant channel gain of the packet from sensor to relay. In this chapter, 3 values of channel gain threshold $h_T = \{-95, -86, -76\}$ dB are used. These values are chosen to reflect likely thresholds according to the specifications of the IEEE 802.15.6 BAN draft standard in the 2360–2483.5 MHz band.

Constant transmit signal power E_s over each packet from sensors and relays is assumed; and constant noise variance (zero-mean additive-white-Gaussian noise) across all devices (N_0); such that the relative performance of the decode-and-forward communications depends on the effective combined channel gain, h_C, as combined channel power $|E_s h_C| = h_C$, and combined channel power $h_C \gg N_0$. Thus the relative performance of the decode-and-forward communications depends on the effective combined channel gain. With decode-and-forward for one-relay, the first types of coherent maximum-ratio combining (MRC) and coherent selection combining (SC) are introduced, where packets are forwarded according to the received channel power at the relay being greater than a particular receive sensitivity, respectively

$$h_{MRC_1} = \begin{cases} \sqrt{|h_{RD}|^2 + |h_{SD}|^2}, & |h_{SR}| > h_T \\ |h_{SD}| & \text{otherwise.} \end{cases} \tag{2.3}$$

$$h_{SC_1} = \begin{cases} \max\left[|h_{RD}|, |h_{SD}|\right], & |h_{SR}| > h_T \\ |h_{SD}| & \text{otherwise.} \end{cases} \tag{2.4}$$

and for two relays, R_1 and R_2, where $k = 1, 2$,

$$\mathbf{h}_{MRC_2} = \begin{cases} \sqrt{|\mathbf{h}_{R_1D}|^2 + |\mathbf{h}_{R_2D}|^2 + |\mathbf{h}_{SD}|^2}, & |\mathbf{h}_{SR_k}| > h_T \\ \sqrt{|\mathbf{h}_{R_1D}|^2 + |\mathbf{h}_{SD}|^2}, & |\mathbf{h}_{SR_1}| > h_T \text{ only} \\ \sqrt{|\mathbf{h}_{R_2D}|^2 + |\mathbf{h}_{SD}|^2}, & |\mathbf{h}_{SR_2}| > h_T \text{ only} \\ |\mathbf{h}_{SD}| & \text{otherwise.} \end{cases} \tag{2.5}$$

$$\mathbf{h}_{SC_2} = \begin{cases} \max\left[|\mathbf{h}_{R_1D}|, |\mathbf{h}_{R_2D}|, |\mathbf{h}_{SD}|\right], & |\mathbf{h}_{SR_k}| > h_T \\ \max\left[|\mathbf{h}_{R_1D}|, |\mathbf{h}_{SD}|\right], & |\mathbf{h}_{SR_1}| > h_T \text{ only} \\ \max\left[|\mathbf{h}_{R_2D}|, |\mathbf{h}_{SD}|\right], & |\mathbf{h}_{SR_2}| > h_T \text{ only} \\ |\mathbf{h}_{SD}| & \text{otherwise.} \end{cases} \tag{2.6}$$

2.8.1   Single-link fading statistics

For all of the single-links between sensors, relays and hubs, the channel gain, normalised to the root-mean-square power of each particular link, is best described by a gamma distribution. The gamma distribution gives the best characterisation of the small-scale fading characteristics in terms of negative log-likelihood of maximum-likelihood (ML) parameter estimates when compared with five other common distributions: lognormal, Weibull, Rayleigh, Nakagami-m and normal. In Reference 51, an argument is made for a gamma distribution to model a slow fading process dominated by shadowing, of which this on-body channel is an example, where deep fades can be introduced at a slow rate with a change of body position. The cumulative distribution function, which gives the outage probability, is

$$P(x|a, b) = \frac{\gamma(a, x/b)}{\Gamma(a)}, \tag{2.7}$$

where $\Gamma(\cdot)$ is the gamma function, and $\gamma(\cdot, \cdot)$ is the lower incomplete gamma function. For normalised channel gain magnitude, the on-body channel ML parameter estimates from all sensors are: (i) to all hubs the shape parameter $a = 1.74$, and the scale $b = 0.459$; (ii) to a hub at the chest ($a = 1.85, b = 0.438$); (iii) to a hub at the left hip ($a = 1.63, b = 0.483$) and (iv) to a hub at the right hip ($a = 1.77, b = 0.454$).

An important second-order statistic for the BAN channel is the AFD. The AFD is the average duration which the received signal remains below a given threshold. Hence, the shorter the AFD, the more time there is for successful packet transmission. Following from Reference 52, where derivations for fading according to a generalised gamma distribution are given, the AFD with gamma fading is given as

$$\text{AFD}(x) = \frac{P(x)}{\text{LCR}(x)} = \frac{b^{a-0.5}\gamma(a, x/b)}{\sqrt{2\pi}f_D x^{a-0.5}} \exp(x/b), \tag{2.8}$$

where f_D is the Doppler spread of the process, and $\text{LCR}(x)$ is the level-crossing rate. In this case f_D is an average of Doppler spreads of hubs relative to all sensors; due to the relative movement of hubs with respect to sensors and also scatterers around the body.

2.8.2 Performance analysis

2.8.2.1 Analysis of outage probability

Here the outage probability performance is presented following from Reference 50, where first the overall outage probability is compared for maximum-ratio combining (MRC) and selection combining (SC) for both one and two relays, i.e., from all sensors to relays to hubs (see Figure 2.5). All three positions of hubs at chest, left hip and right hip are considered, and the direct links all emanate from the seven sensor positions. A channel gain threshold for the source-to-relay link, $h_T = -95$ dB, is considered. It is shown that the outage probability of the direct link, as well as the theoretical outage probability, based on the gamma ML parameter estimated of the direct-link for comparison, where it is noted that theory provides an excellent fit to the empirical outage probability. To make a fair comparison of all cases, a plot against signal strength (ρ_{rms}) normalised to the root-mean-square power of each direct sensor-to-hub link is made, with the outages agglomerated. Overall, there are large improvements

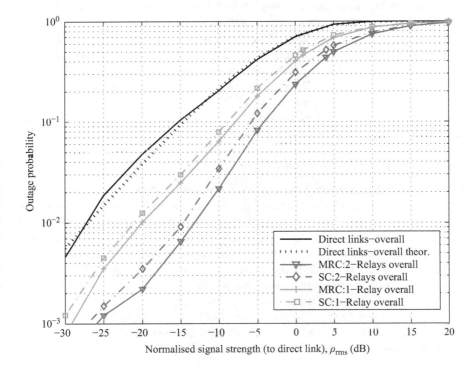

Figure 2.5 *Outage probability, cooperative receive diversity, $h_T = -95$ dB, Selection Combining (SC) and maximum-ratio combining (MRC), 1 or 2 relays, overall; based on Reference 50*

Table 2.2   *Performance gains, over direct-links at 10% outage*
            probability for $h_T = -95\,dB$ *from Reference 50*

Hub location	Relay/s	MRC gain (dB)	SC gain
Chest	Left + right hip	10	9
Chest	Left hip	8	7
Chest	Right hip	6	5
Left hip	Chest + right hip	15	13
Left hip	Chest	13	12
Left hip	Right hip	7	6

using cooperative receive diversity. At 10% outage probability, the use of two relays provides 4 dB performance improvement over that of one relay when using MRC and provides 3.5 dB performance improvement when using SC.[2] There is a further 7 dB performance improvement of MRC for one relay over that of a single, or direct, link. For two relays, there is a 1 dB performance degradation of SC compared with MRC, and a 0.5 dB performance degradation when comparing MRC with SC for one relay. At 1% outage probability, the performance improvement is even more substantial: 7 dB for two relays over one-relay and 7.5 dB for one relay over the direct link.

Next, the two separate scenarios of a hub at the chest and then a hub at the left hip are considered, with $h_T = -95\,dB$. The performance improvements at 10% outage probability are summarised in Table 2.2. With the hub at the chest and the hub at the left hip, there are significant performance gains in all cases of decode-and-forward cooperative communication. With the use of two relays, the hub at the left hip (with relays at chest and right hip) outperforms the hub at the chest (with relays at left hip and right hip). Further, the best performing one relay and hub positions are with the relay at the chest and the hub at the left hip.

2.8.3   *Analysis of second-order statistics*

The average outage (or fade) durations for cooperative decode-and-forward BAN communication with selection combining as described in Section 2.9 is presented here. The AFD (or AOD) is normalised by the different average Doppler spreads of each process. The maximum measured level crossing rates are used as estimates of the Doppler spread of the process, these vary between 1 and 1.2 Hz (which is typical of the "slow movement" that is to be expected as the majority of the experiments were done in an office). This variation depends upon the placement of the hub, the relays being used and, only very marginally, on sensors-to-relay channel gain thresholds h_T.

The normalised AFD by Doppler spread (AFD.f_D), in Figure 2.6 is investigated, with the same range of h_T used previously. Overall, considering all hub positions, SC results are shown in Figure 2.6(a). The SC results for the hub at the left hip are shown

---

[2]10% outage probability corresponds to a guideline for maximum packet error rates according to the draft IEEE 802.15.6 BAN standard [8].

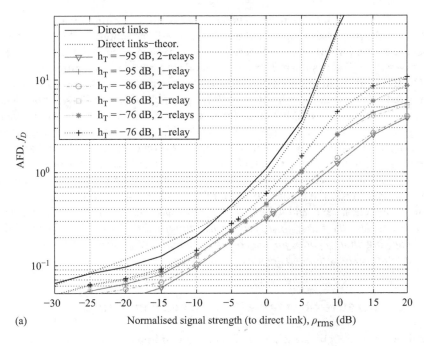

(a)

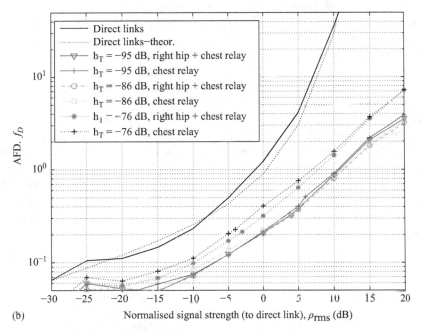

(b)

Figure 2.6   Normalised AFD. f_D, Various h_T, Selection Combining (SC), 1 or 2 relays based on Reference 50

in Figure 2.6(b). Each curve represents one process with only minor variation, if any, in f_D from the processes represented by other curves. The empirical AFDs for direct links are shown for reference in Figure 2.6, with the theoretical AFDs for the direct-link given by (2.8) also shown. These are in good agreement with empirical direct link AFDs. It is found in Reference 50 that the AFD of two relays with $h_T = -95$ dB when compared with two relays for $h_T = -86$ dB is almost identical, similarly for one relay. There is some increase of AFD at the highest threshold, $h_T = -76$ dB, for one relay in Figure 2.6(a). For Figure 2.6(b), with SC and hub at the left hip, the performance in terms of AFD is identical when comparing two relays (at chest and right hip) and also when comparing one relay (at chest) for $h_T = \{-96, -86\}$ dB, and the one relay and two relay performances are also identical to each other. With $h_T = -76$ dB, there is some increase in AFD in Figure 2.6(b) for relay at chest case and also for the relays at right hip and chest case, and hub at left hip. *Most importantly in Figure 2.6 the AFD with relays, over the range of ρ_{rms}, is always significantly lower than for the direct link; implying shorter fades for the combined signal and longer times for successful packet transmissions than for the direct link.*

2.9   Cooperative diversity with switched combining

The diversity technique switched combining can be implemented for both multiple branch co-located diversity and multiple branch cooperative diversity communications. Unlike selection combining, which requires channel estimation for all branches, switched combining requires only a single end-to-end channel estimation for each transmission, thus reducing the overall complexity while still achieving good outage performance [53, 54].

Once again we employ the experimental results described in Section 2.5. Here switch-and-examine combining is presented for WBANs, as a summary of the work in Reference 55. For two-branches when the channel gain for any branch is below (rather than crosses below [56]) a given threshold over the communications for any packet, a switch occurs to the alternate branch and communications occur across that branch [53] for that packet. However, for three branches, a switch to another branch occurs when the channel gain on the current branch is below a certain threshold. But if the channel gain on this alternate branch is also below that threshold then the last-branch to try is switched-to, and this branch becomes the chosen branch, for that packet, whether or not the channel gain is above-or-below the threshold at that time instant; although at the next time instant switching may again occur if the chosen branch is below the threshold [57]. There is switching whenever the channel gain is below this threshold.

In Reference 55, it is demonstrated that there is significant benefit for cooperative traditional switch-and-examine combining for the general BAN channel. It has relatively good performance when compared to the less-simple cooperative selection combining, and significantly better outage performance than that presented in Reference 56. Three-branch is somewhat more effective than two-branch SwC for cooperative communications. The rate of switching between branches is still much lower for this SwC than SC for CoBAN, typically by a factor for SwC of between 10% and 15% of that of SC for any switching threshold.

2.9.1 Switched combining – implementation

The same assumptions are made as in previous sections with respect to receive signal energy, noise power and combined channel gain – with analysis on a per-packet or per-codeword basis. This is also a block-fading channel. Here two forms of switched diversity with decode-and-forward communications are investigated and compared with cooperative selection diversity with decode-and-forward communications.

For the work in Reference 55, a switched combining method similar to that in Reference 53 is proposed. For the two-branch switch-and-examine combining here, $L = 2$, when the channel gain for any branch is below a given threshold h_{ST} at time (for transmission of one packer) τ, a switch occurs to the alternate branch and communications occur across that branch for that packet as described in (2.10),

$$h_C(\tau) = h_k(\tau) \text{ where } k \in \{0, 1\}, \text{ iff}$$
$$\{h_{Sw}(\tau - 1) = h_k(\tau - 1) \wedge h_k(\tau) \geq h_{ST}\} \vee$$
$$\{h_{Sw}(\tau - 1) = h_{(k-1)_2}(\tau - 1) \wedge h_{(k-1)_2}(\tau) < h_{ST}\}, \tag{2.9}$$

where $(x)_2$ implies x modulo 2, i.e., $L = 2$, and $\wedge$ and $\vee$ are the "and" and "or" operations, respectively.

Here in three branch switch-and-examine combining, $L = 3$, a switch to another branch occurs when the channel gain on the current branch is below a certain threshold h_{ST} during packet transmission time τ. But if the channel gain on this alternate branch is also below h_{ST} then the last-branch to try is switched-to and, whether or not this channel gain is above or below h_{ST}, this branch becomes the chosen branch. For switch-and-examine combining (SwC) employed in this chapter, to check whether to switch to an alternate branch, the switching occurs based on the current branch channel gain only (as well as what branch was last chosen). This 3-branch SwC is described in (2.10) in the following, where $(x)_3$ implies x modulo 3.

$$h_{Sw}(\tau) = h_k(\tau) \text{ where } k \in \{0, 1, 2\}, \text{ iff}$$
$$\{h_{Sw}(\tau - 1) = h_k(\tau - 1) \wedge h_k(\tau) \geq h_{ST}\} \vee$$
$$\{h_{Sw}(\tau - 1) = h_{(k-1)_3}(\tau - 1) \wedge h_k(\tau) \geq h_{ST} \wedge h_{(k-1)_3}(\tau) < h_{ST}\} \vee$$
$$\{h_{Sw}(\tau - 1) = h_{(k-2)_3}(\tau - 1) \wedge h_{(k-1)_3}(\tau) < h_{ST} \wedge h_{(k-2)_3}(\tau) < h_{ST}\}. \tag{2.10}$$

2.9.2 Theoretical performance

The theoretical outage probability and switching rates for the proposed switched combining in (2.9) and (2.10) depend on the steady-state branch activation probabilities, which are for two-branches from Reference 58,

$$\rho_k = \frac{q_{(k-1)_2}}{q_0 + q_1}, \quad k = 0, 1, \tag{2.11}$$

where $(x)_2$ is x modulo 2 and q_k is the cumulative distribution function for channel gain h_{ST} on branch k, i.e., probability that $h_k < h_{ST}$.[3] The analysis of Reference 58 is expanded to three branches in Reference 55 such that

$$\rho_k = \frac{q_{(k-1)_3} q_{(k-2)_3}}{q_1 q_0 + q_2 q_0 + q_1 q_2}, \quad k = 0, 1, 2, \tag{2.12}$$

where again $(x)_3$ implies x modulo 3.

Hence the outage probability, $P_C(h) = Pr(h_C < h)$, for switched combining $L = 2$ branches is, from Reference 58,

$$P_C(h) = \sum_{k=0}^{1} \left\{ \rho_k \times \begin{pmatrix} F_k(h) - q_k + q_k F_{(k-1)_2}(h), h \geq h_{ST} \\ q_k F_{(k-1)_2}(h), h < h_{ST} \end{pmatrix} \right\}. \tag{2.13}$$

And following from the method for $L = 2$, an expression for $P_C(h)$ for $L = 3$ is provided in Reference 55:

$$P_C(h) = \sum_{k=0}^{2} \left\{ \rho_k \begin{pmatrix} F_k(h) - q_k + q_k(F_{(k-1)_3}(h) - q_{(k-1)_3}) \\ + q_k q_{(k-1)_3} F_{(k-2)_3}(h), h \geq h_{ST} \\ q_k q_{(k-1)_3} F_{(k-2)_3}(h), h < h_{ST} \end{pmatrix} \right\}. \tag{2.14}$$

An approximation to theoretical switching rate for traditional switch-and-examine combining can be found from the dwell time according to average non-fade duration *ANFD* for each branch at level h_{ST}, i.e., the expected contiguous time for which the channel gain $h \geq h_{ST}$,[4] and the steady-state relay activation probabilities given by (2.11) and (2.12). The average non-fade duration for each branch k is

$$\text{ANFD}_k(h_{ST}) = \frac{1 - q_k}{\text{LCR}_k(h_{ST})}, \tag{2.15}$$

where $\text{LCR}_k(h_{ST})$ is the level crossing rate at threshold h_{ST} (i.e., the rate at which h_C crosses above h_{ST}).

For the particular CoBAN scenarios here all dual-hop and single-hop links, $\text{LCR}_k(h_{ST})$, as well as q_k and outage probabilities $P_C(h)$, can be approximated by closed form expressions, as all small-scale fading links (when normalised by mean path loss) can be well described by a gamma, Weibull, and sometimes lognormal distribution. $\text{LCR}_k(h_{ST})$ for gamma, Weibull and lognormal fading can be found in References 59, [60] and [61], respectively.

Hence the overall branch switching rate is

$$SR_L \simeq \frac{1}{\left\{ \sum_{k=0}^{L-1} \text{ANFD}_k \rho_k \right\}}, \tag{2.16}$$

which is an alternate, i.e., different, formulation to that described in Reference 62.

---

[3]where with respect to the form in Reference 53, this is the same as probability that $h_k^2 < h_{ST}^2$.
[4]Please note this is an approximation due to the possibility of switching to a branch with $h < h_{ST}$.

2.9.3   Analysis of outage probability

Here the outage probability performance is presented of decode-and-forward cooperative communications according to the mode of operation described by equations (2.9) and (2.10) for switched combining (Sw$_1$C) as in Reference 55, the switched combining described in Reference 56 (Sw$_2$C) and selection combining (SC). To make a fair comparison of all cases, signal strength, ρ_{rms} is plotted, normalised to the root-mean-square power of each direct sensor-to-hub link, with outages agglomerated.

2.9.3.1   Optimum h$_{ST}$

The relative performance of the switched combining (Sw$_1$C) used here will be dependent on the choice of switching threshold h$_{ST}$. Here, it is also attempted to reduce the switching rate, which is lower with a lower switching threshold h$_{ST}$, as the contiguous time for a channel gain for any branch to be above this threshold is larger for smaller h$_{ST}$. Accordingly, there is similar performance, empirically determined, in terms of outage probability for h$_{ST} \geq -93$ dB for switched combining here, thus h$_{ST} = -93$ dB is chosen as a switching threshold.

2.9.3.2   Overall outage probability performance

The overall outage probability performance, considering all hub positions for both methods of cooperative switched combining (Sw$_1$C) and (Sw$_2$C) and selection combining (SC), is shown in Figure 2.7. At 10% outage probability, there is 2 dB performance improvement of the three branch Sw$_1$C over two-branch Sw$_2$C, with these SwC modes having 5 dB and 3 dB performance improvement, respectively, over single-link communications. Importantly, at 10% outage probability, two-branch SC only has 2 dB performance improvement over two-branch Sw$_1$C, with the same 2 dB performance for three-branch SC over three-branch Sw$_1$C. Importantly, Sw$_1$C, proposed here, has 2-dB performance improvement over Sw$_2$C in Reference 56 for both two and three branches; and 2-branch, with 1-relay, Sw$_1$C performs the same as 3-branch, with 2-relays, Sw$_2$C.

2.9.3.3   Outage probability performance at particular hub positions

Next considered is the two separate scenarios for Sw$_1$C and SC of a hub at the left hip and then a hub at the chest, with h$_{ST} = -93$ dB. Performance improvements for Sw$_1$C with respect to those for Sw$_2$C, with greater improvement for Sw$_1$C, over single-link communications for 10% outage probability are summarised in Table 2.3. With the hub at the left hip, there are significant performance gains for both cooperative Sw$_1$C and cooperative SC with only 2 dB performance degradation of Sw$_1$C for all relay positions (compared to a further 2 dB performance degradation for Sw$_2$C with respect to Sw$_1$C). Considering the case of the hub at the chest, there is 2.5 dB gain at 10% outage probability with Sw$_1$C for using the left-hip and right-hip relays over just using right hip. Further, the best performing one relay and hub positions are with the relay at the chest and the hub at the left hip.

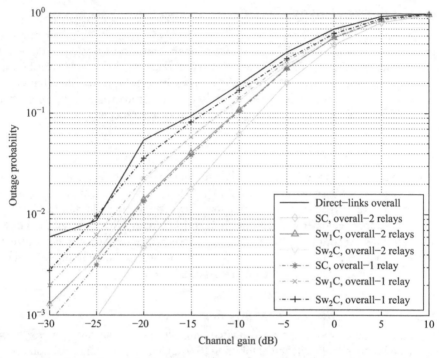

Figure 2.7   Outage probability, cooperative switched combining (SwC), Sw₁C (2.9)
and (2.10), Sw₂C from Reference 56 and cooperative selection
combining (SC), 1 or 2 relays, overall based on Reference 55

Table 2.3   Performance gains, over direct-links at 10% outage probability for
$h_{ST} = -93\,dB$ *[55]*

Hub Loc.	Relay/s	Sw₁C gain (dB)	Sw₂C gain (dB)	SC gain (dB)
Left hip	Chest + right hip	8	6	10
Left hip	Chest	6.5	6	8.5
Left hip	Right hip	4	2	5.5
Chest	Left + right hip	4	2.5	6
Chest	Left hip	1.5	−0.5	3.5
Chest	Right hip	2.5	2	4.5

2.9.3.4   Outage probability performance – empirical compared with theoretical

Here the empirical performance with the theoretical outage probability according
to (2.13) and (2.14) with respect to two and three branch combining, respectively,
is compared, where the theoretical cdf value q_k for any sensor-to-hub or sensor-
to-relay-to-hub branch, with respect to switching threshold h_{ST} varies according to

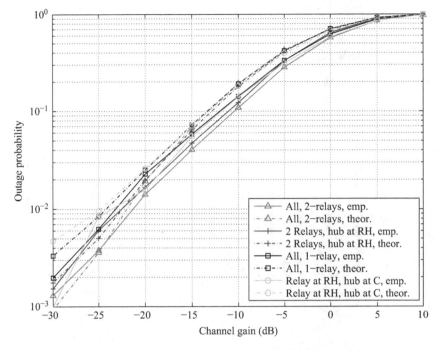

Figure 2.8   Empirical and theoretical outage probability, cooperative switched combining (SwC), Sw₁C, for 2 and 3-branches overall, for 2 relays, with hub at right hip, and for one relay at right hip and hub at chest [55]

sensor location, measurement scenario and subject (as well as being drawn from different statistical distributions; as the mean path loss varies significantly given different scenarios, subjects and sensor locations). Accounting for this variation, as shown in Figure 2.8, two and three branch cooperative switched combining proposed here has around 1.5–2.5 dB better performance at 10% outage according to empirical measurements than theoretical two and three branch outage from (2.13) and (2.14), overall. And also for one relay at right hip and hub at chest there is 1.5 dB improvement of performance over analysis, and with hub at right hip, and relays at chest and left hip this is also 1.5 dB performance improvement in Figure 2.8. According to the theory and in Figure 2.8, at 1% outage probability there is a 2 dB performance gain overall from using two relays to using one-relay, which is similar to the 1% empirical outage probability improvement from using 1 to 2 relays. Importantly, from theory and Figure 2.8, the gain of two relays of 0.5 dB, over one relay, somewhat underestimates the empirical gain of 2 dB from the empirical implementation of switched combining according to (2.9) and (2.10).

2.9.4   Switching rate analysis

The switching rate is the rate at which the selection of any branch changes at the destination; i.e., the hub device. Switching rates for diversity combining play an

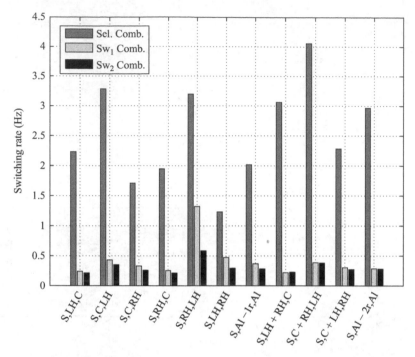

Figure 2.9   Switching rates, $h_{ST} = -93\,dB$, cooperative switched combining (SwC), Sw_1C from Reference 55, Sw_2C from Reference 56, and selection combining, with 2 and 3-branches, considering all three relay placements and hub positions with order "S,R,H" indication source, relay location/s, hub location. Overall-2 relays is A12r, Overall-1 relay is A11r, LH – left hip, S – sensor, RH – right hip, C – chest. Based on Reference 55

important role in assessing receiver outages due to switching transients [63]. Frequent relay switchings can cause synchronisation problems because the system needs to repeat an initialisation process each time the active relay changes, in order to re-adapt to the channel conditions of the new branch [53]. Further, readjustments of frequency synchronisation leads to increased implementation complexity, as well as potential delays and outages.

The switching rate is plotted in Figure 2.9 for all scenarios with a switching threshold $h_{ST} = -93$ dB, where the performance of selection combining is compared, the switched combining method of Reference 56, Sw_2, and the enhanced switched combining proposed here, Sw_1, presented as described for 2 and 3 branches combining in (2.9) and (2.10), respectively. The switching rates for both methods of switched combining are in general shown to be around 10%–15% of that for Selection Combining. As shown the Sw_1 method proposed here has significantly better outage performance than Sw_2 of Reference 56 as demonstrated, and switching between branches occurs more often for the proposed Sw_1 method, due to stricter conditions for a branch remaining selected than Sw_2 in Reference 56. However, there is only a marginal increase in switching rate in ten cases, for all three-branch combining and nearly all 2-branch combining, the increase in switching rate is minimal (for nine cases increase is under 0.05 Hz and an increase of 5%, and for one relay S, LH, RH case there is a 0.2 Hz increase). The only significant increased switching rate from Sw_2 to Sw_1 is for the two-branch, one relay case S, RH, LH, with a relay at right hip and a hub at left hip where the switching rate doubles by 0.6 Hz from Sw_2 to Sw_1.

The theoretical switching rate is determined according to (2.20), where the branch activation probabilities ρ_k are determined with respect to $h_{ST} = -93$ dB from Figure 2.9. From the theoretical expression for two-branch combining according to (2.20), using branch activation probabilities ρ_k in (2.11) for all six combinations and also overall (seven cases), there is a mean error comparing with switching rates presented in Figure 2.9 of 0.15 Hz, with minimum error of 0.002 Hz to a maximum of 0.7 Hz (second largest error is 0.19 Hz). From the theoretical expression for three-branch combining according to (2.20), using branch activation probabilities ρ_k in (2.12) for all three combinations and also overall (four cases), there is a mean error comparing with switching rates presented in Figure 2.9 of 0.14 Hz, with minimum error of 0.05 Hz to a maximum of 0.25 Hz comparing from theory to empirical analysis. This error suggests the switching rate of (2.20) is a reasonable first-order approximation to that of empirical analysis.

2.10 Cooperative switched diversity with power control

The description in this section follows from work in Reference 64, where again we used the channel data captured as described in Section 2.5. Simple transmit power control is described, using a "Sample-and-Hold" channel prediction method following from Reference 65, to be incorporated with simple cooperative switch-and-examine combining following from Reference 55. Using extensive empirical "open-access" multi-link data, the capture of which is as described before in Section 2.5 it is shown that the combination of these two mechanisms *significantly improves communications reliability, in terms of reduced outages, and increases energy efficiency, while maintaining low radio complexity and operational efficiency, further evidenced by second-order statistics such as switching rate.*

As previously, $L = 3$-branch cooperative diversity switch-and-examine combining is investigated as described in the previous section, where the threshold is now modified to a power value, rather than a channel gain. The switch-and-examine choice of diversity branch is with respect to a certain threshold $h_T = h_{ST}/\sqrt{P_{t,k}(\tau)}$, where h_{ST} is the root channel power switching threshold and $P_{t,k}(\tau)$ is the (known) transmit (Tx) power during packet transmission time τ, which can now vary between transmission times.

2.10.1 *Transmit power control using "sample-and-hold" prediction*

According to IEEE 802.15.6 the maximum radiated transmit (Tx) power is 0 dBm (1 mW), accordingly transmit power control is applied where the channel gain of the previous received packet, at $\tau - 1$ is used to set the sensor (and relay, if active) transmit power at time τ. The range of output Tx power is from -30 dBm to 0 dBm (0.001 mW to 1 mW), where 2 dB spacing with 16 discrete power levels is chosen such that $Tx_{range} = \{-30, -28, \ldots, -2, 0\}$ dBm. The transmit power control properly applied uses the knowledge of receiver power sensitivity rx_{sens}. Accordingly, in the linear domain

$$P_t(\tau) = \min \left\{ \left[1, Tx_{range} \geq \frac{rx_{sens}}{h_p(\tau - 1)^2} P_t(\tau - 1) \times \text{offset} \right] \right\}, \tag{2.17}$$

where $h_p(\tau - 1)^2$ is received channel power (in mW) at $\tau - 1$, and a power offset is applied to mitigate for possible reduction in channel gain at the next time sample τ. Across all measurements, a suitable value for offset is found as 6 dB (a multiplier of 3.98 in the linear domain).

2.10.2   First- and second-order statistics

For the three-branch switch-and-examine combining here described in (2.10), incorporating power control described in (2.17), of particular concern is branch activation probabilities, outage probability with respect to non-normalised channel gains and overall-branch switching rate as previously. The switch-and-examine combining is dependent on the branch activation probabilities, for the current packet, inst., and overall, these follow from (2.12).

The outage probability is $P_{out} \approx Pr(h_C < h)$, where h is any given channel gain, and h_C is combined SwC channel gain. Both q_k and $P_C(h)$ found in (2.14) can be estimated in closed-form due to a good fit of a lognormal distribution to all of the non-normalised branch channel gain (as opposed, as previously described in the chapter of the fit of the gamma distribution, to normalised channel gain). Distribution log-means vary from -9.5 to -8.75, and log-standard deviations vary from 0.71 to 1.

The switching rate can be found from the dwell time according to the average non-fade duration ANFD for each branch at root power switching threshold level h_{ST}, with respect to the maximum transmit power required for that threshold level $\sqrt{\max(P_{t,k,inst.})}$, $h_{Tn} = h_{ST} / \sqrt{\max(P_{t,k,inst.})}$, when that branch is chosen,

$$\text{ANFD}_k(h_{Tn}) = \frac{1 - q_k(h_{Tn})}{\text{LCR}_k(h_{Tn})}, \tag{2.18}$$

where $\text{LCR}_k(x)$ is the branch level crossing rate, the rate at which the channel gain crosses above (or below) a given threshold x, which has a closed form for LCR, with the lognormal distribution of x, of

$$\text{LCR}_k(x) = f_D \exp \left(\frac{-(\ln(x) - \mu_l)^2}{2\sigma_l^2} \right), \tag{2.19}$$

where f_D is the Doppler spread, $\ln(\cdot)$ is the natural logarithm, μ_l is the log-mean, and σ_l is the log-standard deviation; a good approximation for f_D according to the empirical data used here is 0.8 Hz.

The switching rate SR_L then can be expressed in terms of the average non-fade duration for any branch, the contiguous duration for channel gain to be above channel gain threshold h_{Tn}, and the branch activation probabilities, giving

$$SR_L \approx \frac{1}{\left\{ \sum_{k=0}^{L-1} \left(\overline{\mathrm{ANFD}_k(h_{Tn})} \overline{p}_{k,\mathrm{inst.}} \right) \right\}}. \tag{2.20}$$

2.10.3   Performance analysis

Here performance analysis is provided, based on the empirical data, and the switched combining and power control outlined for power control, PC, alone (simply using the direct link), switched combining, SwC alone (where a suitable fixed transmission power is used) and SwC + PC, the combination of switched combining and power control recommended here. First, the outage probability with respect to these three cases is presented with respect to four typical BAN receiver sensitivities, for narrowband communications at 2.4 GHz and as specified by IEEE 802.15.6 [8], of $\{-95, -93, -90, -86\}$ dBm. For the least sensitive receiver at -86 dBm for SwC alone, as a good tradeoff of power consumption and reliability, a fixed Tx power of -2 dBm is chosen, for more sensitive receivers a Tx power of -5 dBm is set for SwC alone.

The outage probability with respect to the receiver sensitivity will determine the packet error rate, so results for hub locations at the C, and the LH, and all three choices of hub locations, are given in Figure 2.10. The cases of the hub at the chest perform better than the hub at the LH. For any hub location, there are significant performance gains of using SwC plus power control (PC), over using SwC alone (such that, e.g., performance at -90 dBm receiver sensitivity for SwC and PC, is similar to performance at -95 dBm receiver sensitivity for SwC alone) in Figure 2.10. SwC and PC used together give over an order of magnitude performance improvement over PC alone in Figure 2.10.

Next in Figure 2.11, sensor circuit power consumption is compared, with the cases of SwC with PC, and direct-link PC ("no coop"), for each hub location, and all hub locations. Transmit power is mapped to circuit power consumption accordingly for a low-power radio for use in BAN health-care as in Reference 65. It is clear from pairs of bars that in all cases in Figure 2.11, for the sensor, using PC and SwC reduces power consumption. Further, Figure 2.11 shows SwC plus PC, gives significant power consumption reduction over suitable fixed transmission powers (two marked horizontal lines on graph) of -2 dBm for -86 dBm receiver sensitivity, and -5 dBm for all the lower receive sensitivities.

It is found in Reference 64 that the theoretical approximation for switching rate of (2.20) is reasonable, with a typical (median) error of 0.12 Hz between this approximation for switching rates (for receive sensitivities lower than -86 dBm) and empirical results. It is also found in Reference 64 that the switching rates are all significantly less than 1 Hz, making switched combining with power control an attractive option for BANs.

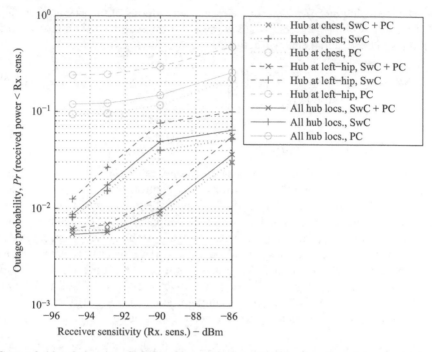

Figure 2.10   Outage probability from Reference 64 in terms of receiver sensitivity, for two particular hub locations, and considering all three hub positions, SwC – switched combining, PC – power control

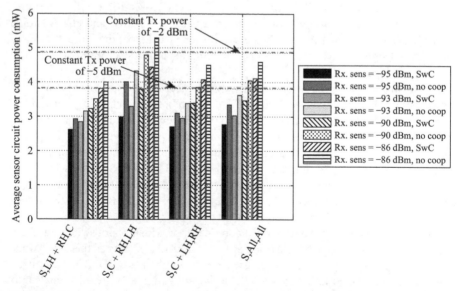

Figure 2.11   Average sensor circuit power consumption from [64] with transmit power control for four receive sensitivities at 2.4 GHz, S – source/sensor, C – chest, LH – left hip, RH – right hip, SwC – switched combining; no coop – direct-link only

2.11   Conclusion

In this chapter, many of the benefits for BANs of diversity and cooperative communications, in particular cooperative diversity, have been explained and demonstrated. These benefits have principally been shown with respect to the physical (PHY) layer – but there has been some discussion of link (MAC), network and application layer issues appropriate to BAN, particularly in a brief survey of state-of-the-art. The demonstrated performance gains over single-link star topologies, particularly with respect to appropriate cooperative combining, motivate the use of cooperative receive diversity to achieve required reliability, which is potentially life-saving for BANs, as well as reduce energy consumption, vital for long BAN life-time. Selection combining gives some performance benefits, but a simple, yet practical, form of combining, namely switched combining, may give sufficient performance improvements in terms of both first- and second-order statistics. It is further demonstrated that such cooperative communications can be enhanced by transmit power control. There are further challenges for cooperative communications in BANs, such as its use to mitigate interference among closely located BANs, and potential cooperative body-to-body communications, where there is some coordination between separate closely located BANs.

References

[1]   Smith DB, Hanlen LW. Channel modeling for wireless body area networks. Ultra-Low-Power Short-Range Radios. 2015; pp. 25–55.

[2]   Lewis D. IEEE 802.15.6 Call for Applications – Summary ID: 802.15-05-0407-05. 2008 July.

[3]   Movassaghi S, Abolhasan M, Lipman J, Smith D, Jamalipour A. Wireless body area networks: a survey. Communications Surveys Tutorials, IEEE. 2014 Third;16(3):1658–1686.

[4]   Miniutti D, Hanlen L, Smith D, *et al.* Dynamic narrowband channel measurements around 2.4 GHz for body area networks. 2008 June;IEEE 802.15.6 technical contribution, document ID:15-08-0354-01-0006-dynamic-narrowband-channel-measurements-around-2-4-ghz-for-body-area-networks.

[5]   Yazdandoost KY, Sayrafian-Pour K. TG6 Channel Model ID: 802.15-08-0780-12-0006. 2010 IEEE submission, Nov.

[6]   Smith DB, Miniutti D, Hanlen LW, Rodda D, Gilbert B. Dynamic narrowband body area communications: link-margin based performance analysis and second-order temporal statistics. In: Wireless Communications and Networking Conference (WCNC), 2010 IEEE, 2010. pp. 1–6.

[7]   Zhen B, Patel M, Lee SH, Won ET. Body Area Network (BAN) Technical Requirements. 2008 July;IEEE 802.15.6 technical contribution, document ID:15-08-0037-01-0006-ieee-802-15-6-technical-requirements-document-v-4-0.

[8]   IEEE Standard for Local and metropolitan area networks Part 15.6: Wireless Body Area Networks. IEEE Std 802156-2012. 2012 Feb;pp. 1–271.

[9]   Hanlen L, Chaganti V, Gilbert B, Rodda D, Lamahewa T, Smith D. Open-source Testbed for Body Area Networks: 200 sample/sec, 12 hrs Continuous Measurement. In: 2010 IEEE International Symposium on Personal, Indoor and Mobile Radio Communications (PIMRC), Istanbul, Turkey, 2010. pp. 66–71.

[10]   Smith DB, Hanlen LW, Rodda D, Gilbert B, Dong J, Chaganti V. Body Area Network Radio Channel Measurement Set, 2012. Available from: http://opennicta .com/datasets.

[11]   Smith DB, Miniutti D, Hanlen LW. Characterization of the body-area propagation channel for monitoring a subject sleeping. IEEE Transactions on Antennas and Propagation, 2011 Nov;59(11):4388–4392.

[12]   Cotton SL, D'Errico R, Oestges C. A review of radio channel models for body centric communications. Radio Science. 2014;49(6):371–388.

[13]   Smith DB, Miniutti D, Lamahewa TA, Hanlen LW. Propagation models for body-area networks: a survey and new outlook. Antennas and Propagation Magazine, IEEE. 2013 Oct;55(5):97–117.

[14]   Ferrand P, Maman M, Goursaud C, Gorce JM, Ouvry L. Performance evaluation of direct and cooperative transmissions in body area networks. Annals of Telecommunications. 2011;66:213–228. 10.1007/s12243-011-0238-y. Available from: http://dx.doi.org/10.1007/s12243-011-0238-y.

[15]   Boulis A, Smith D, Miniutti D, Libman L, Tselishchev Y. Challenges in body area networks for healthcare: the MAC. Communications Magazine, IEEE. 2012;50(5):100–106.

[16]   Nosratinia A, Hunter TE, Hedayat A. Cooperative communication in wireless networks. Communications Magazine, IEEE. 2004;42(10):74–80.

[17]   Sendonaris A, Erkip E, Aazhang B. User cooperation diversity. Part I. System description. IEEE Transactions on Communications. 2003 Nov;51(11):1927–1938.

[18]   Nabar RU, Kneubühler FW, Bölcskei H. Performance limits of amplify-and-forward based fading relay channels. In: IEEE International Conference on Acoustics, Speech, and Signal Processing, 2004. Proceedings (ICASSP'04), vol. 4. IEEE, 2004. pp. iv-565–iv-568.

[19]   Dong J. Wireless body area networks: enabling robust coexistence and interference management. Ph.D. dissertation, The Australian National University; 2015.

[20]   Li Q, Hu RQ, Qian Y, Wu G. Cooperative communications for wireless networks: techniques and applications in LTE-advanced systems. Wireless Communications, IEEE. 2012 Apr;19(2): pp. 22–29.

[21]   Shalmashi S, Ben Slimane S. Cooperative device-to-device communications in the downlink of cellular networks. In: Wireless Communications and Networking Conference (WCNC), IEEE, 2014. pp. 2265–2270.

[22]   Chu S, Wang X, Yang Y. Exploiting cooperative relay for high performance communications in MIMO ad hoc networks. IEEE Transactions on Computers. 2013 Apr;62(4):716–729.

[23] Guo B, Guan Q, Yu FR, Jiang S, Leung VCM. Energy-efficient topology control with selective diversity in cooperative wireless ad hoc networks. In: 2014 IEEE International Conference on Communications (ICC), 2014. pp. 197–202.

[24] Lin J, Jung H, Chang YJ, Jung JW, Weitnauer MA. On cooperative transmission range extension in multi-hop wireless ad-hoc and sensor networks: a review. Ad Hoc Networks. 2015;(29):117–134. Available from: http://www.sciencedirect.com/science/article/pii/S1570870515000281.

[25] Zhu C, Zheng C, Shu L, Han G. A survey on coverage and connectivity issues in wireless sensor networks. Journal of Network and Computer Applications. 2012;35(2):619–632. Simulation and Testbeds. Available from: http://www.sciencedirect.com/science/article/pii/S1084804511002323.

[26] Sharma S, Shi Y, Hou YT, Kompella S. An optimal algorithm for relay node assignment in cooperative ad hoc networks. IEEE/ACM Trans-actions on Networking. 2011 June;19(3):879–892. Available from: http://dx.doi.org.virtual.anu.edu.au/10.1109/TNET.2010.2091148.

[27] Valentin S, Lichte HS, Karl H, Aad I, Loyola L, Widmer J. Opportunistic relay-ing vs. selective cooperation: analyzing the occurrence-conditioned outage capacity. In: Proceedings of the 11th International Symposium on Modeling, Analysis and Simulation of Wireless and Mobile Systems. MSWiM '08. New York, NY, USA: ACM, 2008. pp. 193–202. Available from: http://doi.acm.org/10.1145/1454503.1454539.

[28] Stamatiou K, Chiarotto D, Librino F, Zorzi M. Performance analysis of an opportunistic relay selection protocol for multi-hop networks. Communications Letters, IEEE. 2012 Nov;16(11):1752–1755.

[29] Hwang KS, Chai Ko Y, Alouini MS. Outage probability of cooperative diversity systems with opportunistic relaying based on decode-and-forward. IEEE Transactions on Wireless Communications. 2008 Dec;7(12):5100–5107.

[30] Bletsas A, Khisti A, Reed DP, Lippman A. A simple cooperative diversity method based on network path selection. IEEE Journal on Selected Areas in Communications. 2006;24(3):659–672.

[31] Cui S, Haimovich AM, Somekh O, Poor HV. Opportunistic relaying in wireless networks. IEEE Transactions on Information Theory. 2009 Nov;55(11):5121–5137.

[32] Conway GA, Cotton SL, Scanlon WG. An antennas and propagation approach to improving physical layer performance in wireless body area networks. IEEE Journal on Selected Areas in Communications. 2009 Jan;27(1):27–36.

[33] Gorce JM, Goursaud C, Villemaud G, Errico RD, Ouvry L. Opportunistic relaying protocols for human monitoring in BAN. In: 2009 IEEE 20th Inter-national Symposium on Personal, Indoor and Mobile Radio Communications. IEEE, 2009. pp. 732–736.

[34] Chen Y, Teo J, Lai JCY, *et al.* Cooperative communications in ultra-wideband wireless body area networks: channel modeling and system diversity analysis. IEEE Journal on Selected Areas in Communications. 2009 Jan;27(1): 5–16.

[35]   Huang X, Shan H, Shen X. On energy efficiency of cooperative communications in wireless body area network. In: Wireless Communications and Networking Conference (WCNC), 2011 IEEE, 2011. pp. 1097–1101.

[36]   D'Errico R, Rosini R, Maman M. A performance evaluation of cooperative schemes for on-body area networks based on measured time-variant channels. In: 2011 IEEE International Conference on Communications (ICC), Kyoto, Japan, 2011. pp. 1–5.

[37]   SIG B. Bluetooth specification version 4.0 [vol. 0], 2010. Available from: http://www.bluetooth.org.

[38]   Feng H, Liu B, Yan Z, Zhang C, Chen CW. Prediction-based dynamic relay transmission scheme for Wireless Body Area Networks. In: 2013 IEEE 24th International Symposium on Personal Indoor and Mobile Radio Communications (PIMRC), 2013. pp. 2539–2544.

[39]   Ding J, Dutkiewicz E, Huang X. Energy efficient cooperative communication for UWB based in-body area networks. In: Proceedings of the 8th International Conference on Body Area Networks. BodyNets'13. ICST, Brussels, Belgium, Belgium: ICST (Institute for Computer Sciences, Social-Informatics and Telecommunications Engineering), 2013. pp. 29–34. Available from: http://dx.doi.org.virtual.anu.edu.au/10.4108/icst.bodynets.2013.253567.

[40]   Momoda M, Hara S. A cooperative relaying scheme for real-time vital data gathering in a wearable wireless body area network. In: 2013 7th International Symposium on Medical Information and Communication Technology (ISMICT), 2013. pp. 38–41.

[41]   Cotton SL, Scanlon WG. Channel characterization for single- and multiple-antenna wearable systems used for indoor body-to-body communications. IEEE Transactions on Antennas and Propagation. 2009 Apr;57(4): 980–990.

[42]   Deepak K, Babu A. Improving energy efficiency of incremental relay based cooperative communications in wireless body area networks. International Journal of Communication Systems. 2015;28(1):91–111.

[43]   Smith DB, Miniutti D, Lamahewa TA, Hanlen LW. Propagation models for body-area networks: a survey and new outlook. Antennas and Propagation Magazine, IEEE. 2013 Oct;55(5):97–117.

[44]   Shaban H, Abou El-Nasr M. Amplify-and-forward cooperative diversity for Green UWB-based WBSNs. The Scientific World Journal. 2013. pp. 1–6.

[45]   Maskooki A, Soh CB, Gunawan E, Low KS. Opportunistic routing for body area network. In: Consumer Communications and Networking Conference (CCNC), 2011 IEEE, 2011. pp. 237–241.

[46]   Abbasi UF, Awang A, Hamid NH. Performance investigation of opportunistic routing using log-normal and IEEE 802.15.6 CM 3A path loss models in WBANs. In: 2013 IEEE Malaysia International Conference on Communications (MICC), 2013. pp. 325–329.

[47]   Hanlen L, Chaganti V, Gilbert B, Rodda D, Lamahewa T, Smith D. Open-source testbed for body area networks: 200 sample/sec, 12 hrs continuous

measurement. In: 2010 IEEE 20th International Symposium on Personal Indoor and Mobile Radio Communication, Istanbul, Turkey, 2010. pp. 1–5.

[48] Smith DB, Zhang J, Hanlen LW, Miniutti D, Rodda D, Gilbert B. Temporal correlation of dynamic on-body area radio channel. Electronics Letters. 2009 Nov;45(24):1212–1213.

[49] Dong J, Smith DB. Cooperative Receive Diversity for Coded GFSK Body-Area Communications. IET Electronics Letters. 2011 Sep;47(19):1098–1100.

[50] Smith D, Miniutti D. Cooperative body-area-communications: first and second-order statistics with decode-and-forward. In: Proceedings of IEEE Wireless Communications and Networking Conference (WCNC), Paris, France, 2012. pp. 1–5.

[51] Abdi A, Kaveh M. On the utility of gamma PDF in modeling shadow fading (slow fading). In: 1999 IEEE 49th Vehicular Technology Conference, vol. 3, 1999. pp. 2308–2312.

[52] Primak S, Kontorovich V. On the second order statistics of generalized gamma process. IEEE Transactions on Communications. 2009 Apr;57(4): 910–914.

[53] Michalopoulos DS, Lioumpas AS, Karagiannidis GK, Schober R. Selective cooperative relaying over time-varying channels. IEEE Transactions on Communications. 2010 Aug;58(8):2402–2412.

[54] Abu-Dayya AA, Beaulieu NC. Analysis of switched diversity systems on generalized-fading channels. IEEE Transactions on Communications. 1994 Nov;42(11):2959–2966.

[55] Smith DB. Improved switched combining with cooperative diversity for wireless body area networks: empirical analysis and theory. In: 2014 IEEE International Conference on Communications (ICC), 2014. pp. 5682–5687.

[56] Smith DB. Cooperative switched combining for wireless body area networks. In: 2012 IEEE 23rd International Symposium on Personal Indoor and Mobile Radio Communications (PIMRC), 2012. pp. 2275–2280.

[57] Yang HC, Alouini MS. Performance analysis of multibranch switched diversity systems. IEEE Transactions on Communications. 2003 May;51(5): 782–794.

[58] Yang HC, Alouini MS. Markov chains and performance comparison of switched diversity systems. IEEE Transactions on Communications. 2004; 52(7):1113–1125.

[59] Smith DB, Miniutti D. Cooperative selection combining in body area networks: switching rates in gamma fading. Wireless Communications Letters, IEEE. 2012 Aug;1(4):284–287.

[60] Sagias NC, Zogas DA, Karagiannidis GK, Tombras GS. Channel capacity and second-order statistics in Weibull fading. IEEE Communications Letters. 2004;8(6):377–379.

[61] Cotton SL, Scanlon WG. Higher order statistics for lognormal small-scale fading in mobile radio channels. IEEE Antennas and Wireless Propagation Letters. 2007;6:540–543.

[62]   Xiao C, Beaulieu NC. Node switching rates of opportunistic relaying and switch-and-examine relaying in Rician and Nakagami-m fading. IEEE Transactions on Communications. 2012 Feb;60(2):488–498.

[63]   Wang X, Beaulieu NC. Switching rates of two-branch selection diversity in $\kappa - \mu$ and $\alpha - \mu$ distributed fadings. IEEE Transactions on Wireless Communications. 2009 Apr;8(4):1667–1671.

[64]   Liang T, Smith DB. Energy-efficient, reliable wireless body area networks: cooperative diversity switched combining with transmit power control. Electronics Letters. 2014;50(22):1641–1643.

[65]   Smith DB, Lamahewa T, Hanlen LW, Miniutti D. Simple prediction-based power control for the on-body area communications channel. In: 2011 IEEE International Conference on Communications (ICC), 2011. pp. 1–5.

Chapter 3

Ultra wideband radio channel characterisation for body-centric wireless communication

Qammer H. Abbasi*, Richa Bharadwaj**, Khalid Qaraqe[†], Erchin Serpedin[††] and Akram Alomainy[‡]

This chapter discusses the various experimental investigations undertaken to thoroughly understand the Ultra wideband (UWB) on/off-body radio propagation channels. These characterisation measurement campaigns were performed in both the anechoic chamber and a typical indoor environment (cluttered laboratory). Effect of human body movements on the channel parameters is evaluated. Apart from measurements in an anechoic chamber and in an indoor environment, when body parts were moving, measurements were also taken on a treadmill machine in order to mimic the scenario of UWB body-centric system applied in performance monitoring for sport and exercise medicine. Radio channel parameters are extracted from the measurement data and statistically analysed to provide a preliminary radio propagation model with the inclusion of pseudo-dynamic body movements.

3.1 Analysis methodology applied for body-centric radio channel modelling

To investigate and analyse the performance of single and multiple antennas for body-centric wireless communication channels, various approaches can be adopted. It can either be predicted through detailed simulations using numerical digital phantom, by real-time measurements or by using a statistical channel model, which completely characterises the channels and the environment. The simulation approach is computationally intensive and becomes even more complex, when UWB technology is considered, because of frequency-dependence characteristic of human body and larger bandwidth of UWB. It seems almost unrealistic if random body movement is introduced in the simulations, and seems much less valuable for static postures with no

*Department of Electrical and Computer Engineering, Texas A & M University at Qatar, Qatar and School of Electronic Engineering and Computer Science, Queen Mary University of London, UK
**Department of Electronics and Communication Engineering Thapar University, Patiala, Punjab, India
[†]Department of Electrical and Computer Engineering, Texas A & M University at Qatar, Qatar
[††]Texas A & M University, college station, USA
[‡]School of Electronic Engineering and Computer Science, Queen Mary University of London, UK

movements at all, as system design is typically based on statistical channel models. The statistical channel approach cannot be adopted due to the absence of a standard statistical channel model for on/off-body radio channels. Hence, on/off-body radio propagation channel measurements seems useful in real environments with naturalistic movements of a real human body to quantify the significance of single and multiple antenna systems for body-worn devices. To do so in this work, antennas were placed on a human subject to perform single and multiple antenna channel characterisation and system modelling, while taking safety limits in consideration.

Different propagation measurement set-ups have been presented in the open literature to characterise the radio channel. The measurement techniques can be generally characterised as time domain measurements and frequency domain measurements [1–5]. These two techniques are the basis for many other radio propagation sounders commonly used in characterising wideband radio channels for indoor, outdoor and Body area network (BAN) scenarios [6–8].

In time domain measurements, a digital oscilloscope is used to receive the signal, and is relatively easy to detect multipath components, if their delay with respect to the direct component is greater than the UWB pulse duration [9]. Channel measurements can also be performed using a vector network analyser (VNA) in the frequency domain. In this case, antennas are connected to the ports of the analyser and a sweep of discrete frequency tones is performed and S_{21} is measured from one antenna to the other which represents the channel frequency response ($H(\omega)$). Analysing both, magnitude and phase of S_{21}, enables a transition to the time domain by simply applying an inverse discrete Fourier transform (IDFT). However, one of the problem with frequency domain method is the restriction applied on measurement area freedom since both transmit and receive antennas are connected to the same VNA; nevertheless, such problem can be avoided by the use of ultra low-loss long cables and applying advance calibration techniques [10]. This problem can also be overcome by applications of VNA measurement set-up using radio frequency on fibre-optic connection in scenarios, where electrically small antennas are used [6].

3.2 UWB antennas for body-centric radio propagation measurements

For UWB-Body centric wireless networks (BCWN), the antenna design becomes more complicated due to the presence of the human body. In this work, a miniaturised coplanar waveguide (CPW) fed tapered slot antenna (TSA) for ultra wideband applications [11] is used for all measurements. While designing UWB antenna, a broadband impedance matching network is needed and it can be achieved by employing two tapered radiating slots at the end of the CPW feeding line [12] or by gradually varying the feed gap [13] or with the help of a pair of tapered radiating slots [14]. The TSA used in this chapter is using a similar approach presented in Reference 12. However, the difference is that here the waveguide and radiating slot are inseparable. The antenna is fabricated on RT/Duroid board (with thickness $h = 1.524$ mm, relative permittivity $\varepsilon_r = 3$ and loss tangent $\tan(\delta) = 0.0013$). The total antenna size is 27 mm $\times$ 16 mm which is around $0.27\lambda_0 \times 0.16\lambda_0$ in electrical length, where λ_0 is

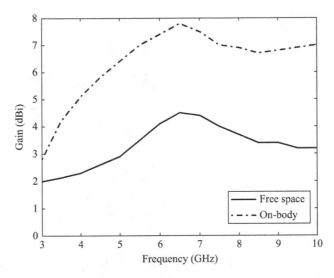

Figure 3.1 Free space and on-body gain as a function of the frequency

the free space wavelength at 3 GHz. Unlike the traditional CPW-fed antenna the TSA is designed to allow for the smooth transition of line impedance. The length of the semi-major and semi-minor axes of the bigger ellipse is 18 and 6.6 mm and smaller ellipse is 12 and 6.1 mm, respectively. The gap between the patch and the ground plane is 0.28 mm. The ratio of semi-major axis to semi-minor axis within the design is the most significant parameter to affect the impedance matching [12]. The return loss of the TSA is below −10 dB in the 3–11.5 GHz band. The antenna preserves good impedance bandwidth even when placed on the human body with slight detuning in lower frequency band due to changes in the effective permittivity and hence the electric length of the antenna. The free space radiation patterns are expected to be omnidirectional and monopole-like performance. The on-body radiation characteristics of the TSA are comparable to the vertical over ground antenna (presented in Reference 9).

Figure 3.1 shows the realised gain of the TSA as a function of the frequency. In free space, the gain ranges from 2.0 to 4.2 dBi, while for on-body case, it varies from 2.8 to 7.5 dBi for a frequency range of 3–10 GHz.

3.2.1 Pulse fidelity

To investigate both frequency domain and transient antenna characteristics, the free space channel between two identical antennas are set side-by-side (most appropriate setting for wireless body area networks (WBAN) applications of the proposed printed antenna) and face-to-face is measured at different angular orientations, namely 0°, 45° and 90°, when the distance between the antennas is 0.5 m as shown in Figure 3.2.

The antennas are connected to the two ports of VNA (Hewlett Packard 8720ES-VNA) to measure the transmission response (S_{21}) using each cable of length 5 m. The measured insertion loss of two long 5 m cables is around 1.5 dB and is obtained by averaging over the whole UWB band, and the changes on the phase are almost

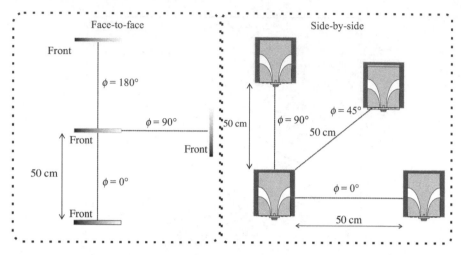

Figure 3.2   Antenna transfer functions measurement set-up in free space with distance of 50 cm between the antennas with different orientations

negligible. Measurements have been done in the anechoic chamber to eliminate multipath reflections from the surrounding scatterers. In this study, the chosen distance between the antennas is 0.5 m which is five times the wavelength at the lower frequency in the 3 GHz band. Thus, the interaction between the antennas is minimal and most of the distortion in frequency channel responses is due to impedance mismatch and inherent radiation properties of the antenna. The time domain responses of the free space radio channel with the antennas are obtained by direct application of inverse fast Fourier transform (IFFT) on the measured real frequency responses S_{21} [15].

Figure 3.3(a) shows the impulse responses in the 3–10 GHz band when the antennas are set side-by-side. The fidelity values for the measured 3–10 GHz band are 96.04% and 72.43% for 45° and 90°, respectively (the pulses used here are from the VNA, and not the Gaussian pulses). The fidelity of the impulse response at different directions (with reference to the response at 0° when the antennas are facing each other) is 91.06% and 95.35% for 90° and 180°, respectively. Fidelity of 77.1% is obtained when comparing side-by-side and face-to-face scenarios (Figure 3.3(c)).

Fidelity studies of commonly used antennas such as monopoles and resistively loaded dipoles presented in References 16–20 have shown that spatially averaged fidelity factor as low as 70% is often deduced. In WBAN specific study presented in Reference 16, a value of 76%–99% is derived for fidelity when numerically comparing input pulse to a transmitted pulses for various antenna types. This indicates that the acceptable minimum value for fidelity is application and environment specific. In the study presented here, the average values obtained for TSA (around 86%) are considered sufficient for the indoor body-centric wireless communication application when considering 99% of energy windowed pulse [16, 18]. In order to determine a specific threshold fidelity factor for generic application (or even for body-centric networks), further evaluations and more in depth system-level analysis are required.

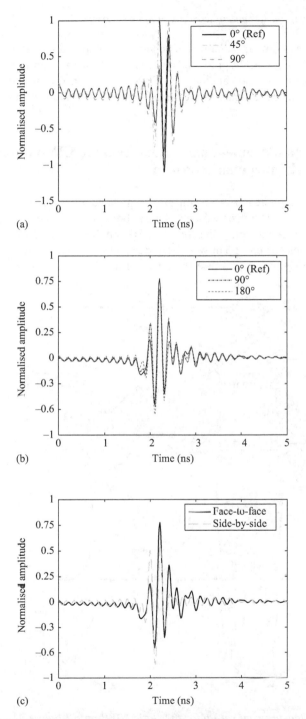

Figure 3.3   Normalised impulse responses of the measured channel of two tapered
slot antennas in the 3–10 GHz band at different angular orientations.
(a) Side-by-side, (b) face-to-face and (c) face-to-side

The TSA presents similar radiation characteristics compared with the planar inverted cone antenna (PICA) (PICA presented in Reference 9), with a significative size reduction that makes it an ideal candidate for BANs. However, for on-body applications, the behaviour of the antenna as a part of the on-body radio channel needs to be investigated.

3.3   Antenna placement and orientation for UWB on-body radio channel characterisation

All measurements were performed on a male candidate of age 24 years, weight 62 kg and height 1.79 m. The distance between the body and the mounted antennas was kept to about 7–10 mm including the distance variation caused by loose clothing. The coaxial cables each of 5 m were used during the measurement and were firmly strapped to the body to minimise the effect of moving cables over the duration of

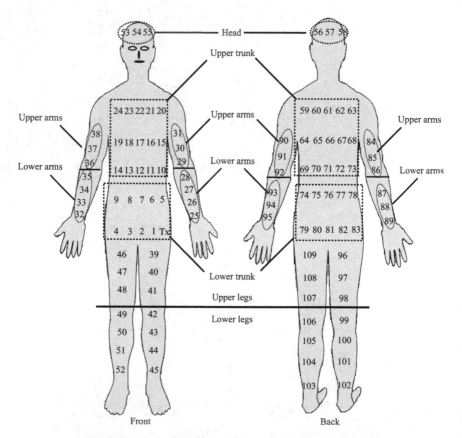

Figure 3.4   Measurement set-up for the static UWB on/off-body radio channel characterisation showing transmit and receive antenna locations as applied in the measurement campaign. ©2011 IEEE, reprinted with permission from Reference 21

the channel measurement. Two sets of measurements have been performed; when the subject was stationary and when the subject was in pseudo-dynamic motion. The antenna used in the measurements is vertically polarised and the radiating element is placed parallel to the human body. Both transmitter and receiver have the same polarisation and orientation, when placed on the human body. For the first set of on-body measurements (when the subject was stationary), the transmitting antenna (Tx) was placed at the waist (belt) position on the left side of the body, about 200 mm away from the body centre line and the receiving antennas (Rx) were placed at 109 different locations as shown in Figure 3.4. For the second set of measurements, the effect of pseudo-dynamic body movements on the UWB on-body radio channels was considered. Two measurement scenarios for Tx were considered for pseudo-dynamic on-body radio channel characterisation:

- when Tx was static with respect to Rx (Tx on waist);
- when Tx was moving in pseudo-dynamic manner with respect to Rx (Tx on wrist).

The receiving antenna was placed at five different positions: on the right chest; right wrist; right ankle; on the centre of the back and on the right side of the head, thus forming five on-body channels as shown in Figure 3.5 (named belt-chest, belt-back, belt-head, belt-wrist and belt-ankle, respectively). The belt-chest channel represents the line-of-sight (LOS) scenario. The belt-back channel is a good representation of the non-line-of-sight (NLOS) scenario. Both of these channels (i.e., belt-chest and belt-back) are static channels in which the distance between the transmitting and

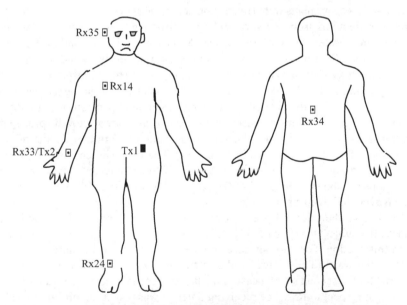

Figure 3.5 Measurement set-up for the pseudo-dynamic UWB on-body radio channel characterisation showing transmit and receive antenna locations as applied in the measurement campaign. ©2010 IEEE, reprinted with permission from Reference 22

receiving antennas is almost constant. To mimic a dynamic channel, in which the path length varies randomly with movement of the body, the belt-wrist channel was selected. Most often, there are scenarios where there is partial LOS or a transition of LOS and NLOS. The belt-head and belt-ankle channels are good examples of this.

An Agilent two-port VNA (Hewlett Packard 8720ES-VNA) was used for UWB on/off-body radio propagation channel measurements. The two antennas (Tx and Rx) were connected to the VNA by a pair of low-loss coaxial cables. For all UWB on/off-body radio propagation channel measurements, a two-port VNA was used to measure the transmission response (S_{21}). A Labview programme written by the author is used to control the VNA remotely. The data measured by the VNA were stored in a computer hard disk by the software in the form of a text file containing the magnitude (in decibel) and the phase (in degrees) of the transmission response (S_{21}). During the measurements, the VNA was always calibrated to exclude the losses that incurred in the cables and thus the measured data reflect the signal measured at the ports of the antenna. The calibration also ensured that a total power of 0 dBm is transmitted by the transmitting antenna. Measurements were performed in the frequency range of 3–10 GHz at a sampling rate of 1601 separate frequency points and sweep time of 800 ms.

3.4   Measurement procedure for UWB on-body radio channel characterisation

In this work, measurements were first performed in the anechoic chamber to eliminate multipath reflections from the surrounding environment, and then repeated in the Body-Centric Wireless Sensor Lab at Queen Mary University of London (Figure 3.6) to consider the effect of the indoor environment on the on-body radio propagation channel. The Body-Centric Wireless Sensor Lab provides a mock hospital room and all necessary equipment to simulate real-life scenarios when investigating wireless sensor networks. The lab has the capability to enable research expansion into implantable devices measurement, compact sensor manufacturing and extensive radio propagation characterisation and modelling. The sensor lab height is 3 m and details of walls, windows and furniture are shown in Figure 3.6. The three-dimensional view of the lab is shown in Figure 3.7. For static on-body radio channel characterisation, Tx and Rx locations are shown in Figure 3.4. For each Rx location, ten sweeps were collected and then averaged to ensure acceptably stable channels.

To observe the effect of pseudo-dynamic movements on UWB on-body radio channels, the subject was in carefully designed motion scenarios, which is explained later in this section. For on-body measurements, the transmitting antenna was first placed on the waist (Tx is considered as static with respect to Rx) and then on the right wrist for the second set of measurements (Tx is considered as in pseudo-dynamic motion with respect to Rx). The receiving antenna was placed on different locations on the body as shown in Figure 3.5.

Due to the short communication distance, the electromagnetic energy propagates from the transmitter to the receiver in a few nanoseconds and hence the channel can

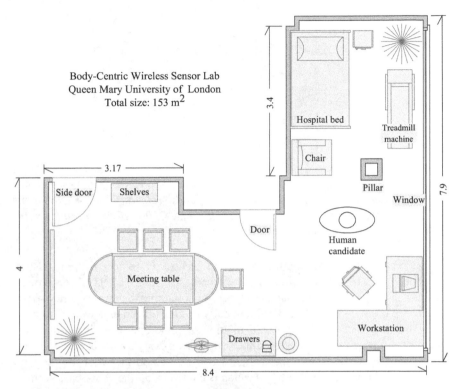

Body-Centric Wireless Sensor Lab
Queen Mary University of London
Total size: 153 m^2

Figure 3.6 Dimensions and geometry of the Body-Centric Wireless Sensor Lab (housed within the Department of Electronic Engineering, Queen Mary University of London, UK). The sensor lab height is 3 m

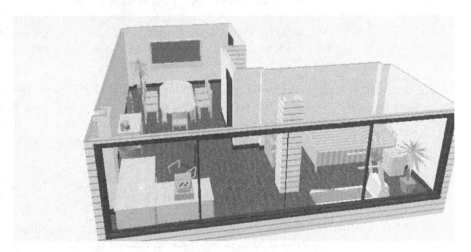

Figure 3.7 Three-dimensional view of Body-Centric Wireless Sensor Lab shown in Figure 3.6

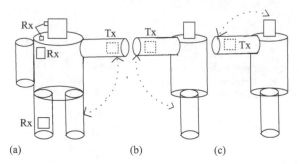

Figure 3.8   (a) Tx movements to the side of body; (b) side view: Tx movements to the front of body and back to side of body; (c) side view: Tx movements from front to right side and back to front. ©2009 IEEE, reprinted with permission from Reference 23

be considered stationary during this time, and it is therefore acceptable to assume "snapshots" of propagation channel measurements to capture effect of movements. This was achieved by ensuring that the subject is maintaining the same position for the entire sweep duration of 800 ms. Different daily routine movements were performed such as bending, leaning forward and rotation of torso and arms including some random movements (when Tx was on the waist for on-body case). In addition, four specific body movements were performed when Tx was on the wrist to gain more insight into the arm movement effect (which is considerably the worst-case scenario for on-body communications as highlighted by previous studies) on the radio channel:

- Arm along the body, moving to the side to form 90° with the body trunk and returning back to the initial position (as shown in Figure 3.8(a)).
- Arm along the body, moving forward to the front so that the arm forms 90° with the body trunk and returning back to the initial position (as shown in Figure 3.8(b)).
- Arm is placed straight in front of the body, moving from the left to right in the front of the body and returning back to the initial position (as shown in Figure 3.8(c)).
- Random arm movements.

For each location and measurement scenarios, more than 100 sweeps were captured to ensure sufficient data points for acceptable statistical analysis. For each measured scenario, four different body movements were performed. Measurements were also made on a treadmill machine as well to incorporate paced walking steps at a certain speed of 1.1 km/h.

3.5   UWB on-body propagation channel analysis

In order to design an efficient radio system for body-centric wireless communications, it is important to provide reliable models of propagation channel. The case is even more complex when it comes to the on/off-body channel characterisation due to the unpredictable and dynamic nature of such a radio channel.

3.5.1 On-body radio channel characterisation for static subjects

The path loss (PL), which is given by the ratio between transmitted and received power, is directly calculated from the measured data by averaging the measured frequency transfers at each frequency point [24]. PL can be represented as a function of distance between Tx and Rx using the following relation [25]:

$$PL_{dB}(d) = PL_{dB}(d_0) + 10\gamma \log\left(\frac{d}{d_0}\right) + X_\sigma \qquad (3.1)$$

where d is the distance between Tx and Rx, d_0 is the reference distance and $PL(d_0)$ is the PL at reference distance. For on-body measurements, $d_0 = d_1 = 10$ cm. The exponent γ is known as PL exponent. It is useful to understand how fast the received power decays with the distance. From Friis formula, it is well known that PL exponent is equal to two for free space propagation; however, for on/off-body communications, exponent is higher due to the factors including losses in tissues, creeping waves and surface wave propagation, and reflections from different parts of human body. In this study, a least-square fit is performed on the PL data and the slope of the curve gives the γ. Where X_σ is a zero-mean Gaussian distributed random variable with standard deviation σ, both values in decibel.

Figure 3.9 shows the variation of PL with the logarithmic distance in an anechoic chamber and in an indoor environment. The slope of the fitted curve is equal to the PL exponent (γ), which is 2.96 and 2.48 for chamber and indoor environment, respectively. When measurements are performed in the indoor environment, the reflections from the surrounding scatterers increase the received power, causing reduction of the PL exponent. A reduction of 13% is observed in this study. The values of γ agree with

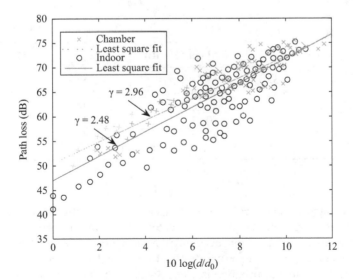

Figure 3.9 *On-body path loss model for the measurements in an anechoic chamber and in an indoor environment for static subject*

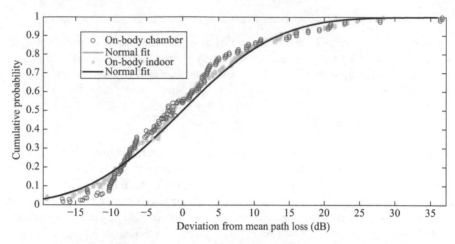

Figure 3.10   Deviation of on-body measurements from the average PL fitted to a normal distribution in an anechoic chamber and in an indoor environment, respectively (for the all, 109 Rx locations)

Table 3.1   On-body path loss exponent (calculated in similar manner using empirical model as shown in Figure 3.9) and mean path loss for different sectors of body. ©2011 IEEE, reprinted with permission from Reference 26

Body parts	On-body			
	LOS		NLOS	
	γ	Mean PL (dB)	γ	Mean PL (dB)
Trunk	3.39	59.41	2.64	62.56
Arms	4.31	59.10	3.34	67.98
Legs	1.16	59.76	2.35	69.78
Head	1.17	70.92	2.57	69.83
All parts	2.8	60.05	1.34	66.04
LOS + NLOS	$\gamma = 2.48$ and mean PL $= 63.02$ dB			

the ones presented in Reference 9, where it was found $\gamma = 3$ in free space, and $\gamma = 2.6$ in the office environment. The shadowing factor is a zero mean, normally distributed statistical variable and it takes into account of the deviation of the measurements from the calculated average PL (see Figure 3.10). In the anechoic chamber, the standard deviation of the normal distribution is $\sigma = 8.34$ and 5.88 for an indoor environment.

Table 3.1 shows the PL exponent and mean PL for different sectors of body for both LOS (front side of body) and NLOS (back side of body) scenarios. Lowest γ

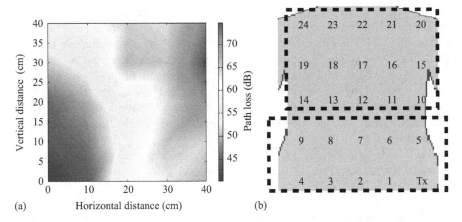

Figure 3.11   Radiograph for on-body PL in an indoor environment for the front side of trunk with Tx at origin of coordinate plane, the Tx and Rx locations on the body are shown in panel (b)

is obtained for legs case for front side of body (i.e., LOS scenario). This is because reflections from the ground increase the received power, which results in reduction of γ for legs case. Same results are obtained for legs case for back side of body (i.e., NLOS scenario). The low value of PL exponent for head case as compared to trunk case is due to less variations in mean PL among different receiver locations for head case as shown in Table 3.1. Variations in γ among different parts of human body for NLOS are very small as compared to LOS because for NLOS case all propagation takes place through multipath components; hence, the received signal strength is approximately close to each other for different sectors of body. Figure 3.11 shows radiograph for on-body PL for the front side of trunk (for Rx1–Rx9 [lower trunk] and Rx16–Rx24 [upper trunk] as shown in Figure 3.4) with Tx at origin (i.e., when both horizontal and vertical distances are equal to zero). It shows the distribution of PL, with respect to the horizontal and vertical distances, which increases with increasing both distances.

3.5.2   Transient characterisation of UWB on-body radio channel

Time-delay analysis provides information about the amount of signal spreading caused by channel and it is well described by mean excess delay (τ_m) and root mean square (τ_{RMS}) delay spread, which are calculated from the first and second central moment of the derived power delay profile (PDP), respectively [10]. Since the time of arrival (including multipath components) of the signal restricts transmitted data rates and also limits system capacity [10], it is commonly used to characterise the transient behaviour and hence the system capacity limits for radio propagation.

PDP can be easily obtained by averaging the obtained channel impulse responses, which are calculated from the measured frequency transfer functions, applying windowing and IFFT. The time domain results for both measurement sets are compared for evaluation. The time domain window can detect multipath signals separated up to 228 ns, with a resolution of 50 ps. PDP is produced simply by averaging all impulse responses by considering samples with the signal level higher than a selected threshold and observing their delay with respect to the peak sample (the direct pulse). In this study, three different threshold levels are considered: 20, 25 and 30 dB below the peak power.

The Akaike information criterion (AIC) is a method widely used to evaluate the goodness of a statistical fit [27]. The second-order AIC (AIC_c) is defined as:

$$AIC_c = -2\log_e(L) + 2K + \frac{2K(K+1)}{n-K-1} \tag{3.2}$$

where L is the maximised likelihood, K is the number of parameters estimated for that distribution and n is the number of samples of the experiment. The criterion is applied to evaluate the goodness of five different distributions commonly used in wireless communications that seem to provide the best fitting for under study measurements (Rayleigh, normal, log-normal, Weibull and Nakagami). All these distributions have two parameters ($K = 2$), except for the Rayleigh ($K = 1$). Smaller value of AIC_c means better statistical model fit, and the criterion is used to classify the models from the best to the worst. To facilitate this process, the relative AIC_c is considered and results are normalised to the lowest value obtained:

$$\Delta_i = AIC_{c,i} - \min(AIC_c) \tag{3.3}$$

A zero value indicates the best fitness. In this analysis, the effect of the receiver sensitivity (the threshold applied to calculate the PDP) on the statistical model is considered. Result shows that the best case ($\Delta_i = 0$) is found in the anechoic chamber adopting a less sensitive receiver (threshold -20 dB) for both RMS and mean excess delay. Higher value of Δ_i is obtained as compared to chamber making statistical model less accurate. This is due to the fact of considering more scattered components in case of indoor environment.

Figures 3.12 and 3.13 show the cumulative distribution of RMS and mean excess delay fitted to log-normal distribution, respectively, based on Akakie criterion. Tables 3.2 and 3.3 show an average value and standard deviation (respectively, μ and σ) of the log-normal distribution for each case for different threshold levels, respectively. When measurements are performed in an indoor environment, the multipath effect produces higher mean and standard deviation for both RMS and mean spread delay. Furthermore, using a more sensitive receiver, a higher number of secondary components are considered, and the average value of the spread delay and standard deviation is higher.

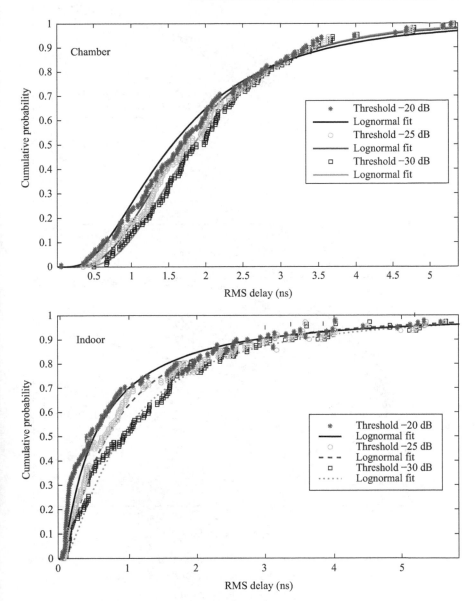

Figure 3.12   On-body RMS delay distribution fitting for the measurements in an anechoic chamber and in an indoor environment

3.5.3   Pulse fidelity

Figure 3.14 shows the fidelity value obtained for more than 100 receiver locations considering both the front and back sides of body in different environments as shown in Figure 3.4. Results show that the pulse shape is better preserved in case of chamber.

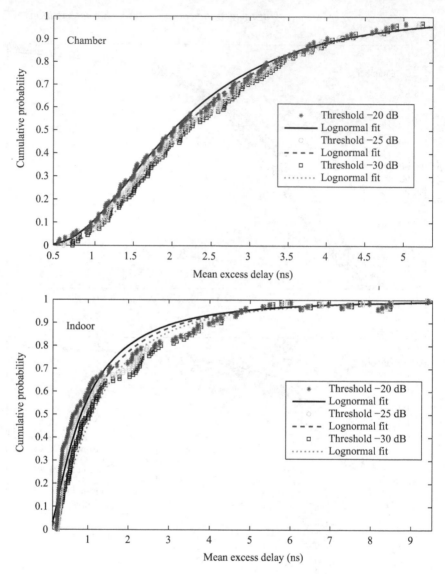

Figure 3.13 On-body mean excess delay distribution fitting for the measurements in an anechoic chamber and in an indoor environment

Table 3.2 Average value and standard deviation of log-normal distribution applied to RMS delay for on-body communications with respect to different threshold levels in the chamber and indoor environment

Threshold	Chamber		Indoor	
	μ	σ	μ	σ
−20 dB	1.22	1.47	1.94	2.92
−25 dB	1.31	1.23	1.99	21.9
−30 dB	1.56	1.12	2.08	2.27

Table 3.3   *Average value and standard deviation of log-normal distribution applied to mean excess delay for on-body communications with respect to different threshold levels in the chamber and indoor environment*

Threshold	Chamber		Indoor	
	μ	σ	μ	σ
−20 dB	1.43	1.46	2.35	2.03
−25 dB	1.55	1.42	2.43	1.98
−30 dB	1.69	1.37	2.48	1.96

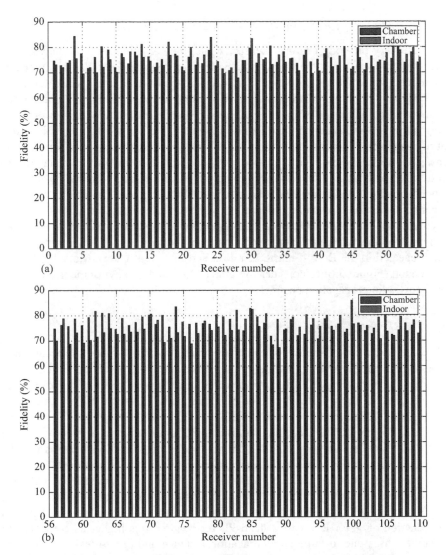

Figure 3.14   *Calculated pulse fidelity for different antenna locations in the anechoic chamber and in the indoor environment for on-body channels. (a) Front side of body and (b) back side of body*

The mean fidelity is 76.45% in the anechoic chamber (77.16% for front side and 75.75% for back side of body, respectively), as compared to 74.69% (74.79% for front side and 74.59% for back side of body, respectively) in an indoor environment. As for both the front and back sides of body, i.e., LOS and NLOS scenarios, the average fidelity is above 70%.

3.6   UWB off-body radio propagation channel characterisation

3.6.1   Antenna placement and measurement procedure

For stationary off-body channel measurements, the communication is between the access point away from the body and devices worn on the body. Same antenna as mentioned above for on-body case is used for off-body case. The distance between the body and the mounted antennas was again kept to about 7–10 mm including the distance variation caused by loose clothing. The coaxial cables each of 5 m were used during the measurement and were firmly strapped to the body to minimise the effect of moving cables over the duration of the channel measurement as done for on-body case. The radiating element is placed parallel to the human body. Similar to on-body case, both transmitter and receiver have the same polarisation and orientation, when placed on the human body. For static off-body channel characterisation, the receiving antennas (Rx) were placed at the same 109 different locations as shown in Figure 3.4 as for the on-body case. Measurements were first performed in the anechoic chamber to eliminate multipath reflections from the surrounding environment, and then repeated in the Body-Centric Wireless Sensor Lab at Queen Mary University of London (Figure 3.6) to consider the effect of the indoor environment on the on/off-body radio propagation channel.

3.6.2   PL characterisation

PL calculations for off-body communications are similar to those in the on-body case as presented above but for off-body measurements, the reference distance is $d_0 = d_1 = 100$ cm. Figure 3.15 shows the variation of PL with the logarithmic distance in an indoor environment for back side of human body. The slope of the fitted curve is equal to PL exponent (γ), which is 1.09.

Table 3.4 shows the PL exponent and mean PL for different parts of body for both LOS (the front side of body) and NLOS (the back side of body) scenarios for off-body communications at the distance of 100 cm from Tx (which is mounted on wall at waist height from the ground) in an indoor environment. Lowest PL is obtained for legs case, for both LOS and NLOS cases (similar to on-body communication case), because reflections from the ground increase the received power, which results in reduction of γ for legs case. Mean PL for head case is higher as compared to the other cases due to large communication distance and due to less variations in PL among different receiver locations. The overall γ is 3.79, this higher value is

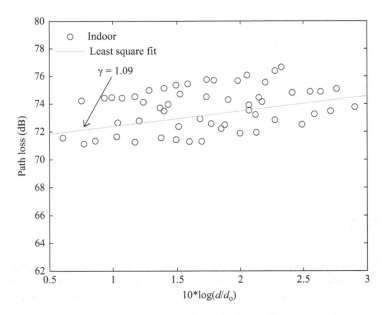

Figure 3.15   Off-body PL for the measurements in an indoor environment for back side of body (i.e., NLOS scenario)

Table 3.4   Off-body PL exponent (calculated in similar manner using empirical model as shown in Figure 3.15) and mean PL for different sectors of body (at distance of 100 cm). ©2012 IEEE, reprinted with permission from Reference 29

Body parts	Off-body (100 cm)			
	LOS		NLOS	
	γ	Mean PL (dB)	γ	Mean PL (dB)
Trunk	1.89	58.63	2.18	73.37
Arms	1.17	54.14	0.99	70.96
Legs	0.26	55.09	0.97	73.50
Head	1.08	60.53	1.64	77.88
All parts	1.07	56.76	1.09	73.18
LOS + NLOS	$\gamma = 3.79$ and mean PL $= 64.97$ dB			

due to large variation between mean PL for LOS and NLOS scenarios, as shown in Table 3.4.

Like for on-body communications, γ for off-body communication in an anechoic chamber is higher ($\gamma = 3.98$) than in an indoor environment as the reflections from the surrounding scatterers increase the received power, causing reduction of the PL

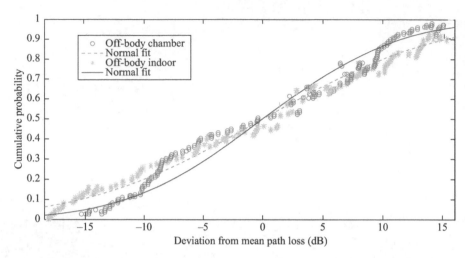

Figure 3.16   Deviation of off-body measurements from the average PL fitted to a normal distribution in an anechoic chamber and in an indoor environment, respectively (when Tx was at a distance of 1 m from the subject carrying 109 Rx locations)

exponent in an indoor environment. A reduction of 4.77% is observed for PL exponent in this study. As mentioned in Section 3.5.1, the shadowing factor is a zero mean, normally distributed statistical variable and it takes into account the deviation of the measurements from the calculated average PL (see Figure 3.16). In the anechoic chamber, the standard deviation of the normal distribution is $\sigma = 11.87$ and 8.99 for an indoor environment.

Figure 3.17 shows variation in the mean PL for both the front side (LOS scenario) and back side (NLOS scenario) of body with respect to different distances (from 10 to 100 cm with step of 10 cm) between on-body receivers and off-body transmitters (which is mounted on wall). Mean PL is calculated by averaging the PL of all receivers for the front and back sides of the body individually. Figure 3.17 shows that mean PL for the front side of body (i.e., LOS scenario) increases almost linearly with increasing the separation distance, which shows that PL is directly proportional to distance. For the back side of the body (i.e., NLOS scenario), the mean PL (as shown in Figure 3.17) is not linear, as in this case, the attenuation through the body is quite high and the main propagation paths are creeping (surface) waves and multipath components with the latter being the dominant contributor due to the dense indoor environment presence. Based on work by Alomainy *et al.* in Reference 28, the contribution due to creeping waves is very small in comparison to the multipath components.

Figure 3.18 shows sliced radiograph for front side of trunk (for Rx1–Rx9 [lower trunk] and Rx16–Rx24 [upper trunk] for front side of body as shown in Figure 3.4) with Tx at origin (when both horizontal and vertical distances are equal to zero) and off-body distance is varied from 10 to 50 cm with step of 10 cm. It shows the

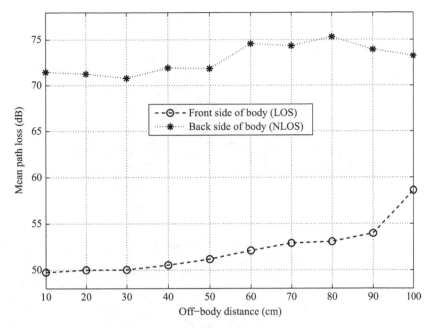

Figure 3.17   Variations of mean PL with respect to change of spacing between off-body Tx (on wall) and on-body receivers on the trunk for the front and back sides of body (as shown in Figure 3.4)

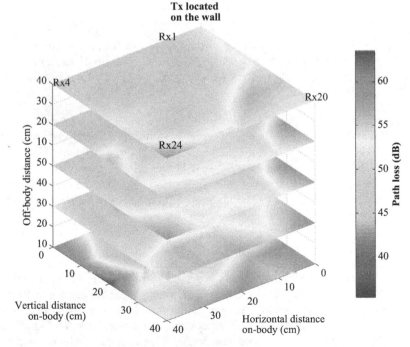

Figure 3.18   Sliced radiograph for off-body PL on the front side of trunk (off-body distance is varied from 10 to 50 cm) with Tx on wall (for Rx1–Rx9 [lower trunk] and Rx16–Rx24 [upper trunk] as shown in Figure 3.4)

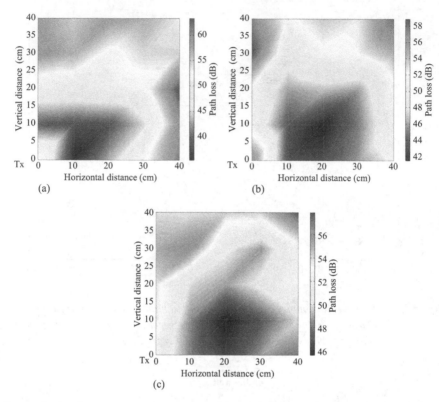

Figure 3.19 Radiographs for different off-body distances (i.e., (a) 10 cm,
(b) 30 cm and (c) 50 cm taken from sliced radiograph as shown in
Figure 3.18). ©2012 IEEE, reprinted with permission from
Reference 29

distribution of PL and its variation with respect to varying off-body distance. Three slices are taken randomly from this sliced radiograph at three different off-body distances (i.e., 10, 30 and 50 cm) and are shown in Figure 3.19. Figure 3.19 clearly shows that as human body is moving away from the Tx (i.e., on wall), the region for lower concentration of PL goes on increasing, possibly because of reduction of near-field effects of antenna as off-body distance increases.

3.6.3 Transient characterisation

The time domain dispersion of the received signal strongly affects the capacity of UWB systems [1]. This effect is characterised by mean excess delay and RMS of the PDP as explained in Section 3.5.2. On the basis of Akaike criterion like for on-body communications, different distributions are being tested for RMS and mean excess delay for different threshold levels (20, 25 and 30 dB). Figures 3.20 and 3.21 show the cumulative distribution of the RMS and mean excess delay fitted

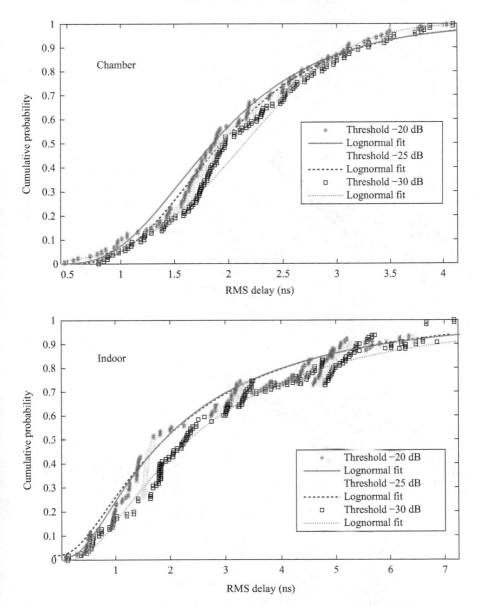

Figure 3.20   Off-body RMS delay distribution fitting for the measurements in an anechoic chamber and in an indoor environment, respectively. ©2012 IEEE, reprinted with permission from Reference 29

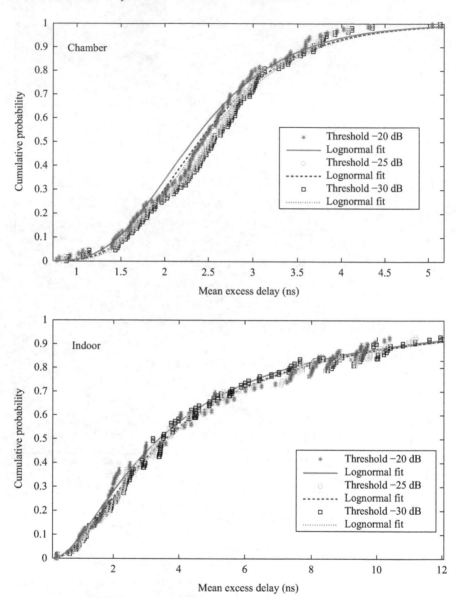

Figure 3.21   Off-body mean excess delay distribution fitting for the measurements in an anechoic chamber and in an indoor environment, respectively. ©2012 IEEE, reprinted with permission from Reference 29

Table 3.5 *Average value and standard deviation of log-normal*
distribution applied to RMS delay for off-body
communications with respect to different threshold
levels in the chamber and indoor environment

Threshold	Chamber		Indoor	
	μ	σ	μ	σ
−20 dB	2.03	0.90	2.93	3.42
−25 dB	2.09	0.87	3.13	3.38
−30 dB	2.15	0.83	3.34	3.42

Table 3.6 *Average value and standard deviation of log-normal*
distribution applied to mean excess delay for off-body
communications with respect to different threshold levels
in the chamber and indoor environment

Threshold	Chamber		Indoor	
	μ	σ	μ	σ
−20 dB	2.46	0.89	5.17	5.95
−25 dB	2.53	0.79	5.35	5.91
−30 dB	2.57	0.89	5.39	5.80

to the log-normal distribution, respectively, based on Akakie criterion for the off-body communications. Tables 3.5 and 3.6 show average value and standard deviation (respectively, μ and σ) of the log-normal distribution for each case for different threshold levels. Same conclusions can be drawn as for the on-body case when measurements are performed in the indoor environment, the multipath effect produces higher mean and standard deviations for both RMS and the mean spread delay. Like for the on-body case, using a more sensitive receiver, more multipath components are considered, and the average value of the spread delay and the standard deviation is higher.

3.6.4 Pulse fidelity

Figure 3.22 shows the fidelity value obtained for more than 100 receiver locations considering both the front and back sides of body in different environments for off-body case. Results show that the pulse shape is better preserved in case of the chamber as that for the on-body case. The mean fidelity is 78.01% in the anechoic chamber (78.84% for the front side and 77.23% for the back side of the body, respectively), as compared to 77.26% (77.24% for the front side and 75.10% for the back side of the body, respectively) in an indoor environment. Again, results similar to the on-body

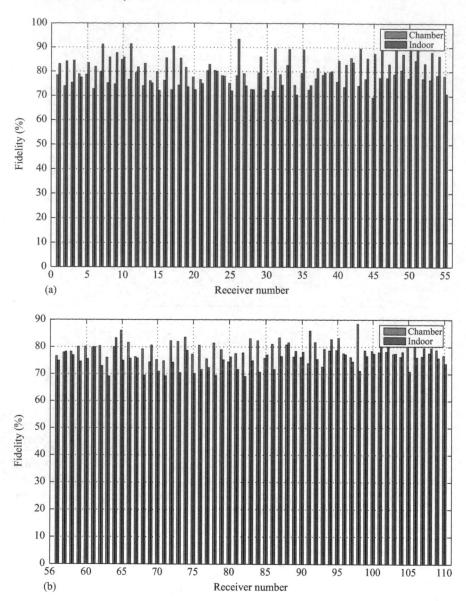

Figure 3.22   Calculated pulse fidelity for different antenna locations in an anechoic chamber and in an indoor environment for off-body channels. (a) Front side of body and (b) back side of body

communications are found and for both the front and back sides of body, i.e., for LOS and NLOS scenarios, the average fidelity is above 70%, and hence acceptable for short-range communications.

3.7 UWB on-body radio channel characterisation for pseudo-dynamic motion

In the analysis presented so far, the radio channel in all measurements was considered to be static. However, the effect on radio channel due to the change in body postures during normal activities is not considered. While performing movements, the transmitter and receiver may alter their position from LOS to NLOS, and also the distance between transmitter and receiver and the relative orientation of the antennas can be modified. Moreover, the antenna input impedance and even its radiation characteristics can be affected by the movements. All these changes on the radio channel introduce a significant fading on the received signal, which, if not accurately considered, could lead to a marginal loss of communication. In this work, a set of measurements is performed while the human candidate is carrying out several pre-defined movements, as explained in Section 3.4.

3.7.1 Channel PL variations as a function of link and movements

The PL, which is given by the ratio between the transmitted and received power, is directly calculated from the measured data, by averaging over the measured frequency transfers at each frequency points [24]. When the receiver is moving with respect to the transmitter, the changes in their relative distance and orientation lead to a variation of the signal strength.

Figure 3.23 shows the PL variation for measured data in an anechoic chamber and in an indoor environment with the receiver placed at different locations. Fluctuations in PL are observed due to a relative change of the distance between transmitter and receiver. The average PL obtained for anechoic chamber is 77 dB, which is higher as compared to the average PL, when the subject was stationary. PL of the on-body channel in an indoor environment is lower than that for the anechoic chamber as predicted due to the contribution of multipath components from obstacles and also from the walls, ceiling and floor.

The cumulative distribution function of the PL variations is compared to well-known distributions and on the basis of Akaike criterion [27], a normal distribution provides the best fit for these measured results (Figure 3.24). Table 3.7 shows the average value (μ) and standard deviation (σ) of the fitted normal distribution that is applied to model PL variation for the on-body channel including movements effect. Table 3.7 demonstrates that the highest value of PL is obtained for the back case when there is a NLOS communication and the main propagation mechanism is multipath reflections within the indoor environment and creeping waves along the body surface for the anechoic chamber scenario. The highest standard deviation is observed for the chest case where there is a large relative distance change between Tx and Rx,

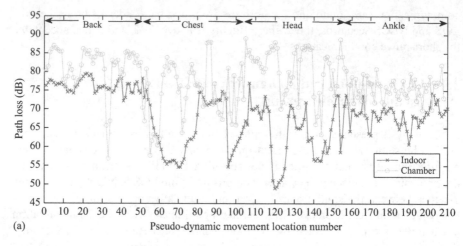

(a)

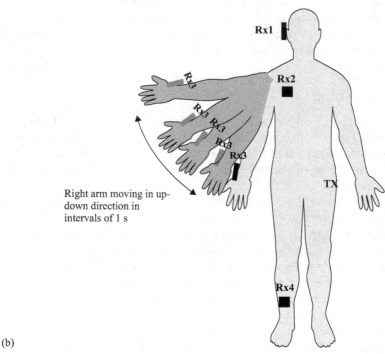

(b)

Figure 3.23   *PL variations (shown in panel (a)), ©2010 IEEE, reprinted with*
permission from Reference 21, when Tx is moving in pseudo-dynamic
motion (as shown in panel (b)) and Rx is placed at different locations,
©2011 IEEE, reprinted with permission from Reference 30

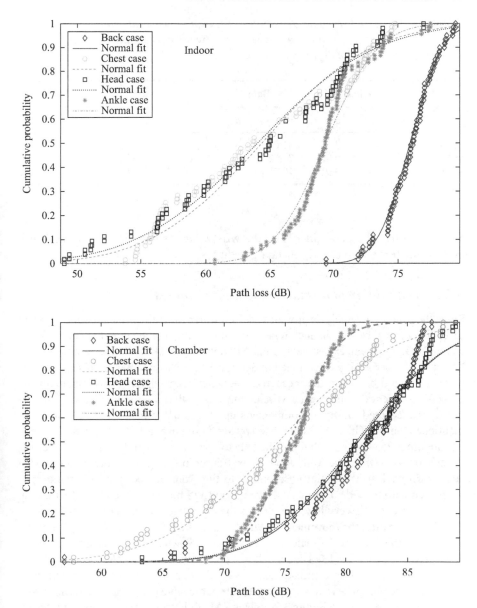

Figure 3.24   Cumulative distribution plot for PL in an indoor environment and chamber, when Tx is moving in pseudo-dynamic motion and Rx is placed at different locations. ©2011 IEEE, reprinted with permission from Reference 30

Table 3.7   Mean and standard deviation of PL using normal distribution for different channels and environments (whereas, running is performed in an indoor environment)

	Chest		Back		Head		Ankle	
	μ	σ	μ	σ	μ	σ	μ	σ
Chamber	74.43	7.42	81.03	6.07	80.85	6.15	75.37	3.06
Indoor	64.32	6.82	75.92	2.02	63.98	6.37	69.36	3.25
Running	68.64	9.91	70.78	8.98	68.32	7.66	72.21	4.72

considering that the transmitter is on the wrist which is moving rapidly and hence path parameters are constantly changing in these cases.

3.7.2   Time-delay and small-scale fading analysis

Since the time-delay analysis provides useful information about the amount of signal spreading caused by the channel, it is important to take into account the delays of the channel, i.e., the mean excess delay and RMS delay spread [10]. These parameters can be obtained from PDP, as mentioned in Section 3.5.2. The channel impulse responses were calculated based on the measured frequency transfer functions which consist of 1601 frequency points using windowing and IDFT. The applied time window can detect received multipath components up to 228 ns with 50 ps of resolution, as mentioned before. The PDP is calculated by averaging over all the measured postures.

Figure 3.25 shows the PDP for the belt-to-wrist link, when the arm moves along the side of body (0° to the front of body) to straight in front (90° to the front of body) and return back to the initial position (0° to the front of body). At 0°, the arm is connected with the side of trunk and there is no LOS between Tx and Rx, that is why the signal strength is very low; the same is the case when the arm comes back to the initial position after the movement. At 90°, the arm is straight in the front of the body; in this case, the distance is greater as compared to 45° where strongest signal strength is observed due to some LOS and small distance between Tx and Rx.

The root mean square spread delay (τ_{RMS}) is a crucial parameter for multipath channels since it imposes a limit to the data rate achievable [1]. Mean excess delay and RMS delay are calculated. Figure 3.26 shows RMS delay with respect to different body movements and Rx locations for both anechoic chamber and indoor measurement scenarios. RMS delay spread in the indoor scenario is higher due to the reflection from the surrounding scatterers, whereas for the chamber case, only reflection from body parts is considered. The goodness of different statistical distributions in fitting the data have been evaluated. For the case of the study, the log-normal distribution provides the highest likelihood among a wide set of distributions using the Akaike criterion. Figure 3.27 presents fitted probability distributions of calculated RMS delay

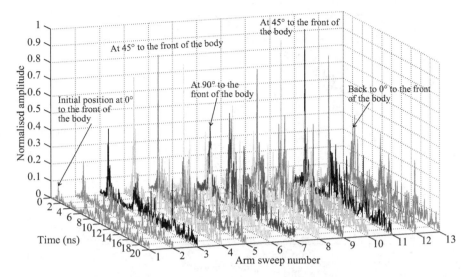

Figure 3.25 *Power delay profile for chest-to-wrist link, when arm moves along the side of body (0° to the front of body) to straight in front (90° to the front of body) and return back to initial position (0° to the front of body). ©2011 IEEE, reprinted with permission from Reference 30*

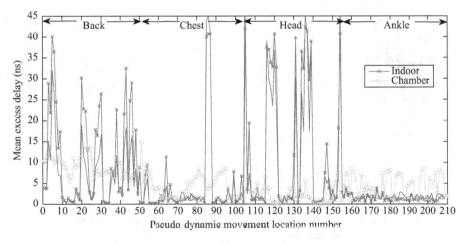

Figure 3.26 *RMS delay when Tx is moving in pseudo-dynamic motion and Rx is placed at different locations. ©2010 IEEE, reprinted with permission from Reference 22*

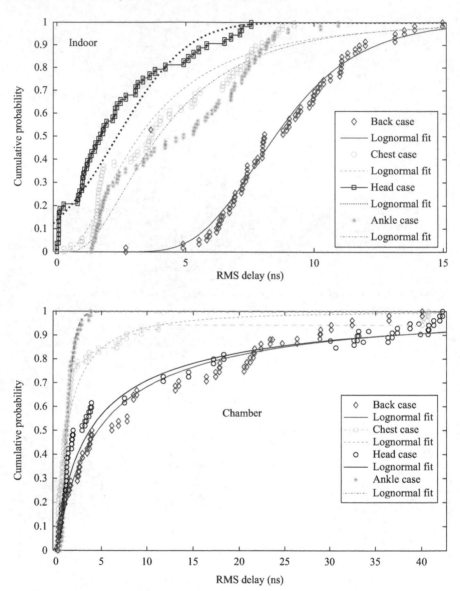

Figure 3.27   Cumulative distribution plot for RMS delay in an indoor and chamber environment, when Tx is moving in pseudo-dynamic motion and Rx is placed at different locations. ©2011 IEEE, reprinted with permission from Reference 30

Table 3.8   *Mean and standard deviation of RMS delay using*
log-normal distribution for different channels and
environments (ns) (whereas, running is performed in
an indoor environment)

	Chest		Back		Head		Ankle	
	μ	σ	μ	σ	μ	σ	μ	σ
Chamber	3.30	2.76	4.73	2.0	1.78	2.68	1.30	1.98
Indoor	4.34	1.24	8.84	2.70	2.36	2.16	4.90	1.14
Running	4.65	3.17	9.68	3.06	5.17	3.05	4.96	2.12

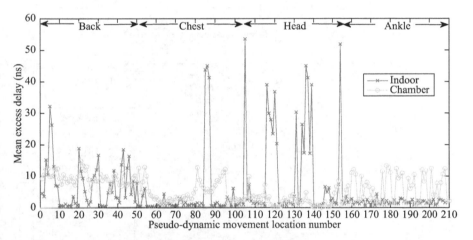

Figure 3.28   *Mean excess delay when Tx is moving in pseudo-dynamic motion and*
Rx is placed at different locations

spread values, and the fitted parameters (μ and σ) for the log-normal distribution are
listed in Table 3.8.

Figure 3.28 shows the mean excess delay with respect to different Rx locations.
Similar interpretation can be obtained for this case as for RMS delay. Again log-
normal distribution provides the best fit for measured data (Figure 3.29) as the case
for RMS delay spread. Table 3.9 lists μ and σ for log-normal distribution used to
model delay.

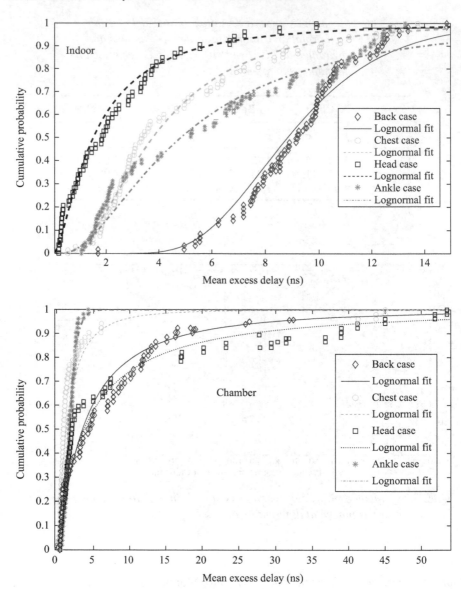

*Figure 3.29   Cumulative distribution plot for mean excess delay in an indoor and
chamber environment, when Tx is moving in pseudo-dynamic motion
and Rx is placed at different locations*

Table 3.9   Mean and standard deviation of mean excess delay using log-normal distribution for different channels and environments (ns) (whereas, running is performed in an indoor environment)

	Chest		Back		Head		Ankle	
	μ	σ	μ	σ	μ	σ	μ	σ
Chamber	2.42	1.24	8.08	1.68	1.79	2.62	1.16	1.11
Indoor	4.44	2.19	9.28	3.03	6.79	2.13	2.64	1.21
Running	6.51	2.84	9.44	4.02	7.72	2.98	5.89	1.58

3.8 Summary

This chapter presents the channel characterisation for on- and off-body communications in different environments (i.e., anechoic chamber and indoor environment) for both static and pseudo-dynamic motion scenarios. For the static subject, PL characterisation is presented for different sectors of the body to get a better insight into the PL variations. Statistical analysis is performed on the measured data. Variety of empirical statistical models has been tested to find the best fitting for the measurement data. On the basis of the Akaike criterion, it is observed that the PL is best modelled by using normal distribution; whereas, the log-normal distribution provides the best fit for the time-delay parameters for both on/off-body communications.

The effect of the body movements on the UWB on-body radio channel has also been analysed in this chapter. To enable prediction and modelling, the variation of PL and time-delay parameters with the body movements has also been compared to statistical models. It was concluded that for the movement case, the log-normal distribution provides the best fit for the RMS and mean excess delay, while the normal distribution is best for modelling PL, as was the case for static scenario. From the analysis of several on-body radio links, it was also concluded that amplitude and delay spread of the received signal can vary significantly with the changes in the posture and position of antennas on the body.

Acknowledgements

This publication was made possible by NPRP grant # 6-415-3-111 from the Qatar National Research Fund (a member of Qatar Foundation). The statements made herein are solely the responsibility of the authors.

References

[1]   T. S. Rappaport. *Wireless Communications Principles and Practice*. Englewood Cliffs, NJ: Prentice Hall Inc., 1996.

[2]   H. Hashemi. The indoor radio propagation channel. *Proceedings of IEEE*, 81(7):943–968, 1993.

[3]   A. A. Saleh and R. A. Valenzuela. A statistical model for indoor multipath propagation. *IEEE Journal on Selected Areas in Communications*, 5(2): 128–137, February 1987.

[4]   H. Suzuki. A statistical model for urban radio propagation. *IEEE Transactions on Communications*, 25(7):673–680, July 1977.

[5]   M. Z. Win and R. A. Scholtz. Impulse radio: how it works. *IEEE Communications Letters*, 2(2):36–38, February 1998.

[6]   I. S. Kovacs, G. Pedersen, P. Eggers, and K. Olesen. Ultra wideband radio propagation in body area network scenarios. In *IEEE Eighth International Symposium on Spread Spectrum Techniques and Applications*, Sydney, Australia, pages 102–106, 30 August–2 September 2004.

[7]   S. L. Cotton and W. G. Scanlon. A statistical analysis of indoor multipath fading for a narrowband wireless body area network. In *IEEE 17th International Symposium on Personal, Indoor and Mobile Radio Communications*, Helsinki, pages 1–5, September 2006.

[8]   D. Neirynck. *Channel Characterisation and Physical Layer Analysis for Body and Personal Area Network Development*. Ph.D. Thesis, University of Bristol, Bristol, UK, November 2006.

[9]   A. Sani. *Modeling and Characterisation of Antenna and Propagation for Body-Centric Wireless Communications*. Ph.D. Thesis, Queen Mary University of London, London, UK, 2010.

[10] J. H. Reed. *An Introduction to Ultra Wideband Communication Systems*. Englewood Cliffs, NJ: Prentice Hall Inc., 2005.

[11] A. Rahman, A. Alomainy, and Y. Hao. Compact body-worn coplanar waveguide fed antenna for UWB body-centric wireless communications. In *The Second European Conference on Antennas and Propagation (EuCAP 2007)*, Edinburg:IET, pages 1–4, November 2007.

[12] T.-G. Ma and C.-H. Tseng. An ultra wideband coplanar waveguide-fed tapered slot ring antenna. *IEEE Transactions on Antenna and Propagation*, 54(4), pages 359–362, April 2006.

[13] L. Liang, L. Guo, C. C. Chiau, X. Chen, and C. G. Parini. Study of CPW fed circular disc monopole antenna for ultra wideband application. *IEE Proceedings on Microwave, Antennas and Propagation*, 152(6):520–526, November 2005.

[14] H. G. Schantz and M. Barnes. The COTAB UWB magnetic slot antenna. In *IEEE AP-S International Symposium*, Boston, MA, pages 104–107, 2001.

[15] T. Yang, S.-Y. Suh, R. Nealy, W. Davis, and W. L. Stutzman. Compact antennas for UWB applications. In *2003 IEEE Conference on Ultra Wideband Systems and Technologies*, IEEE, Virginia, USA, pages 205–208, November 2003.

[16] M. Klemm, I. Z. Kovacs, F. G. Pedersen, and G. Troester. Novel small-size directional antenna for UWB WBAN/WPAN applications. *IEEE Transactions on Antennas and Propagation*, 53(12):3884–3896, 2005.

[17] D. Lamensdorf and L. Susman. Baseband-pulse-antenna techniques. *IEEE Antennas and Propagation Magazine*, 36(1), pages 20–30, February 1994.

[18] J. S. McLean, H. Foltz, and R. Sutton. Pattern descriptors for UWB antennas. *IEEE Transactions on Antennas and Propagation*, 53(1 Part 2):553–559, 2005.

[19] T. Dissanayake and K. P. Esselle. Correlation-based pattern stability analysis and a figure of merit for UWB antennas. *IEEE Transactions on Antennas and Propagation*, 54(11 Part 1):3184–3191, 2006.

[20] M. Kanda. A relatively short cylindrical broadband antenna with tapered resistive loading for picosecond pulse measurements. *IEEE Transactions on Antennas and Propagation*, 26(3):439–447, 1978.

[21] Q. H. Abbasi, A. Alomainy, and Y. Hao. Characterisation of MB-OFDM based ultra wideband systems for body-centric wireless communications. *IEEE Antenna and Wireless Propagation Letter*, 10:1401–1404, 2011.

[22] Q. H. Abbasi, A. Sani, A. Alomainy, and Y. Hao. On-body radio channel characterisation and system-level modelling for multiband OFDM ultra wideband body-centric wireless network. *IEEE Transactions on Microwave Theory and Techniques*, 58(12):3485–3492, 2010.

[23] Q. H. Abbasi, A. Sani, A. Alomainy, and Y. Hao. Arm movements effects on ultra wideband on-body propagation channel and radio systems. In *Proceedings of the 2009 IEEE Loughborough Antennas and Propagation Conference (LAPC)*, Loughborough, UK, pages 261–264, 2009.

[24] J. A. Dabin, N. Ni, A. M. Haimovich, E. Niver, and H. Grebel. The effects of antenna directivity on path loss and multipath propagation in UWB indoor wireless channels. In *Proceeding of the IEEE Conference on Ultra Wideband Systems and Technologies*, Newark, NJ, pages 305–309, 2003.

[25] S. S. Ghassemzadeh, R. Jana, C. W. Rice, W. Turin, and V. Tarokh. A statistical path loss model for in-home UWB channels. In *IEEE Conference on Ultra Wideband Systems and Technologies (UWBST)*, IEEE, Baltimore, MD, USA, pages 59–64, May 2002.

[26] Q. H. Abbasi, M. M. Khan, A. Alomainy, and Y. Hao. Characterization and modelling of ultra wideband radio links for optimum performance of body area network in health care applications. In *Proceedings of the 2011 IEEE International Workshop on Antenna Technology (IWAT)*, Hong Kong, pages 206–209, 2011.

[27] K. P. Burnham and D. R. Anderson. *Model Selection and Multimodel Inference: A Practical Information-Theoretic Approach*. New York, NY: Springer-Verlag, 2002.

[28] A. Alomainy, Y. Hao, and W. F. Pasveer. Numerical and experimental evaluation of a compact sensor antenna performance for healthcare devices. *IEEE Transactions on Biomedical Circuits and Systems*, IEEE, 1(4):242–249, December 2007.

[29] Q. H. Abbasi, M. M. Khan, S. Liaqat, A. Alomainy, and Y. Hao. Ultra wideband off-body radio channel characterisation for different environments. In *Proceedings of the Seventh International Conference on Electrical and Computer Engineering (ICECE)*, Dhaka, Bangladesh, 2012.

[30] Q. H. Abbasi, A. Sani, A. Alomainy, and Y. Hao. Experimental characterisation and statistical analysis of the pseudo-dynamic ultra wideband on-body radio channel. *IEEE Antenna and Wireless Propagation Letter*, 10:748–751, 2011.

Chapter 4

Sparse characterization of body-centric radio channels

Xiaodong Yang*, Aifeng Ren*, Zhiya Zhang*,
Qammer Hussain Abbasi**, Erchin Serpedin[†], Wei Zhao[††],
Shuyuan Yang*, and Akram Alomainy[‡]

4.1 Introduction

Body-centric wireless communications systems (BWCS) will be a focal point for future communications from end-to-end user's perspective [1]. In BWCS, one of the key issues is to characterize BWCS channels effectively and accurately. Some researchers have investigated the BWCS for various scenarios. In Reference 2, on-body ultra-wideband (UWB) channels were characterized, and the optimal model for root-mean-square (RMS) delays is obtained. In Reference 3, different types of antennas were used as transceivers; a typical statistical model, the log-normal model, was used to fit mean excess delay and RMS delay data. Alomainy *et al.* [4] obtained the fitting parameters for microstrip patch antennas. In Reference 5, the authors demonstrated that Rayleigh distribution can fit the indoor and outdoor radio channels data well. In Reference 6, a Gaussian distribution was used as the reference model for mean excess delay and RMS delay of BWCS. For UWB BWCS channels, they are usually measured by time domain and frequency domain techniques [7]. In frequency domain measurement, to ensure the accuracy and range of the measurement, people usually use high-end vector network analyser (VNA); on the other hand, a high-rate analogue-to-digital converter (ADC), which is very expensive, is a key device for the whole system [8]. Through deep thinking of the state of the art of BWCS radio channels [1–8], we find that there are two issues deserve further discussion. In the following two paragraphs, we would like to further discuss these two issues.

First of all, it is important to realize that for most current BWCS channel characterization methods, some fitting parameters from available statistical models are obtained; then, these parametric statistical models are selected as appropriate models for radio propagation. It is worth noting that these propagation models are all for

*School of Electronic Engineering, Xidian University, Xi'an, Shaanxi, China
**Department of Electrical and Computer Engineering, Texas A&M University at Qatar, Doha, Qatar
†Electrical and Computer Engineering, Texas A&M University, College Station, USA
††School of Electro-Mechanical Engineering, Xidian University, Xi'an, Shaanxi, China
‡School of Electronic Engineering and Computer Science, Queen Mary, University of London, London, UK

specific propagation scenarios, including specific posture, individual, surrounding environment, etc. It is therefore common for people to propose such a problem: is there any generalized BWCS channel model which can be used in different scenarios? On the other hand, it is noticeable that for BWCS channel measurement, a large number of samples are necessary to guarantee the accuracy of the results [9]. Considering these two factors, a small sample based non-parametric propagation model is needed for BWCS.

Then, we continue to look into the on-body UWB communications. It is found that for on-body radio propagation, the receiving antenna get most of the energy through direct path [3]. This propagation characteristic implies that the on-body UWB channel is sparse, and this exciting fact encourages us to explore novel approach to characterize such kind of radio channels.

In general, a small-sample non-parametric model and sparse on-body UWB channel characterization method are important supplements to the current BWCS research; since they share the common character of sparsity, the authors would like to discuss these interesting topics in this chapter.

The chapter is organized as follows: Section 4.2 gives the basics of sparse non-parametric technique and compressive sensing; Section 4.3 presents some results regarding non-parametric modelling and on-body impulse response estimation; Section 4.4 uses statistical technique to explore obesity's effect on the on-body narrowband channels; Section 4.5 draws a conclusion.

4.2   Basics of sparse non-parametric technique and compressive sensing

4.2.1   Sparse non-parametric technique

The approaches in machine learning can be roughly divided into supervised learning and unsupervised learning. For supervised learning, an optimal model is obtained via training available samples; then, the inputs can be mapped to corresponding outputs by applying the model. It is worth mentioning that the model belongs to the set of some function, and optimum is for certain criterion. Finally, the unknown data are classified. On the other hand, the model needs to be built without any training samples, the approach is known as unsupervised learning. The typical example in unsupervised learning is clustering. The sparse non-parametric technique presented in this paper is a supervised learning approach. In Reference 10, it is proposed that to solve the problem of probability estimation, the support vector technique can be introduced. Following Weston's train of thoughts, in this chapter, the support vector technique is used to construct the probability model for BWCS.

4.2.1.1   Empirical distribution function

In the definition of distribution function, the probability can be obtained directly. However, in BWCS, the distribution function is unknown; it should be inferred from known channel data. To achieve this step, the empirical distribution function is introduced. For BWCS, one-dimensional empirical distribution function is used to approximate the distribution function. To express probability models mentioned at

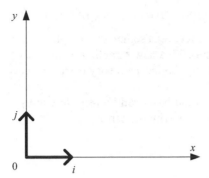

Figure 4.1 Two-dimensional (2D) Euclidean space formed by orthonormal basis © 2015 (Reproduced by permission of the Institute of Electrical and Electronics Engineers [9])

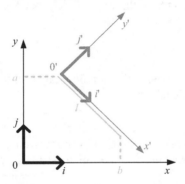

Figure 4.2 Random curve in the 2D Euclidean space © 2015 (Reproduced by permission of the Institute of Electrical and Electronics Engineers [9])

the beginning of the chapter, the basis of the regression function should be selected properly (Figures 4.1 and 4.2).

4.2.1.2 The coefficients for the probability approximation

Vapnik and Mukherjee [11] represented the regression problem as:

$$f(x_j) = \sum_{i=1}^{N} \alpha_i k(x_j, x_i) + b, \tag{4.1}$$

where f is the linear combination of the basis, and k is the basis. Equation (4.1) is then be used to fit the empirical distribution function mentioned in Section 4.2.1.1. To find the coefficients for the approximation, first, the standard [11] should be considered; then, to use the data in the image space, the linear programming technique should be applied. The specific mathematical procedure for solving coefficients can be found in Reference 9.

In summary, wireless channels for BWCS are affected by many uncertainties; therefore, the propagation model should be determined entirely by the data. The sparse non-parametric technique presented in the chapter can fulfil the task well.

4.2.2   Basics of compressive sensing framework

It is known that in order to recover a signal, the sampling rate must satisfy the sampling theorem [12]. In Reference 13, a new sampling theory was developed; it is known as compressive sensing theory. So far the theory has been successfully applied to many fields of science.

Suppose an orthonormal basis can be used to express discrete time signal $x \in R^{N \times 1}$; then, the signal will be with the form:

$$x = \sum_{i=1}^{N} \psi_i \theta_i, \qquad (4.2)$$

where θ is a sparse vector. If we use matrix to express the signal, we have

$$x = \Psi \theta. \qquad (4.3)$$

We define that the sparsity of the vector is $K(K \ll N)$, x then becomes

$$x = \sum_{k=1}^{K} \psi_{i_k} \theta_{i_k}. \qquad (4.4)$$

In compressive sensing theory, the sampling rate is reduced significantly. The M linear measurements for signal x are

$$s = \Phi x, \quad \Phi \in R^{M \times N}. \qquad (4.5)$$

The linear measurement can be understood better from sensor point of view. Then, the next step is to reconstruct the original signal. The basic equation for compressive sampling is

$$s = \Phi x = \Phi \Psi \theta, \qquad (4.6)$$

where Ψ is a matrix, θ is a sparse vector, s is the measurement signal, and Φ is the measurement matrix. Next, by considering l_1 optimization problem [14], the reconstruction can be achieved.

4.3   Results and discussions regarding non-parametric modelling and on-body impulse response estimation

4.3.1   Establishing sparse non-parametric propagation models and their evaluation

4.3.1.1   Measurement setup

To measure the on-body narrowband channels, printed monopole antennas were used [7]. The resonance frequency for the antenna is 2.4 GHz. An HP8720ES VNA was used to obtain the transmission response between two printed monopole antennas. Proper distance was selected to consider the effect of human body on the antennas. The position for transmitting antenna was fixed while the receiving antenna was placed on representative positions on the body. The experiment took place at Queen Mary, University of London.

To obtain the time domain characteristics of the on-body channels, a pair of miniaturized coplanar wave-guide-fed tapered slot antennas [15] were used. The frequency range for the measurement is 3–10 GHz. By applying inverse discrete Fourier transform, the channel impulse response can be obtained. The sampling time is 50 ps, and the sampling rate is 1601 frequency points. The measurement was also performed at Queen Mary, University of London.

4.3.1.2 Kernel functions

It is known that the linear classifier can only process linearly separable samples. If the samples are not linearly separable, we cannot get the solution from linear classifier. Therefore, the scope of application of the linear classifier narrows down significantly, and it is difficult to make full use of its advantages. The basic train of thought for solving linearly inseparable problems is to convert them to high dimensions, making them linearly separable. The key issue in the conversion is to find the corresponding mapping method. Unfortunately, there is no mature method to follow. However, what we concern is the value of inner product in high-dimensional space; once we know the value, the classification results will be obtained. In theory, x' is the function of x; w' is a constant, it is transformed from a constant w in low-dimensional space. Therefore, it is common to consider the existence of a kind of function $K(w,x)$, which accepts input values in low-dimensional space but gets the inner product values in high-dimensional space. Once we have such kind of functions, it is not necessary to find out the mapping relationship any more. Luckily, such kind of functions do exist and are known as kernel functions. Actually, as long as a function satisfies Mercer conditions, it can be seen as a kernel function. The basic function of kernel function is to accept two vectors in low-dimensional space and to obtain the inner product of vectors in high-dimensional space after certain transformation. In this chapter, the sigmoid function $k(x_i, x_j) = \tanh(ax_i^T x_j + r)$ is selected as the kernel function. The kernel function is with the form $k = \frac{1}{1+e^{-t(x_i - x_j)}}$ [10]. The kernel function is given in Figure 4.3.

The whole procedure of building sparse non-parametric model for BWCS can be divided into four steps and is given in Figure 4.4.

4.3.1.3 Sparse non-parametric model characterization results and discussions

It is known that for BWCS, radio channels are usually characterized by existing log-distance parametric models. However, one may note that there is still some room for the improvement of the model. Such room comes from two aspects: one is limited number of samples; the other is the uncertainty of propagation around the human body. Therefore, it is necessary to establish novel propagation model which is sparse and independent of prior knowledge. A typical non-parametric model is given in Figure 4.5. By defining related coefficients, the final regression function can be established. The independent coefficient ε and regularization parameter C can be

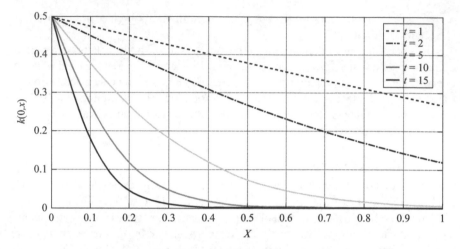

Figure 4.3   *Kernel functions with specific parameters © 2015 (Reproduced by permission of the Institute of Electrical and Electronics Engineers [9])*

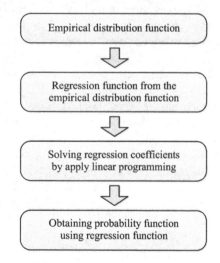

Figure 4.4   *Flow chart for establishing sparse non-parametric propagation models for BWCS (Replotted from Reference [16], Institute of Electrical and Electronics Engineers)*

adjusted so that the error of the model can be controlled. In Figure 4.5, the specific parameter combination is $C = \frac{\max\left(|\bar{y}+3\sigma_y|,|\bar{y}-3\sigma_y|\right)}{l}$, $\varepsilon = 0.1$. One should avoid over-fitting situation when finding appropriate parameters for the model; actually, the fitting error and the sparsity should be considered simultaneously. Take Figure 4.5 as an example, although smaller errors, compared with available parametric models, can be obtained; sparsity of the model should not be sacrificed. Thus, the coefficients presented in Figure 4.5 are considered to be the initial parameters for on-body large-scale fading.

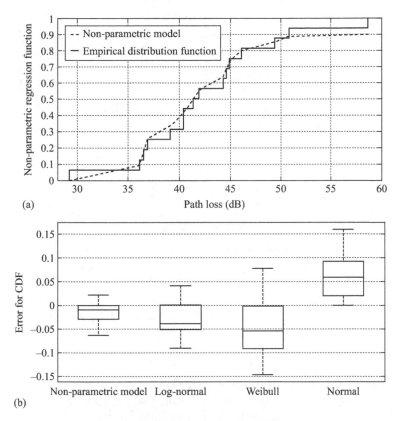

Figure 4.5　A non-parametric model for BWCS © 2015 (Reproduced by permission of the Institute of Electrical and Electronics Engineers [9])

To evaluate the performance of the proposed model completely, time domain delays are considered as well. The RMS delays are considered to be the characterization object. It is found that, by carefully tuning the independent coefficient and regularization parameter, the sparsity and the fitting error of the model can be balanced (Figure 4.6), demonstrating the feasibility of the model.

It is common to propose such a question: how do we know the propagation mechanism from the novel model presented in this chapter? It is important and necessary that the propagation model should contain mechanism information in it. In the following content, these issues are discussed. It is interesting to note that the number of support vectors is closely related to the specificity of the propagation. Figure 4.7 gives the sparsity of the model for large-scale fading under specific body posture. It can be seen that the novel model is stable.

4.3.2　Sparse on-body UWB channel estimation

It is known that the on-body UWB channel satisfies sparsity condition [3,17], such propagation characteristic creates conditions for applying compressive sensing strategy. The aim of this part of the chapter is to recover the impulse response of the

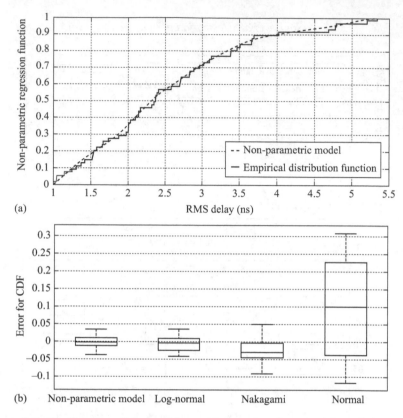

(a)

(b)

Figure 4.6   Verification of the model considering time domain delays © 2015 (Reproduced by permission of the Institute of Electrical and Electronics Engineers [9])

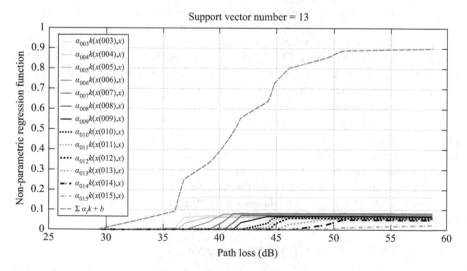

Figure 4.7   Sparsity of the novel non-parametric model © 2015 (Reproduced by permission of the Institute of Electrical and Electronics Engineers [9])

corresponding UWB channel. It is well known that the transmitting signal for a UWB system is [18]:

$$s(t) = d(t) \otimes p(t) \quad t \in [0, T_s), \tag{4.7}$$

where the pulse excitation and the monopulse are all contained. To better illustrate the channel recovery process, we use two steps to accomplish channel estimation.

For the first step of channel reconstruction, we do not consider actual radio channels. The premise of sparse channel estimation is that the Finite Impulse Response (FIR) filter system is similar to a Compressive Sensing (CS) framework. By comparing the convolution process of FIR filter with basic framework of CS presented in Section 4.2.2, the orthonormal basis, the sparse vector, and the measurement matrix can be obtained. The variable for CS framework is given in Figure 4.8.

In Figure 4.9, the output signal of ADC is characterized. Due to lower speed of sampling, the time interval has been changed. Finally, the result of l_1 reconstruction is shown in Figure 4.10. At this point, we have finished step one of on-body UWB channel estimation.

For the second step of UWB on-body channel estimation, the real impulse response is taken into consideration. The experiment process is similar to the measurement setup presented in Section 4.3.1.1. The reconstruction result is given in Figure 4.11. Due to the effect of human body, disorder of the impulse response can be clearly observed.

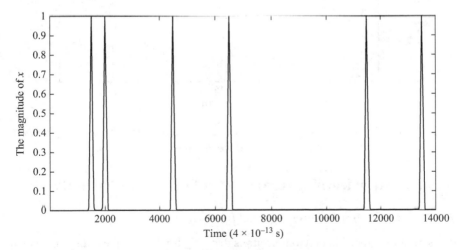

Figure 4.8   Variable for CS framework © 2015 (Reproduced by permission of the Institution of Engineering & Technology [8])

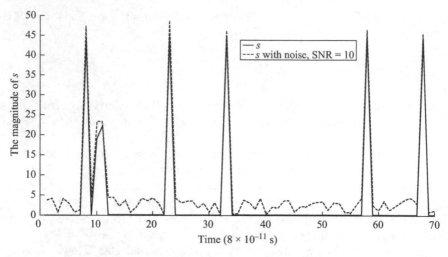

*Figure 4.9   The output of ADC © 2015 (Reproduced by permission of the
Institution of Engineering & Technology [8])*

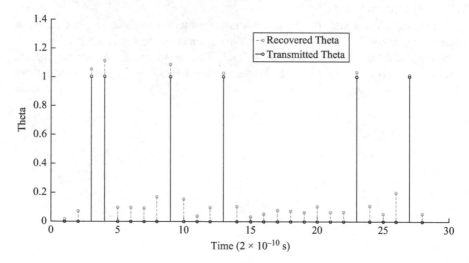

*Figure 4.10   Results for l_1 reconstruction © 2015 (Reproduced by permission of the
Institution of Engineering & Technology [8])*

4.4   Statistical learning technique and its application in BWCS

4.4.1   Small-sample learning and background

It has been mentioned in previous sections that only limited samples can be collected
for BWCS. One may easily realize that it will increase the difficulty of extracting
useful information from available results. Fortunately, it has been demonstrated that
such a problem can be safely solved by applying statistical learning technique, which

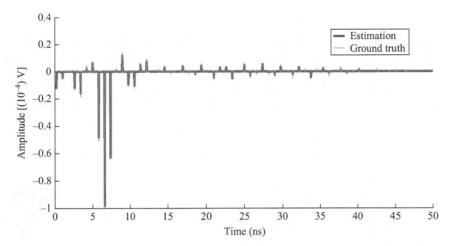

Figure 4.11   *Impulse response reconstruction results considering actual on-body channels © 2015 (Reproduced by permission of the Institution of Engineering & Technology [8])*

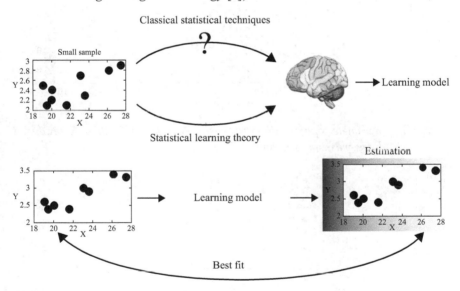

Figure 4.12   *Small-sample problems in BWCS and their solutions*

stresses small-sample-size problems [19] (Figure 4.12). In traditional channel model acquisition methods, least square fitting technique is usually utilized to obtain environmental factors. However, when sample size is small, support vector technique should be introduced to enhance the characterization accuracy. In the next section, the authors would like to use small-sample learning technique to characterize the relationship between obesity and on-body narrowband channels.

4.4.2   Example of support vector regression

In this part, we apply small-sample learning technique to Finite Difference Time Domain (FDTD) based on-body channels to explore the relationship between obesity factors and propagation factors. It is worth mentioning that in the process of parameter optimization, ten-folded cross-validation, rather than leave-one-out cross-validation, was applied. The result of parameter optimization using Radial Basis Function (RBF) kernel is given in Figure 4.13.

To optimize the searching result, a standard particle swarm algorithm was used; velocity and position are updated using standard equations (Figures 4.14 and 4.15).

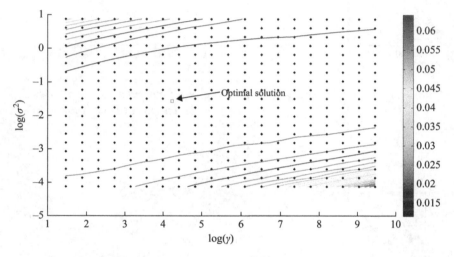

Figure 4.13   Parameter optimization result using RBF kernel © 2014 (Reproduced by permission of Springer [20])

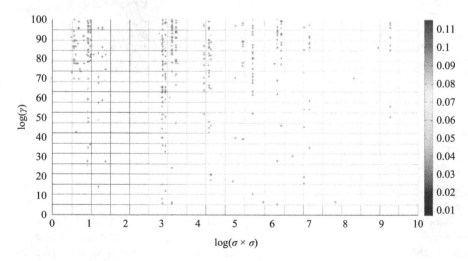

Figure 4.14   2D trajectory of particles © 2014 (Reproduced by permission of Springer [20])

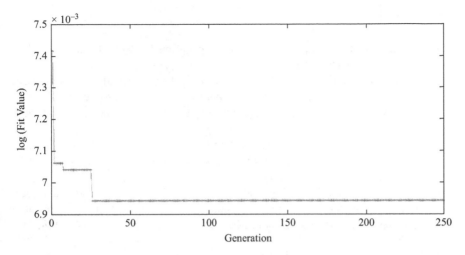

Figure 4.15   Global fitness function © 2014 (Reproduced by permission of Springer [20])

To characterize the obesity's effect more completely, RBF kernel and poly kernel are both used to establish the regression model. It is worth mentioning that different confidence bands are used for the two kernels; and the pointwise confidence band is narrower. Finally, it is found that, for both kernels, the obesity's effect on the narrowband on-body channels is almost the same.

4.5   Conclusion

In this chapter, sparse characterization of BWCS is discussed. First of all, a novel sparse non-parametric model is proposed to characterize BWCS channels, it has been demonstrated that it is an important supplement to the existing parametric models; and then, compressive sensing technique is applied to the on-body UWB channel estimation, the impulse response of the channel is perfectly reconstructed; finally, particle swarm optimization based support vector regression technique is used to explore obesity's effect on the on-body narrowband wireless channels. This chapter provides readers a totally new angle of view of looking at the current channel modelling technique in BWCS; thus will be beneficial to the ones who aim to develop new radio channel models for BWCS.

Acknowledgements

This work was supported in part by the National Natural Science Foundation of China under Grant 61301175, the Fundamental Research Funds for the Central Universities, the Natural Science Basic Research Plan in Shaanxi Province of China under Grant 2014JQ8311, and the Project Funded by China Postdoctoral Science Foundation.

References

[1]   P. S. Hall and Y. Hao, Antennas and Propagation for Body-Centric Wireless Networks. Norwood, MA, USA: Artech House, 2006.

[2]   A. Sani, A. Alomainy, G. Palikaras, *et al.*, "Experimental characterization of UWB on-body radio channel in indoor environment considering different antennas," IEEE Trans. Antennas Propag., vol. 58, no. 1, pp. 238–241, Jan. 2010.

[3]   A. Alomainy, Y. Hao, X. Hu, C. G. Parini, and P. S. Hall, "UWB onbody radio propagation and system modelling for wireless body-centric networks," IEE Proc. – Commun., vol. 153, no. 1, pp. 107–114, Feb. 2006.

[4]   A. Alomainy, Y. Hao, A. Owadally, *et al.*, "Statistical analysis and performance evaluation for on-body radio propagation with microstrip patch antennas," IEEE Trans. Antennas Propag., vol. 55, no. 1, pp. 245–248, Jan. 2007.

[5]   P. S. Hall, Y. Hao, Y. I. Nechayev, *et al.*, "Antennas and propagation for on-body communication systems," IEEE Antennas Propag. Mag., vol. 49, no. 3, pp. 41–58, Jun. 2007.

[6]   A. Alomainy, A. Sani, A. Rahman, J. G. Santas, and Y. Hao, "Transient characteristics of wearable antennas and radio propagation channels for ultra-wideband body-centric wireless communications," IEEE Trans. Antennas Propag., vol. 57, no. 4, pp. 875–884, Apr. 2009.

[7]   A. Alomainy, "Antennas and radio propagation for body-centric wireless networks." PhD Thesis, 2007.

[8]   X. Yang, A. Ren, Z. Zhang, M. U. Rehman, Q. H. Abbasi, and A. Alomainy, "Towards sparse characterization of on-body ultra-wideband wireless channels," IET Healthcare Technol. Lett., vol. 2, no. 3, pp. 74–77, Jun. 2015.

[9]   X. Yang, S. Yang, Q. H. Abbasi, *et al.*, "Sparsity-inspired non-parametric probability characterization for radio propagation in body area networks," IEEE J. Biomed. Health Inform., vol. 19, no. 3, pp. 858–865, May 2015.

[10] J. Weston, A. Gammerman, M. Stitson, V. Vapnik, V. Vovk, and C. Watkins, "Support vector probability estimation," in Advances in Kernel Methods— Support Vector Learning, B. Schölkopf, C. Burges, and A. Smola, Eds. Cambridge, MA, USA: MIT Press, 1999.

[11] V. Vapnik and S. Mukherjee, "Support vector method for multivariate Density estimation," Neu. Inf. Pro., pp. 659–665, 1999. Available: http://papers.nips.cc/paper/1652-support-vector-method-for-multivariate-density-estimation

[12] M. Unser, "Sampling: 50 years after Shannon," Proc. IEEE, vol. 88, no. 4, pp. 569–587, 2000.

[13] E. Candes, J. Romberg, and T. Tao, "Robust uncertainty principles: exact signal reconstruction from highly incomplete frequency information," IEEE Trans. Inf. Theory, vol. 52, no. 2, pp. 489–509, 2006.

[14] S.-J. Kim, K. Koh, M. Lustig, S. Boyd, and D. Gorinevsky, "An interior-point method for large-scale l1-regularized least squares," IEEE J. Sel. Top. Signal Process., vol. 1, no. 4, pp. 606–617, 2007.

[15] A. Rahman, A. Alomainy, and Y. Hao, "Compact body-worn coplanar waveguide fed antenna for UWB body-centric wireless-networks," Proc. 2nd Eur. Conf. Antennas Propag., Edinburgh, UK, pp. 1–4, Nov. 2007.

[16] X. Yang, A. Ren, S. Yang, *et al.*, "On the sparse non-parametric model for body-centric ultra-wideband channel," 2014 XXXIth General Assembly and Scientific Symposium, pp. 1–4, 2014.

[17] A. Alomainy, Y. Hao, C. G. Parini, and P. S. Hall, "Comparison between two different antennas for UWB on-body propagation measurements," IEEE Antennas Wirel. Propag. Lett., vol. 4, pp. 31–34, 2005.

[18] "First report and order," Technical Report, Federal Communications Commission, 2002.

[19] V. Vapnik, The Nature of Statistical Learning Theory, 2nd edn. New York: Springer-Verlag New York Inc., 2000.

[20] X. Yang, Q. Zhang, S. Yang, *et al.*, "Risks posed by obesity to body-surface narrowband wireless communication," Chin. Sci. Bull., vol. 59, nos. 29 and 30, pp. 3949–3954, Oct. 2014.

Chapter 5

Antenna/human body interactions in the 60 GHz band: state of knowledge and recent advances

Maxim Zhadobov*, Carole Leduc*, Anda Guraliuc*,
Nacer Chahat†, and Ronan Sauleau*

5.1 Introduction

Body-centric wireless networks constitute an extremely attractive next-generation wireless technology representing a cognitive interface to higher-level networks (WPAN, WLAN, WMAN, etc.). These emerging systems open new possibilities in the fields of wireless communications, remote monitoring and sensing of the human body activity, and detection and localization for a great number of applications (medical, entertainment, defence, smart homes and cities, sport, etc.). They could also play a key role in wireless sensor networks and internet of things (IoT) whose economic impact is growing exponentially. Body-centric wireless networks have emerged as an alternative or add-on to traditional wired network systems (e.g. in medical environments), and new exciting applications are now under development (e.g. high-data-rate body-to-body streaming, remote monitoring of patients at home).

The upper limit of the spectrum used for wireless networking has been recently progressively shifting towards the millimetre-wave (MMW) band due to the increasing need in network capacity and high data rates [1]. Wireless data traffic is rising exponentially (nearly a tenfold increase in traffic is expected from 2014 to 2019 according to the Cisco global mobile data traffic forecast (Figure 5.1) [2], mainly driven by video streaming applications and cloud computing). Recently, the 60 GHz band has been identified as highly promising for body-centric wireless communications including body-area network (BAN) technologies. One of the main features differentiating the 60 GHz BAN from a lower frequency BAN is confidentiality and low interference with neighbouring networks, which has been demonstrated to be crucial for body-centric and inter-BAN communications [3], for instance in military scenarios

*Institute of Electronics and Telecommunications of Rennes (IETR), UMR CNRS 6164, University of Rennes 1, 11D, 263 Avenue du G. Leclerc, 35042 Rennes, France
†Jet Propulsion Laboratory (JPL), California Institute of Technology (CalTech), 4800 Oak Grove Dr, Pasadena, CA 91109, USA

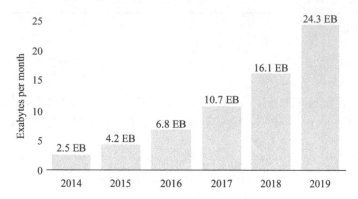

Figure 5.1   Global mobile traffic forecast by 2019 [2]

where communication security is vital [4]. Limited operating range in this band (e.g. shifting the frequency from 70/80 GHz to 60 GHz decreases the operating range from 3 km to 400 m [5]) is mainly related to the strong oxygen-induced atmospheric attenuation (typically 16 dB/km). Besides, high data rates can be achieved [beyond several Gb/s], which is extremely attractive to support the increasing demand in data traffic and high-data-rate transmissions. To address these needs, IEEE 802.11 NG60 group was formed in 2014 aiming at significantly increasing the data rates in the 60 GHz frequency band in a backwards compatible way to IEEE 802.11ad. In addition, the 60 GHz band provides other advantages, such as miniature size of antennas and sensors compared to their counterparts in the lower part of the microwave spectrum. Today, MMW circuits and antennas can be implemented with a high-level integration and reasonable cost. Finally, this band provides high accuracy that can be beneficially used for high-resolution localization (this can be exploited for new applications, such as immersive video games).

The implementation of 60 GHz technologies, including body-centric applications, is an unavoidable evolution of wireless networks, and first commercial solutions are now emerging for WPAN (e.g. Dell Latitude 6430u, first ultrabook using Qualcomm/Wilocity 60 GHz chipset). The 60 GHz band is unlicensed, meaning that no application for license has to be made prior to the deployment of a service operating in this band. Different spectra are allocated depending on countries (e.g. 57–66 GHz in Europe, 57–64 GHz in North America and South Korea, 59.4–62.9 GHz in Australia, and 59–66 GHz in Japan [6]). Note that the available bandwidth is hundreds times higher compared to existing wireless technologies at lower microwave frequencies.

A massive deployment of wireless devices equipped with 60 GHz Tx/Rx modules is foreseen in coming years. The corresponding new usages and services (some examples are provided in the next section) will unavoidably involve coupling of radiating devices with the human body, both in terms of the body impact on wireless device performances and in terms of user exposure. This includes the near-field interactions of wearable and mobile devices operating in the vicinity of the human body. This chapter provides a comprehensive overview of basic features and recent advances in the field of antenna/human body interactions in the 60 GHz band.

5.2   Emerging body-centric applications at millimetre waves

This section summarizes several representative application examples and potential use cases of 60 GHz technologies involving the interaction of radiating structures with the human body in different contexts.

5.2.1   Heterogeneous mobile networks

Next-generation heterogeneous cellular mobile networks (5G) are expected to be implemented by 2020. They will take advantage from the wide unlicensed frequency bands, including the 60 GHz band, to allow flexible spectrum usage as well as peak capacities above 10 Gbit/s, achieving tens to hundreds of times more capacity compared to current 4G cellular networks. The MMW mobile broadband system will be a part of 5G design. It will be used for user access, backhaul and fronthaul applications, meshed relay implementations, potentially sharing the same radio resources. MMW technologies are expected to be integrated within miniature high-data-rate small-cell access points connected to a cellular network through optical fibre or MMW wireless backhaul (Figure 5.2) to support massive data exchanges for mobile users with low latency, low interferences, high quality of service, and low power consumption per bit. Expected user exposure scenarios mainly include phone call (smartphone close to the head), browsing (smartphone in front of the head), and access point (mainly whole-body exposure) (Figure 5.3). In particular, it is expected to be used in dense urban environments meaning that multi-source exposure scenarios are likely. Many R&D projects are now ongoing to support and contribute to the development of heterogeneous 5G networks (e.g. large-scale MiWaves project supported by EU specifically focused on the integration of MMW technologies in 5G and bringing together academics, industry, and mobile operators [7]).

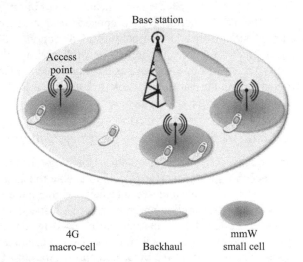

Figure 5.2   60 GHz technologies within the 5G networks architecture – image courtesy of MiWaves project [7]

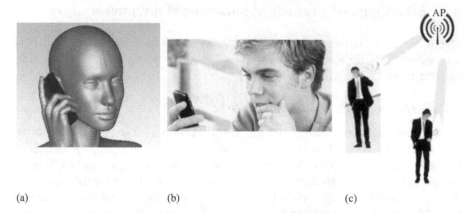

(a)   (b)   (c)

Figure 5.3   Expected user exposure scenarios at 60 GHz within 5G networks.
(a) Phone call, (b) browsing, and (c) access point – image courtesy of
MiWaves project [7]

5.2.2   Body-to-body secured communications

Besides, 60 GHz smart antenna technology is suitable for communications in military context such as relay information on situational awareness, tactical instructions, and covert surveillance in situations where high-speed, short-range, soldier-to-soldier wireless communications are required. The concept of a soldier-to-soldier mobile ad-hoc wireless networks (MANET) was introduced in Reference 4 together with technologies that could be used to provide high-speed wireless networking for dismounted combat personnel. The concept of a short-range soldier-to-soldier MANET is represented in Figure 5.4. Here a small team of co-located infantry troops is wirelessly connected to facilitate high-speed communications within an urban warfare environment. One of the main benefits of the proposed 60 GHz concept is that the communications are secure, with a low probability of detection and interception. Besides, a feasibility study of a V-band wireless network optimized for interconnecting various subsystems worn by a soldier has been presented in Reference 8 for different postures demonstrating that it is possible to establish a BAN with reliable coverage to specified nodes using auxiliary nodes.

5.2.3   Radar-on-chip for gesture and movement recognition

In 2015, Google introduced a 60 GHz radar-on-chip technology for the accurate gesture and movement recognition [9]. A radar is used to enable touchless interactions where the human hand becomes an intuitive interface for manipulating electronic devices. The sensor can track sub-millimetre motions at high speed and accuracy. It fits onto a chip, can be produced at scale, and can be used inside small wearable devices. Two chips have been introduced, a 9 mm^2 one using pulse radar and 11 mm^2 chip using continuous signal radar. In the presented application example, the radar illuminates a hand and tracks the hand and fingers movements by processing the

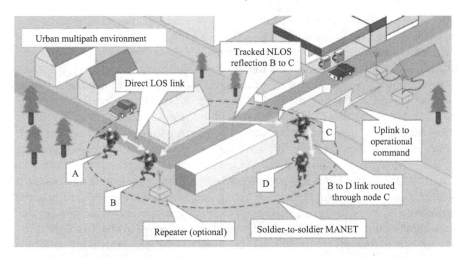

Figure 5.4   Soldier-to-soldier MANET concept [4]. © 2009 IEEE

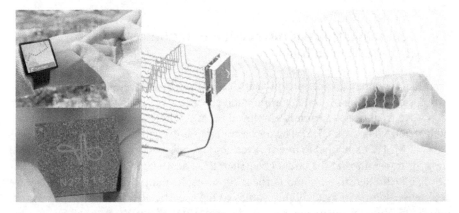

*Figure 5.5   60 GHz radar-on-chip for motion recognition from Google [9] – image
courtesy of Google ATAP*

received signal (Figure 5.5). The target applications include controlling small-screen
wearable devices, for instance by remotely zooming the picture on the screen or
changing the hour of a smartwatch by a motion of fingers in free space.

5.2.4   e-Health monitoring and medical applications

While MMW technologies have not been directly employed for e-Health applications so far, promising potential of this band for high-data-rate diagnostic
systems was discussed in Reference 10 introducing possible architectures of patient-
centric and hospital-centric systems. Through several examples the authors demonstrate that multi-Gb networks integrating 60 GHz technologies will allow real-time
high-speed transfer of patient-related data (e.g. HD video streaming or high-resolution

MRI scans) to local and remote locations. Besides, some concepts of using V-band technologies for medical purposes have been proposed including sensing of vital parameters and thermal therapy. For instance, Glucowise concept has been recently introduced for the fast non-invasive monitoring of the glucose level [11]. It consists in monitoring the through-tissue transmission at locations where the tissue thickness does not exceed several millimetres (e.g. between the thumb and forefinger or at the earlobe), and the received signal is correlated to the glucose level in the blood. The announced operating frequency is around 65 GHz explaining the choice of the application points with an acceptable through-tissue attenuation. The sensors are expected to have integrated nano-composite films enhancing the transmission by reducing the reflections from skin [12]. Measured data are digitally encrypted and then transmitted through Bluetooth to a smartphone or tablet. Another recently introduced potential application is related to the thermal therapy such as hyperthermia or ablation. For such applications, currently used operating frequencies are limited to about 10 GHz, and it has been demonstrated that 10–100 GHz band can be advantageously used to the local selective heating of the skin layers using enforced air cooling of the skin [13]. This opens a new potential, for instance for the local non-invasive treatment of melanoma.

5.3   General features of interaction of millimetre waves with the human body

From the viewpoint of interaction with the human body, the 60 GHz band, originally absent from the natural environmental spectrum, is substantially different from microwave frequencies currently employed in most wireless communication systems (mainly 900 MHz to 10 GHz). The power transmission coefficient in human tissues is higher at MMW compared to microwaves, and therefore more electromagnetic (EM) energy is transmitted to the body. The penetration depth to the body is much smaller and mainly limited to skin and cornea. As a result, the power absorption is more localized, resulting in much higher values of the specific absorption rate (SAR) (i.e. roughly 25 W/kg for the incident power density (IPD) of 1 mW/cm^2 at 60 GHz). Due to these high SAR levels, significant heating (>0.1°C) may appear even for relatively low IPD (>1 mW/cm^2). These fundamental aspects of MMW interactions with the human body are reviewed in detail in this and next sections.

5.3.1   Target organs and tissues

Above 30 GHz, the EM power absorbed by the human body is superficial (penetration depth is 1 mm or less), mainly due to the strong relaxation of free water molecules occurring at these frequencies. In particular, at 60 GHz, typical penetration depth to the biological tissues is around 0.5 mm, and more than 90% of the EM power transmitted to the body is absorbed by skin. Therefore, the main target organs of MMW are eyes and skin. Exposure of eyes leads to the absorption of the EM energy by the cornea with a free water content of 75% and thickness of about 0.5 mm. Hereafter, we will essentially consider the interactions with skin as it covers 95% of the human body surface, and it is the most exposed organ in all above-considered scenarios. From the EM viewpoint, the skin is an anisotropic multi-layer dispersive structure made of four different layers, namely the stratum corneum (SC, consisting of dead cells with a low water content, its thickness ranges from 10 to 40 μm), epidermis (50-μm thick on the eyelids and up to 1.5-mm thick on the palms and soles), dermis (thickness 300 μm on the eyelid up to 3 mm on the back), and subcutaneous fat layer (thickness 1–5.6 mm). The skin also contains capillaries and nerve endings. It is mainly composed of 65.3% of free water, 24.6% of proteins, and 9.4% of lipids [14]. Note that the water content of tissues may vary with age impacting their EM properties.

5.3.2 EM properties of tissues

The knowledge of the EM properties of the human skin is essential to understand the field behaviour close to or inside the human body. Available data for the effective complex permittivity of skin at MMW, in particular above 50 GHz, are very limited due to technical difficulties associated with measurements [15]. Around 60 GHz, the dispersive EM properties of biological tissues are mainly related to the rotational dispersion of free water molecules [16]. Skin permittivity data reported in the literature so far strongly depend on the measurement technique, type of study (*in vivo* or *in vitro*), experimental conditions (e.g. skin temperature, location on the body, thickness of different skin layers, etc.).

The first data on the skin permittivity around 60 GHz were reported by Gandhi and Riazi [17]. An extrapolation of the measured data on the rabbit skin at 23 GHz using a Debye model was performed at 60 GHz. Gabriel *et al.* reported the complex permittivity of the human skin up to 110 GHz, based on an extrapolation of measured data performed up to 20 GHz [18]. Two skin models have been considered: wet skin and dry skin. The dry skin model represented the human skin under normal environmental and physiological conditions ($T \sim 32.5°C$). The corresponding broadband dispersive behaviour of the complex permittivity obtained using the first-order Debye model is illustrated in Figure 5.6. Today, this is the most used model and, according to the authors' experience, provides in most of the cases a good agreement with the human skin.

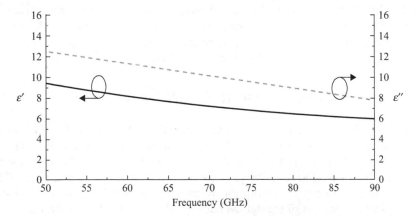

Figure 5.6   Complex permittivity of the dry skin around 60 GHz

Table 5.1   EM properties of skin at 60 GHz

Complex permittivity	Temperature [°C]	Method	Reference
In vivo			
$7.98 - j \cdot 10.90$	32.5	Extrapolation	Gabriel [18] (dry skin)
$8.12 - j \cdot 11.14$	32.5	Measurement	Alekseev [20]
$8.02 - j \cdot 10.05$	32.5	Measurement	Chahat [21]
In vitro			
$8.89 - j \cdot 13.15$	37	Extrapolation	Gandhi [17]
$10.22 - j \cdot 11.84$	37	Extrapolation	Gabriel [18] (wet skin)
$9.90 - j \cdot 9.00$	23	Measurement	Alabaster [19]
$13.2 - j \cdot 10.30$	37	Extrapolation	Alabaster [19]

Since 2000s, a few reports completed these data (Table 5.1). Alabaster performed *in vitro* studies on human skin samples, and the dielectric properties were obtained using a free-space technique in the 60–100 GHz frequency range [19]. Alekseev and Ziskin characterized the EM properties the human forearm and palm skin by reflection measurements with an open-ended waveguide [20]. Homogeneous and multi-layer models were proposed to fit the experimental data. The latest results come from Chahat *et al.* who used an open-ended coaxial probe and a new temperature-based method to measure human skin EM properties [21, 22]. In the later method, forearm skin was exposed at 60.4 GHz using an open-ended waveguide and continuous wave signal. Temperature distribution was recorded using an infrared camera. By fitting the analytical solution of the bioheat transfer equation to the experimental heating kinetics, the values of the power density (PD) and penetration depth were found and used to retrieve the complex permittivity of skin described by Debye equation.

Important variations exist among the reported data. This may be related to: (1) temperature variations (e.g. 23°C [19] and 37°C [17] while the skin temperature under normal environmental conditions is around 32.5°C); (2) water content of *in vitro* skin samples that might vary depending on the measurement protocol. More realistic skin permittivity models are expected from *in vivo* measurements (i.e. Gabriel *et al.* [18] (dry skin), Alekseev and Ziskin [20], and Chahat *et al.* [21]). Note that the best agreement is observed among these three sets of data.

5.4   Plane wave illumination at the air/skin interface

In order to understand the basic phenomena occurring when MMW impinge on the human body, we consider here an incident plane wave at the interface of a flat skin model with the EM properties illustrated in Figure 5.6. If not stated otherwise, a semi-infinite dry skin model is considered. Since the penetration into skin in the 60 GHz band is limited to a fraction of a millimetre and the wavelength is smaller compared to the typical dimensions of the human body, results obtained for the flat model can be extrapolated to the most of practical exposure scenarios.

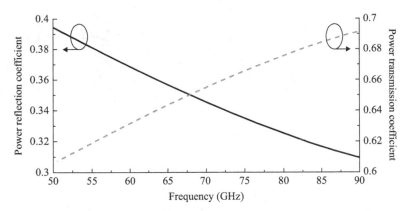

Figure 5.7 Power reflection and transmission coefficients in the 50–90 GHz range for the normal incidence

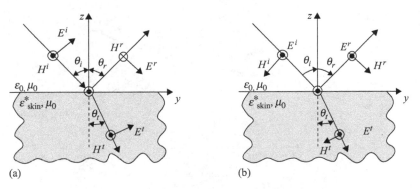

Figure 5.8 Polarization: (a) TM mode and (b) TE mode

5.4.1 Reflection and transmission

The power reflection and transmission coefficients computed for a normally incident plane wave are represented in Figure 5.7 in the 50–90 GHz range. At 60 GHz, roughly 37% of the incident EM power is reflected and 63% penetrate to the body. Note that, in the 50–90 GHz frequency range, the change in power reflection and transmission coefficients is within 10%; for the 57–66 GHz range it is within 2%.

Power transmitted to the body varies significantly depending on the polarization and angle of incidence. Any incidence can be represented as a superposition of two modes illustrated in Figure 5.8: (a) TM mode or parallel polarization (E field is parallel to the plane of incidence), (b) TE mode or perpendicular polarization (E field is perpendicular to the plane of incidence). Figure 5.9 represents the power reflection and transmission coefficients for both polarizations at three different frequencies around 60 GHz, namely 57 GHz, 60 GHz, and 66 GHz. Transmission to the skin is higher for the parallel polarization. While dispersive EM properties were used for

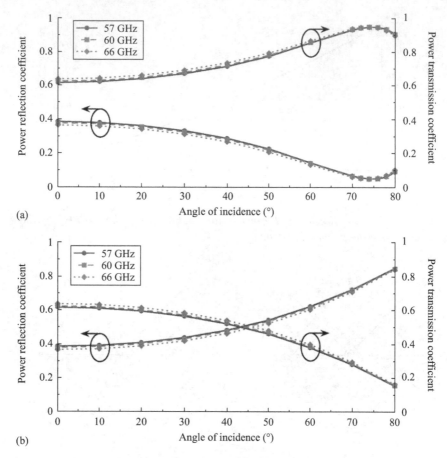

Figure 5.9   Power reflection and transmission coefficients in the 57–66 GHz range: (a) TM mode and (b) TE mode

the computations, the differences within the 57–66 GHz are insignificant. Note that results may change depending on the EM properties of skin (Table 5.1); data for different models are provided and discussed in Reference 23.

To investigate the impact of the multi-layer skin structure on the reflection from and transmission to the skin, two- and three-layer planar skin models were considered. They consist of: (1) SC above homogeneous skin and (2) SC/epidermis (E) and dermis (D)/fat. The typical properties of the layers are summarized in Table 5.2. Both two- and three-layer models demonstrate that the power reflection and transmission coefficients are similar to those of the homogeneous model (Figure 5.10). This suggests that reflection at the air/skin interface, and transmission to skin can be well described by a homogeneous model. Therefore, it can be used to study the impact of the presence of the human body on an on-body antenna performances as well as for the assessment of the body-centric propagation channel.

Table 5.2 Homogeneous vs. multi-layer skin models at 60 GHz

Layers	Permittivity	Conductivity [S/m]	Thickness [mm]
Homogen. skin [18]	7.98	36.4	–
SC [24, 25]	2.96	10^{-4}	0.015
E + D [24]	8.12	1.4	1.45
Fat [18]	3.13	2.8	1

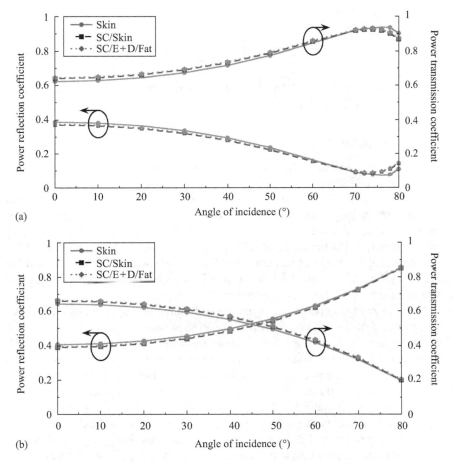

(a)

(b)

Figure 5.10   Power reflection and transmission coefficients at 60 GHz for
(a) TM and (b) TE modes at the air/skin interface – comparison
between homogeneous and multi-layer skin models

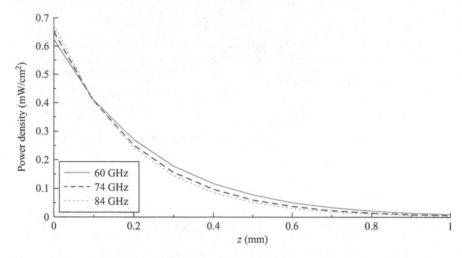

Figure 5.11   Attenuation of PD in the skin at 60 GHz compared to 74 GHz and 84 GHz. IPD is 1 mW/cm²

5.4.2 Absorption

Figure 5.11 shows the attenuation of PD at different frequencies (60 GHz, 74 GHz, and 84 GHz) calculated for the normal incidence. The power transmitted to the body decreases exponentially in skin as a function of depth, and peak PD inside skin increases with frequency as absorption becomes more localized. Assuming that the average epidermis thickness is 0.1 mm, about 40% of the incident power reaches dermis and only 0.1% the fat layer [23]. Figure 5.12 shows the attenuation of PD as a function of the angle of incidence for TM and TE polarizations. Note that the maximum values are obtained for the normal incidence, and the attenuation of IPD is stronger for TE polarization compared to TM.

Alekseev *et al.* compared homogeneous and multi-layer skin models [24]. They demonstrated that the PD and SAR profiles in skin with a thin SC (e.g. 15 μm at forearm) are almost identical for homogeneous and multi-layer models. Indeed, the SC containing little or no free water, and having a low permittivity, can be considered as a lossless very thin dielectric film in contact with the medium of a higher permittivity. For the locations with thinner SC (e.g. at palm the thinness can reach 0.43 mm), this layer may impact PD and SAR distribution. Internal reflections from the fat layer are too small to notably change the PD and SAR distributions in the dermis close to the fat layer. As in most of the BAN-related scenarios, the body regions of interest have much lower SC thickness compared to palm, in majority of the practical cases a homogeneous skin-equivalent phantom can be used for exposure dosimetry.

5.4.3 Impact of clothing

In body-centric applications, it is important to account for the effect of clothing and possible impact on the antenna/body interactions. The impact of various fabrics as

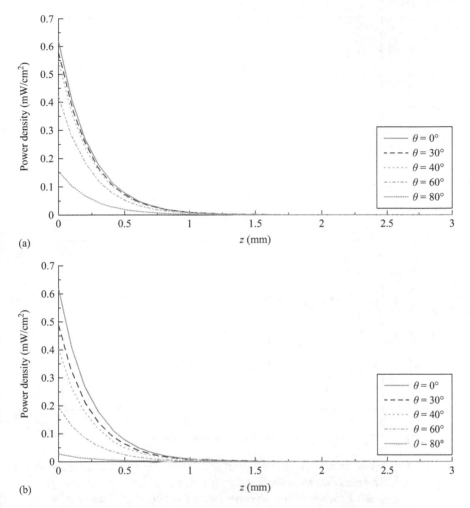

Figure 5.12   PD for (a) TM and (b) TE polarizations at different angles of incidence θ at 60 GHz (θ = 0° corresponds to the normal incidence). IPD is 1 mW/cm²

well as of the clothing–skin air gap on the transmission coefficient was studied for a normally incident plane wave [26]. In particular, a change of transmission coefficient ranging from 10% to 18% depending on the EM properties of textiles was demonstrated (maximum deviation of transmission occurs for the highest value of relative permittivity). In Reference 23, clothing thickness and air gap impacts were assessed analytically for a plane wave impinging on a multi-layer structure, namely a three-layer model (skin/clothing/air) and four-layer model (skin/air gap/clothing/air). Gabriel's data and dry fleece permittivity ($\varepsilon = 1.25 + j \cdot 0.024$) were used to represent skin and textile EM properties, respectively. The transmission coefficient was

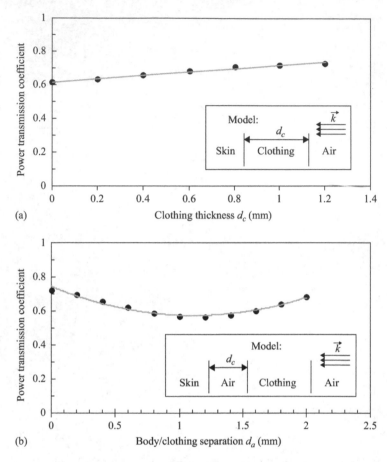

Figure 5.13 *Power transmission coefficient at 60 GHz: (a) depending on the textile*
thickness and (b) as a function of an air gap for $d_c = 1.25\,mm$ [23].
© 2011 Cambridge University Press (reprinted with permission)

computed for a typical range of clothing thickness between 0 mm and 1.2 mm. The
transmission coefficient slightly increases with the clothing thickness (Figure 5.13).
However, the presence of an air gap between clothing and skin results in a decrease
of the power transmitted to the skin.

5.4.4 Heating

At 60 GHz, small penetration depth to biological tissues results in SAR levels significantly higher compared to those obtained at microwaves at identical PD. This
local absorption may result in a significant heating for medium- and high-power
exposures. The steady-state distribution of the temperature increments for IPD of
1 and 5 mW/cm^2 is represented in Figure 5.14. These data correspond to an analytical

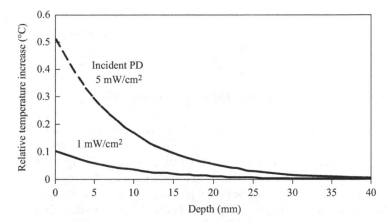

Figure 5.14 Temperature rise for a homogeneous skin model exposed to a plane wave at 60 GHz [23]. © 2011 Cambridge University Press (reprinted with permission)

solution of the 1D bioheat transfer equation [27]. The results suggest that the maximum temperature rise on the body surface due to the exposure at 60 GHz can be roughly estimated as IPD [in mW/cm^2] divided by 10. Note that heating due to local MMW exposure affects not only skin but also subcutaneous tissues including fat and muscles. Therefore, in contrast to the EM dosimetry, for an accurate assessment of thermal effects, a multi-layer model should be used. However, it is important to stress that MMW heating is different from microwaves that penetrate deeper into the tissues and represents a volumetric heating source. MMW induced heating is similar to superficial heating induced by other conventional sources, such as exposure to sun or touching a hot object.

Parametric study performed by Kanezaki *et al.* [28] demonstrated that the temperature distribution induced by a MMW exposure strongly depends on the geometry and thermal properties of the multi-layer model. In particular, they demonstrated that the surface temperature elevation in a three-layer model was 1.3–2.8 times greater than that in single model due to the thermally insulating nature of the fat layer. Furthermore, Alekseev and Ziskin [29, 30] demonstrated that heating is strongly related to the blood flow in skin that depends on the environmental temperature and physiological conditions. It was shown that, depending on these parameters, steady-state temperature increments may vary by a factor of 3. Nelson *et al.* [31] also demonstrated that heating is affected by various environmental conditions (such as perfusion, convection, and sweat rate), but it is not significantly affected by the metabolic rate. In a recently published report, Wu *et al.* [32] also highlighted that the MMW-induced heating can be increased by almost two times due to the presence of clothing.

Finally, it is important to underline that temperature increments (typically below 0.5°C) induced by PD below current international exposure limits (summarized in the next section) are lower than environmental temperature fluctuations. Note also that the average skin temperature is around 32.5°C while the threshold of the pain sensation in humans is around 43°C, and thermal injuries occur only above 43–45°C (depending on the heating duration) [33].

5.5   Exposure limits: guidelines and standards

Wireless devices, including 60 GHz body-centric technologies, should comply with the exposure limits established to protect humans against biological and health effects that might be potentially adverse. To this end, a number of organizations at the international and national levels issued exposure recommendations, guidelines, and standards. These limits are intended to apply to all human exposures except exposure of patients undergoing procedures for diagnostic or treatment. In this section, we first introduce the most appropriate dosimetry metrics at MMW and then discuss ICNIRP and IEEE recommendations used as a reference for most of local regulations worldwide.

5.5.1   Dosimetry metrics

The major dosimetric quantities at MMW are IPD, SAR, and temperature. They are detailed hereafter.

1) *Incident power density (IPD)*. This is the main exposure characteristic adopted by most of international guidelines and standards in the 60 GHz band. The IPD is defined as:

$$\text{IPD} = \frac{P}{S} = \left| \vec{E} \times \vec{H} \right| \tag{5.1}$$

where P is the incident power, S is the exposed surface area, and $\vec{E}$ and $\vec{H}$ are the rms values of electric and magnetic field strengths, respectively. Under far-field exposure conditions it can be determined numerically or based on $\vec{E}$ or $\vec{H}$ field measurements and free-space wave impedance η as:

$$\text{IPD} = \frac{\left| \vec{E} \right|^2}{\eta} = \eta \left| \vec{H} \right|^2 \tag{5.2}$$

Note that, in contrast to SAR or temperature, exposure assessment based on IPD does not rely on knowledge of the distribution of fields or power absorption in the tissues but only on the density of power propagating towards the tissue. For a given transmit antenna, it can be also calculated as:

$$\text{IPD} = \frac{GP_r}{4\pi d^2} \tag{5.3}$$

where G is the antenna gain (linear scale), P_r is the total power radiated by the antenna, and d is the distance from the antenna to the observation point. However, under near-field exposure conditions, especially in the reactive zone where antenna performances in presence of the body are different from those in free space, its determination is more challenging as it is not always possible to distinguish the incident from total field (i.e. incident plus scattered field). In this case, an equivalent plane-wave IPD (eIPD), determined based on the SAR or temperature assessment, can be used. One of possible definitions of eIPD is the following: it is IPD of normally incident plane wave inducing the same SAR or temperature as the considered transmit antenna or device. For completeness, space-averaging issues should be also taken into account together with the exposure duration related aspects in case of the transient temperature.

2) *Specific absorption rate (SAR).* The SAR is a quantitative measure of power absorbed per unit of mass and time. In contrast to IPD, it also takes into account the physical properties of exposed samples:

$$\text{SAR} = \frac{P}{m} = \frac{\sigma\left|\vec{E}\right|^2}{\rho} = C\left.\frac{dT}{dt}\right|_{t=0} \tag{5.4}$$

where m is the tissue mass, σ is its total conductivity, and ρ is its mass density. C is the heat capacity, and T is the temperature. Whole-body averaged or local 10-g averaged SAR are usually used as dosimetric quantities at microwaves. However, at MMW, where most of the energy is absorbed in the few outer millimetres of tissue, using local 10-g or 1-g averaged SAR as a metric is meaningless. However, as mentioned above, SAR can be used as an intermediate parameter to retrieve eIPD.

3) *Steady-state and/or transient temperature (T).* As MMW energy absorption results in heating, especially in case of medium- and high-power exposures, temperature is an important parameter linking the exposure metrics that can be found by post-processing based on temperature recordings with potential impacts at the level of tissues, organs, or whole organism. Some authors suggested that temperature should be used as a dosimetric quantity at MMW (e.g. [32]). However, its use for the low-power exposures dosimetry might be challenging as this would require using very sensitive temperature measurement techniques and/or multi-physics computations.

5.5.2   Exposure limits

Exposure limits recommended for the 60 GHz band by the International Commission on Non-Ionizing Radiation Protection (ICNIRP) [34] and by Institute of Electrical and Electronics Engineers (IEEE) in Std. C95.1 [35] are summarized in Table 5.3. The limits are provided in terms of IPD. Under far-field exposure conditions, the corresponding electric and magnetic field values can be found using (5.2). Additional safety margin of ×5 is applied to the general public compared to the occupational exposure in ICNIRP guidelines, while in IEEE Std. C95.1 it is of ×10 for controlled environments. It is important to stress that limits provided in Table 5.3 should be averaged over a certain area, and for local exposures (averaged over 1 cm^2) very

Table 5.3   ICNIRP and IEEE exposure limits for general public and occupational exposure at 60 GHz

	Exposure	IPD [mW/cm²]	E field [V/m]	H field [A/m]	Averaging	
					Surface [cm²]	Time [min]
ICNIRP	Occupational exposure	5 100	137 –	0.36 –	20 1	0.92 0.92
	General public	1 20	61 –	0.16 –	20 1	0.92 0.92
IEEE Std. C95.1	Controlled environment	10 100	– –	– –	100 1	0.36 0.36
	General public	1 20	– –	– –	100 1	3.6 3.6

high field values are permitted (up to $20\,mW/cm^2$ for the general public). This means that any wireless device with the radiated power below $20\,mW$ automatically complies with these limits. Note that time averaging is also involved meaning that, for a realistic signal, allowable peak radiated power can be even higher.

In relation to the emerging body-centric 60 GHz applications, it is important to underline that current regulations do not provide any recommendations for near-field exposures at MMW. In particular, ICNIRP guidelines only state that "Exposures in the near-field are more difficult to specify, because both E and H fields must be measured and because the field patterns are more complicated". To the best of our knowledge, today there is no any formal guideline providing standard numerical and/or experimental procedure for exposure compliance testing of MMW devices operating in the 60 GHz band and involving exposure of the human body under near-field conditions.

ICNIRP and IEEE limits, used as a basis for regulations in most of European countries and in United States, are science based meaning that they take into account all scientifically demonstrated well-established biological and health effects. Some countries (e.g. Switzerland and Italy) adopted lower exposure limits compared to those discussed above due to the precautionary principle.

5.6   Antennas for body-centric communications

Antennas are a key element in design of a body-centric wireless system. First MMW antennas for body-centric communications in the 60 GHz band were reported in 2011, and since that time several antenna prototypes have been introduced for both on- and off-body communications. Here, these antennas are presented, and their performances are reported both in free space and on body.

On-body antennas should be compact, light-weight, and low profile, possibly conformal to the body surface and allowing the integration with clothes and garments.

They have to be efficient, with minimal power absorption inside the human body, which behaves as a highly lossy dispersive dielectric material at MMW. Radiating structures placed on or close to the human body may experience detuning, radiation pattern distortion and changes in the input impedance and efficiency. To operate robustly close to the human body and minimize the user exposure, the coupling of antennas with the human body has to be minimized whenever possible.

5.6.1 On-body communications

Antennas for on-body communications should radiate along the body with the electric field normal to the body surface to maximize the coupling between body-worn devices. Previous studies showed that a monopole antenna is a very good candidate for on-body communications at lower microwave frequencies [36], but it does not provide high enough gain required in the 60 GHz band. Note that in this band the on-body propagation loss is very high; typical path loss on the human body is of 57–88 dB, depending on the link [37]. High antenna gain (above 12–13 dBi) is therefore required making end-fire antennas appropriate candidates for on-body MMW communications.

Several on-body antenna solutions have been reported in the 60 GHz band. Among available solutions, Yagi-Uda antennas offer a good trade-off in terms of size and gain performance compared to other end-fire antennas, such as tapered-slot antennas. Three Yagi-Uda antenna array designs were proposed (Figure 5.15):

- *Yagi-Uda antenna* on a 0.254-mm-thick RT Duroid 5880 substrate ($\varepsilon_r = 2.2$, $\tan \delta = 0.003$) with 12 dBi gain covering 55–65 GHz range [38]. The design is similar to the one presented in Figure 5.15a.
- *Textile Yagi-Uda antenna*, fabricated on a 0.2-mm-thick fabric substrate with $\varepsilon_r = 1.5$ and $\tan \delta = 0.016$, demonstrating 11.9 dBi gain and covering 57–67 GHz band [39] (Figure 5.15a).
- *SIW Yagi-Uda antenna* consisting of a row of vias, acting as directors, placed in front of the output of a substrate integrated waveguide (SIW), with 12.5 dBi measured gain and 1 GHz bandwidth [6] (Figure 5.15b). RT/Duroid 5880 with a thickness of 0.787 mm is used as a substrate.
- *Four-array Yagi-Uda antenna* printed on a 0.127-mm-thick RT/Duroid 5880 substrate with a 15 dBi gain and 5 GHz bandwidth [40] (Figure 5.15c).

Two other candidates for on-body communications in the 60 GHz band were reported (Figure 5.16). Brizzi *et al.* reported a woodpile antenna consisting in a resonant cavity fed by a monopole-like feeding source [41] (Figure 5.16a). The woodpile structure improves the gain by a more than 5 dB compared to a conventional monopole antenna. Recently, a disc-like antenna covering 59.3–63.4 GHz band has been reported [42] (Figure 5.16b). Its radiation is mainly in the plane of substrate making this structure an appropriate candidate for on-body communications. The gain of the antenna is of 6.7 dBi for the antenna placed 5 mm from the body. In contrast to Yagi-Uda antennas, these two antennas are rather complex to fabricate and bulky.

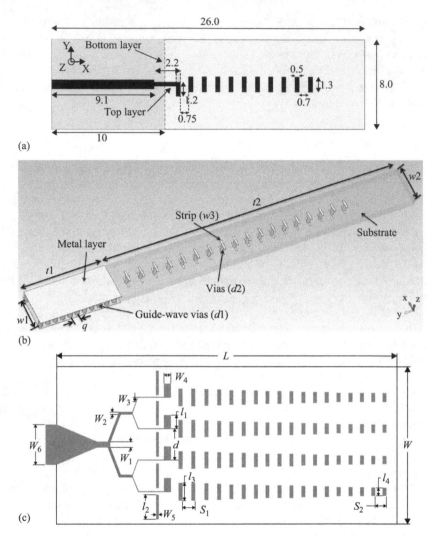

Figure 5.15   Yagi-Uda antennas for on-body communications: (a) textile antenna
[39] © 2012 IEEE; (b) SIW antenna [6] © 2013 IEEE; and (c) four
arrays antenna [40] © 2011 Wiley

5.6.2   Off-body communications

Medium-gain antennas (typically 10–13 dBi) are often required for off-body communications. In controlled environments, line-of-sight channels can be efficiently
exploited using medium-gain passive antennas, whereas directive beam steering
antennas might be desirable for non-line-of-sight adaptive channels to comply with the
power link budgets. Patch antennas have been identified as one of the best solutions for
off-body communications [43]. They are simple, compact, low-profile, light-weight,

(a)

(b)

*Figure 5.16   Other antennas for on-body communications: (a) woodpile antenna
[6] © 2013 IEEE and (b) disc-like antenna [42] © 2014 IEEE*

low-cost, and allow maximizing radiation towards the opposite side of the human body
while reducing radiation towards the body. However, in contrast to lower frequencies,
at MMW the influence of spurious waves due to the feeding lines on radiating patterns
cannot be neglected.

Two 60 GHz patch antennas were introduced in Reference 44, and one of them
was optimized to operate on a textile. A simple microstrip patch antenna was designed
to achieve a maximum gain (i.e. 6 dBi gain in free space) (Figure 5.17a). It is printed
on a 127-μm-thick RT Duroid 5880 substrate and covers 59.4–60.9 GHz band in free
space. The bandwidth is only slightly affected by the human body presence (i.e. it is
shifted by 0.2 GHz towards higher frequencies). The radiation pattern remains stable in

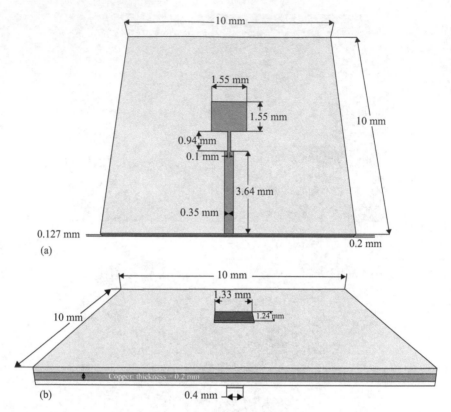

Figure 5.17   Patch antennas for off-body communications [44]: (a) microstrip patch antenna and (b) aperture-coupled patch antenna. © 2014 Chahat et al.

the direction opposite to the human body. Optimized aperture-coupled patch antenna covering 59–61 GHz band with 6.2 dBi gain in free space (6.7 dBi on body) is shown in Figure 5.17b. In this case, the reflection coefficient S_{11} is almost insensitive to the presence of the human body. However, the peak SAR is 40 times higher compared to the microstrip patch antenna due to the location of the feeding line below the ground plane. To avoid high exposure levels in case of the aperture-coupled antenna, the feeding line could be sandwiched between two substrates with top and bottom grounds.

To enhance the gain while preserving acceptably small antenna size, patch antenna arrays may be beneficially used. A 12 dBi gain microstrip-fed 2×2 patch single-layer antenna array printed on 127-µm-thick RT Duroid 5880 substrate was introduced in Reference 45. It covers the 59–65 GHz band both in free space and on body. The antenna radiation performances at broadside are nearly insensitive to the presence of the body, mainly thanks to the ground plane shield.

A similar antenna array was designed and fabricated using a copper foil deposited on a 0.2-mm-thick textile ($\varepsilon_r = 2.0$, $\tan \delta = 0.02$). In figure 5.18, flexible

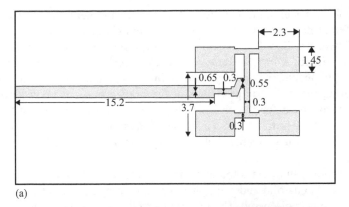

(a)

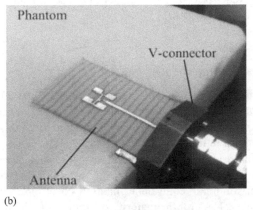

(b)

Figure 5.18   Microstrip antenna array printed on the cotton textile: (a) layout and
(b) antenna prototype on a flat-skin-equivalent phantom [46]. © 2013
IEEE

ShieldIt Super™ electrotextile was used for the ground plane. The announced fabrication accuracy is around 10 μm. While it covers the 57–65 GHz band, its gain
is reduced from 12 dBi to 8 dBi and efficiency drops from 60% to 40% compared
to the antenna array on a classical substrate. Bending has a small impact on the
reflection coefficient and gain of the textile antenna. Under crumpling conditions,
the antenna characteristics also remain satisfactory (the gain drops by 1 dB, and the
antenna remains matched over the 57–65 GHz range).

5.7   Experimental skin-equivalent models

Experimental tissue-equivalent phantoms have been extensively used in research
and compliance testing for the analysis of antenna/human body interactions, but
also to study wave propagation around and inside the human body. Such phantoms

ensure stable and well-controlled reproducible measurement conditions. Over the years, different phantoms have been proposed and classified from different points of view. There are liquid, semi-solid, and solid phantoms that can simulate different body tissues with dielectric properties covering a wide frequency range. Liquid or gel phantoms typically consist in a container filled with a liquid or gel simulating the dielectric properties of the human body or body parts. In particular, they are used for the conservative assessment of the SAR at frequencies ranging from 30 MHz to 6 GHz. This type of phantoms is of a limited interest at MMW because of the container shell and very shallow penetration of the field into the tissues.

This section summarizes the experimental models available up to date in the 60 GHz band. Semi-solid and solid phantoms are compared, and possible range of their applications is discussed. Thermal model of the semi-solid phantom with skin-equivalent properties is presented.

5.7.1 Semi-solid phantom: EM model

Semi-solid phantoms have been widely used for emulating the EM properties of tissues with a high water content (such as muscle, brain, or skin) mainly below 10 GHz and more recently up to 40 GHz [47]. The main advantage of this type of phantoms is easy fabrication both in terms of the mixture preparation and forming a required shape, including multi-layer structures. Here a solidifying agent is used to make the material jelly-like, eliminating the outer shell used in liquid phantoms. The first homogeneous semi-solid skin-equivalent phantom covering 55–65 GHz range was reported in Reference 21. As for the majority of semi-solid phantoms, its main component is deionized water. Being the main component of skin, it mainly determines the dispersive properties of the phantom. Agar is used to jellify the phantom allowing retention of the self-shaping without a container. The contribution of agar to the phantom dielectric properties is negligible for small concentrations (typically below 4%). Polyethylene powder is used to tune the complex permittivity of the water-agar mixture to approach the target skin permittivity values. Finally, TX-151 is used to increase the viscosity. Preservatives can be employed to increase the phantom lifetime from several days to several weeks or even months; however most of them are toxic and require special environmental conditions for manipulation. For short-term measurement campaigns, their use can be omitted.

An example of a semi-solid phantom representing a human arm is shown in Figure 5.19. Its fabrication procedure is described in detail in Reference 21. The EM properties of the phantom are in agreement with those of skin (relative deviation within 10%). Table 5.4 provides the dielectric properties of the phantom measured using an open-ended coaxial probe [21] and heating kinetics technique [22]. This phantom has been successfully used for exposure dosimetry studies, mainly using infrared thermometry, for assessing the impact of the presence of the human body on the antenna performances [45], as well as for body-centric propagation studies [48]. Some application examples are provided in Section 5.8.

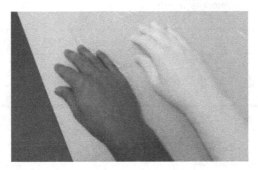

Figure 5.19   *Skin-equivalent phantom representing an arm and a hand: (left) human arm used to make a realistic mould and (right) phantom [6]. © 2013 IEEE*

Table 5.4   *EM properties of the semi-solid phantom compared to those of the dry skin [44]*

	Complex permittivity	Relative deviation	Penetration depth [mm]
Dry skin [18]	$7.98 - j \cdot 10.9$	–	0.48
Phantom (coaxial probe) [21]	$7.4 - j \cdot 11.4$	$7.3\% - j \cdot 4.6\%$	0.45
Phantom (heating kinetics) [22]	$8.3 - j \cdot 10.8$	$4\% - j \cdot 0.9\%$	0.49

5.7.2   Semi-solid phantom: thermal model

When semi-solid phantoms are used for the exposure assessment induced by on-body antennas using thermometry techniques, such as infrared imaging, the SAR and IPD are retrieved by fitting the measured temperature data to the analytical solution of the bioheat transfer equation [21, 45]. In this case, the analytical thermal model used for fitting impacts directly the accuracy of the temperature rise dynamics and dosimetric results. Recently, thermal behaviour of a semi-solid 60 GHz phantom has been analysed in detail [49]. An improved thermal model was proposed taking into account the finite thickness of the phantom and compared to a simplified semi-infinite thermal tissue model previously introduced by Foster [27].

The measured thermo-physical properties of the phantom are presented in Table 5.5 demonstrating a good agreement with skin values. The specific heat capacity and thermal conductivity were measured using a differential scaling calorimeter, and thermal method of guarded hot plate, respectively.

The thermal behaviour of the phantom was studied analytically by solving the 1D heat transfer equation. The impact of the phantom thickness on the temperature rise was investigated in detail, and results obtained for phantom models with three different thicknesses were compared to the analytical semi-infinite model (Figure 5.20). It was demonstrated that the thermal behaviour of the phantom is well

*Table 5.5 Thermo-physical properties of the skin-equivalent phantom
and human skin [49]*

	Human skin [50]	Phantom
Mass density [kg/m^3]	1093–1190	880
Heat capacity [kJ/(kg°C)]	3.15–3.71	3.48
Heat conductivity [W/(m°C)]	0.293–0.385	0.386

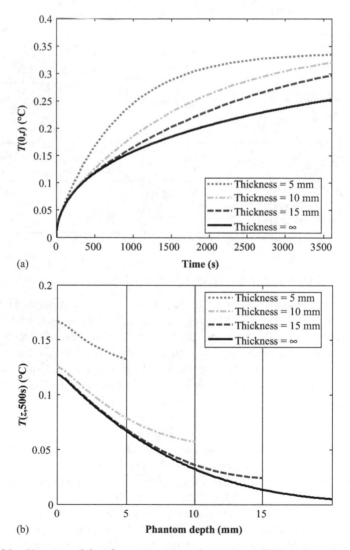

(a)

(b)

*Figure 5.20 Heating of the phantoms exposed to a normally incident plane wave at
IPD of 1 mW/cm^2: (a) heating dynamics on the phantom surface and
(b) temperature distribution in the phantom after 500 sec of exposure*

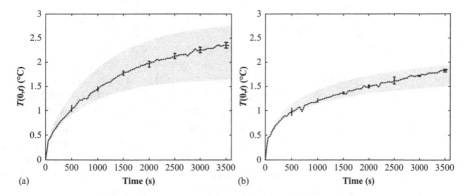

Figure 5.21   Measured temperature rise (dotted line) compared to numerical envelops (highlighted in grey): (a) 10 mm thick and (b) 150 mm thick phantom (representing semi-infinite phantom). The error bars correspond to the standard deviation for three measurements

described by the model with a finite thickness as well as by the semi-infinite model for short exposure durations (shorter than 1 min). However, for exposure durations exceeding 1 min, it is crucial to use the finite-thickness model to accurately assess the temperature rise.

In practice, variations of thermo-physical (i.e. heat conductivity and capacity) and heat transfer parameters at the air/phantom interfaces (i.e. heat transfer coefficient), which cannot be always accurately controlled in measurements, may strongly impact the heating. Figure 5.21 demonstrates the influence of these variations, as well as of typical uncertainty in definition of mass density and variability of the phantom EM properties during the exposure, on the heating of the phantom surface. Numerical results are compared with measurements performed by an infrared camera on a phantom exposed by an open-ended V-band rectangular waveguide placed 5 cm from the phantom. They demonstrate that the heating envelope (1) decreases with the phantom thickness and (2) increases with time. For both thicknesses, a very good agreement is noticed between numerical and experimental results.

Note that the phantom discussed above is skin-equivalent from the EM viewpoint. To improve the accuracy of the model in terms of the thermal equivalence to skin, it is important to create an experimental multi-layer model taking into account thermoregulation, through calibration or incorporating an artificial thermoregulation system.

5.7.3   Solid phantom

While water-based semi-solid phantoms have some important advantages, they also have two important drawbacks: (1) limited lifetime due to evaporation that also results in variations of the phantom EM properties with time and (2) its EM properties are temperature sensitive because of the high water concentration. To overcome these limitations, solid phantoms may be used since they provide stable EM and mechanical properties over a longer time span. Up to recently, existing solid phantoms have been limited to about 6 GHz and have been mainly used for body surface SAR and radiation pattern measurements. Most of them are costly and require specific equipment for fabrication, together with special high-temperature and high-pressure manufacturing procedures. At MMW, due to high loss tangent of skin (tan $\delta \approx 1.3$ at 60 GHz), primarily determined by its high free water content, design of solid phantoms with the same EM properties as skin is very challenging if not impossible. However, for certain applications, such as study of impact of the presence of the human body on the antenna performances or characterization of the body-centric propagation channel, it is sufficient to reproduce the same power reflection coefficient as that of skin without necessarily having the same EM properties (as a result, the absorption inside the phantom and surface wave propagation may be different).

In 2014, Guraliuc *et al.* introduced the first phantom with the power reflection reproducing that of skin in the 58–63 GHz range [51]. It consists in a lossy flexible PDMS (silicon-based organic polymer) dielectric sheet containing carbon powder with a metallic backing (Figure 5.22a). Its preparation is described in detail in Reference 51. It is not as straightforward as for semi-solid phantoms; one of the difficulties consists in obtaining a homogeneous carbon–PDMS mixture without bubbles. Figure 5.22b shows the complex permittivity of the PDMS–carbon composite for

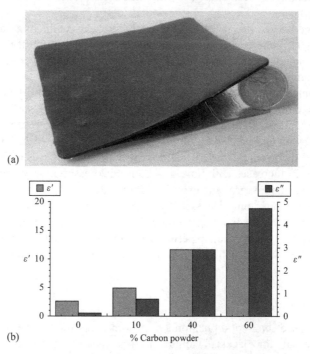

Figure 5.22   *Solid phantom fabricated using a PDMS–carbon composite with a metallic backing: (a) phantom prototype and (b) measured complex permittivity of the PDMS–carbon powder composite at 60 GHz for various concentrations of carbon [52]. © 2014 IEEE*

different concentrations of the carbon black powder. It is determined using a free-space technique with transmission/reflection quasi-optical setup. The thickness of the dielectric layer was optimized to obtain the same reflection coefficient as that of skin for a wide range of incidence angles. For the optimal configuration (i.e. carbon particles concentration of 40% and thickness of the composite 1.3 mm), average relative deviations of the power reflection coefficient with respect to skin are 6% and 3% for TM and TE polarizations, respectively.

Figure 5.23 shows an example of on-body propagation measurements between two V-polarized open-ended waveguides separated by 15 cm placed over: semi-solid

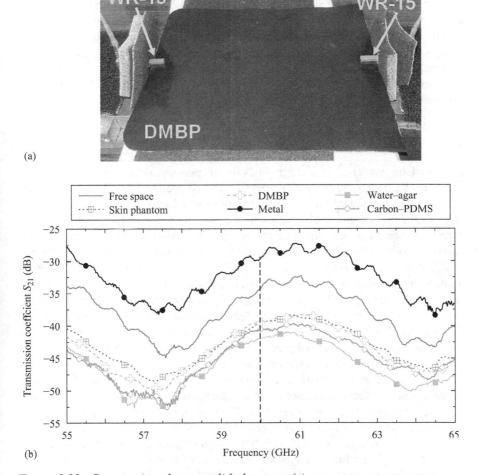

Figure 5.23 *Propagation above a solid phantom: (a) measurement setup: two open-ended waveguides positioned on the surface of a $20 \times 20\,cm^2$ dielectric metal backed phantom (DMBP) [51] © 2014 IEEE and (b) measured transmission coefficient S_{21} between waveguides in free space and above different phantoms/media [52] © 2014 IEEE*

skin-equivalent phantom, dielectric metal-backed solid phantom, metal, water–agar phantom, and carbon–PDMS sheet without metal backing. The results show that only the solid phantom provides the transmission coefficient close to the one obtained with the skin-equivalent phantom. The best agreement is obtained at 58–63 GHz (the maximum deviation is of 0.7 dB). The phantom was also used in propagation studies with realistic wearable antennas (e.g. textile Yagi-Uda antenna [51]) demonstrating a very good accuracy and robustness in different communication scenarios as well as an excellent reproducibility of results.

5.8   Near-field coupling between antennas and human body

In this section, we discuss challenges in terms of the near-field dosimetry at 60 GHz and provide an application example of IPD and SAR retrieval using the infrared thermometry. Furthermore, impact of the feeding type on the antenna/body coupling is considered in detail to demonstrate that the power absorbed in the body can be substantially reduced by appropriately choosing the feeding type. Finally, it is demonstrated that exposure in on-body communication scenarios can be reduced by using electrotextiles, simultaneously enhancing the propagation both in line-of-sight and non-line-of-sight scenarios.

5.8.1   Tools for the exposure assessment

Impact of the presence of the human body on performances of wearable MMW antennas (S_{11}, radiation, and efficiency) can be evaluated numerically and/or experimentally using standard measurement techniques and human body phantoms. The exact evaluation of power absorption in the body is a challenging task at MMW both from numerical and experimental viewpoints. The major challenges for computations are the following: (1) electrically large problems (λ_{skin} varies from 2.5 mm to 1.25 mm in 30–100 GHz range; this implies small mesh cell sizes of numerical models – of the order of 0.1 mm), (2) lack of accurate well-established body models at these frequencies, and (3) multi-scale problems related to the presence of electrically small sub-structures. Furthermore, the EM problem should be coupled with the thermodynamic one to carefully take into account possible heating and permittivity variations related to the thermal gradients. Note that noticeable change of permittivity values of biological tissues typically appears at 60 GHz for temperature gradients $\Delta T > 3°C$. As far as the experimental dosimetry is concerned, the direct field-based MMW dosimetry faces two major problems. First, the gradients of PD and SAR values within biological tissues are very high. This implies that measurements should be performed with a sub-mm accuracy. Second, today there is no commercially available 60 GHz sensors with small enough spatial resolution providing acceptable accuracy and sensitivity. Therefore, this approach has a limited practical interest for assessing the near-field exposure induced by on-body MMW antennas.

An alternative solution consists in measuring remotely or invasively the near-surface thermal dynamics on a phantom or directly on skin. SAR and IPD can then be retrieved by post-processing fitting measured temperature dynamics to an appropriate thermal model (usually a solution of heat or bioheat transfer equation). Some non- or minimally perturbing techniques have been reported for the simultaneous determination of local T, SAR, and IPD, including optical fibre measurements, high-resolution infrared imaging, thermosensitive liquid crystals, and magnetic resonance thermal imaging. The last two techniques provide 3D distribution but suffer from a low sensitivity and are difficult to implement in practice. The first one implies contact measurements only in one point. Infrared thermometry is the most suitable for MMW dosimetry allowing non-invasive surface temperature measurements on a sample.

As an example, SAR and IPD were retrieved based on the measured temperature rise on a semi-solid skin-equivalent phantom for the microstrip patch antenna array printed on Duroid 5880 substrate (see Section 5.6.2 [45]). The experimental setup is represented in Figure 5.24a. A FLIR SC5000 high-resolution infrared camera

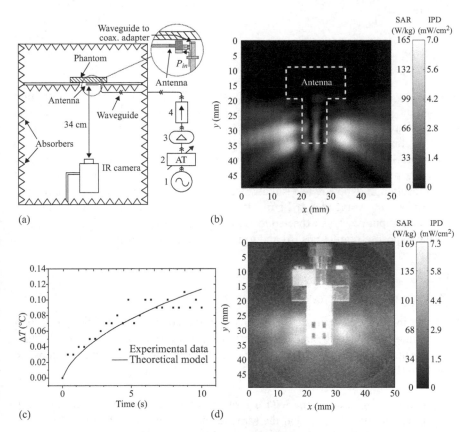

Figure 5.24   *Temperature-based dosimetry applied to the near-field exposure assessment: (a) experimental setup, (b) computed, and (d) measured SAR and IPD distributions, (c) temperature dynamics obtained for the antenna placed on the skin-equivalent phantom (input power 322 mW, antenna/phantom spacing 5.6 mm, peak SAR location), [45]. © 2012 IEEE*

(thermal sensitivity 0.02°C, spatial resolution 640×512 pixels) operating in the 2.5–5.1 μm range is used to record the heating pattern and temperature dynamics on the surface of the phantom. The camera is located on the same side as the antenna, 34 cm away from the phantom. The recorded temperature dynamics was fitted to the theoretical model deduced from the 1D heat transfer equation (Figure 5.24c). The solution of this equation is provided in Reference 27 for semi-infinite tissue-equivalent media. Heat conductivity and capacity of the phantom are reported in Table 5.5. Fitting is performed by minimizing the standard deviation value varying PD, the only unknown parameter. Once the peak PD value corresponding to the minimal standard deviation has been determined, IPD and SAR [17] can be retrieved as follows:

$$IPD = \frac{PD}{1 - R}$$

$$SAR = \frac{2PD}{\rho_{phantom}\,\delta_{phantom}}$$

(5.5)

where $\rho_{phantom}$ is the mass density of the phantom (Table 5.5), and δ the penetration depth to the phantom. A very good agreement was demonstrated between numerical results compared to SAR and IPD data retrieved using the temperature-based technique (Figure 5.24b,d). Note that here IPD is an equivalent of eIPD introduced in Section 5.5.1. It was demonstrated that for the considered scenario, even a relatively high input power of the antenna (up to 550 mW for an antenna/body separation of 1 mm) resulted in exposure levels that were below international exposure guidelines (i.e. 20 mW/cm^2 averaged over 1 cm^2). It is worthwhile to underline that in practice, the antenna input power is expected to be restricted to several tens of mW to comply with the exposure regulations [6] and to reduce the power consumption of wireless devices and sensors.

5.8.2 Impact of the feeding type

Here, we investigate the impact of the feeding type on the antenna/body coupling for antennas similar to the one considered in the previous sub-section. Three standard configurations of 2×2 patch antenna arrays for off-body communications are compared (Figure 5.25).

- *Single-layer antenna array with a ground plane fed by a microstrip line (A_1).* The feeding line is located at the same level as the patches (Figure 5.25a). The patch-to-patch spacing d_s is of 3.7 mm ($0.74 \cdot \lambda_0$ at 60 GHz).
- *Aperture-coupled array excited by a microstrip line (A_2).* The coupling slots are etched in the ground plane and excited by a microstrip beam former printed on the lower substrate (Figure 5.25b). The feeding line and radiating elements are located at different levels allowing to decrease d_s to 2.5 mm ($0.5 \cdot \lambda_0$ at 60 GHz) and thus to reduce the perturbation of the radiation pattern by the feeding.

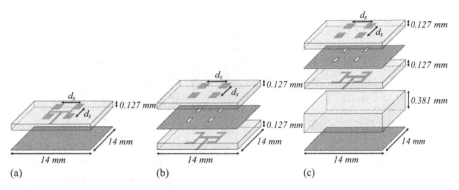

Figure 5.25   *2 × 2 patch antenna arrays layouts: (a) antenna array fed by a*
microstrip line (A₁), (b) aperture-coupled array excited by a
microstrip line (A₂), and (c) aperture-coupled array with a ground
plane (A₃)

- *Aperture-coupled array excited by a stripline (A₃).* It has a similar design with d_s
equal to 2.5 mm, with an additional substrate layer and ground plane added under
the feeding line.

The total size of the three antenna arrays is identical ($14 \times 14\,mm^2$); the
thicknesses are of 161 μm, 305 μm, and 603 μm for A_1, A_2, and A_3, respectively.

Antenna performances in terms of the reflection coefficient S_{11}, radiation pattern,
and gain were computed in free space and on a $60 \times 60 \times 1\,mm^3$ flat skin-equivalent
phantom (antenna/phantom spacing is 1 mm). The S_{11} is similar for the three
structures; the $-10\,dB$ S_{11} bandwidth is equal to 2.9 GHz, 2.7 GHz, and 2.5 GHz
for A_1, A_2, and A_3, respectively (Figure 5.26). As expected, the reflection coefficient
S_{11} of A_1 and A_3 is insensitive to the presence of the phantom, whereas S_{11} of A_2 is
very slightly shifted when the antenna is placed on body. Note that, in contrast to A1
and A3, A2 cannot be in the direct contact with the body.

The radiation patterns in free space and on the phantom are provided in Figure 5.27
(for the sake of brevity only E-plane patterns are shown). In contrast to A_2 and A_3,
in case of A_1, the larger spacing d_s between patches results in relatively high side
lobes levels in E plane. This also leads to a reduction of $-3\,dB$ beamwidth of A_1
(44°) compared to A_2 (53°) and A_3 (46°). Due to the absence of a ground plane,
the strongest backward radiation is observed for A_2 (only $-7\,dB$ compared to the
maximum gain, in contrast to $-23\,dB$ and $-29\,dB$ for A_1 and A_3, respectively).
As a result, the strongest influence of the presence of the phantom on the radiation
performances is observed for A_2, also resulting in a noticeable increase of directivity
(10%) and gain (5%) (Table 5.6).

SAR induced in the skin-equivalent phantom is represented in Figure 5.28. For
the antennas with a ground plane (A_1 and A_3), SAR is mainly due to the presence of
the side lobes. This is clearly seen in Figure 5.28a where two zones of maximum SAR
appear in E plane (parallel to x-axis) close to the antenna. For the antenna without a

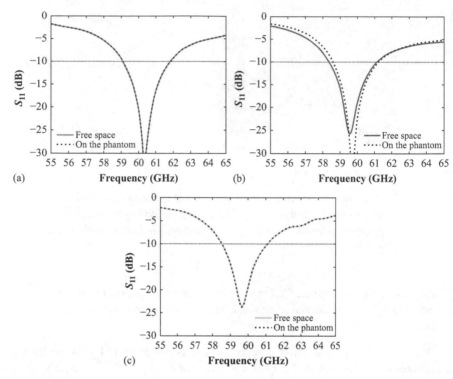

Figure 5.26   Simulated reflection coefficient S_{11} of the antenna arrays (a) A_1,
(b) A_2, and (c) A_3

ground plane (A_2), as expected, the most exposed area is located under the antenna due to the strong backward radiation of the feeding line and slots (Figure 5.27b).

A_3 ensures the lowest exposure. The peak SAR (SAR_{peak}) is 1.4 and 78.5 times lower compared to A_1 and A_2, corresponding to exposure reduction of 30.0% and 98.7%, respectively (Table 5.7). The SAR averaged over a square of 1 cm^2 around the peak SAR is reduced by a factor of 3.9 (74%) and 44.3 (97.7%) compared to A_1 and A_2, respectively. The SAR averaged over 20 cm^2 is reduced by a factor of 2.5 (59%) and 7.9 (87.4%) times for SAR averaged over 20 cm^2. The total power absorbed by the phantom P_{abs} reaches only 4.2%, 9.8%, and 0.5% of the power accepted by A_1, A_2, and A_3, respectively. Note that eIPD can be retrieved based on SAR data from (5.5):

$$\text{eIPD} = \frac{\rho_{\text{phantom}} \, \delta_{\text{phantom}} \, \text{SAR}}{2(1 - R)} \tag{5.6}$$

These results suggest that, by appropriately choosing the feeding type, the antenna/body coupling can be substantially reduced. Using of a ground plane allows decreasing significantly the backward radiation. Reducing the side lobes levels

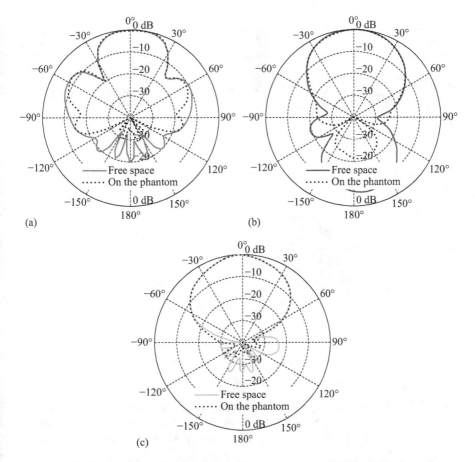

Figure 5.27 Simulated normalized radiation patterns at 60 GHz in E-plane for antenna arrays (a) A_1, (b) A_2, and (c) A_3

Table 5.6 Computed directivity, gain, and efficiency for antenna arrays (a) A_1, (b) A_2, and (c) A_3

	A_1		A_2		A_3	
	Free space	On the phantom	Free space	On the phantom	Free space	On the phantom
Directivity [dBi]	13.3	13.5	10.5	11.6	11.9	11.9
Peak gain [dB]	12.4	12.5	9.7	10.2	10.9	10.9
Efficiency [%]	0.91	0.85	0.91	0.84	0.89	0.9

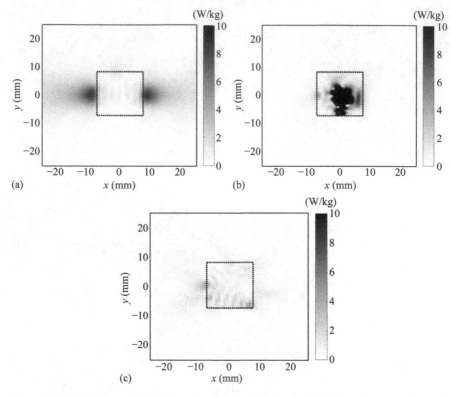

Figure 5.28   *Computed SAR distributions on the skin-equivalent phantom exposed at 60 GHz by the antenna arrays (a) A_1, (b) A_2, and (c) A_3. Input power is 10 mW. Dotted lines schematically represent the antenna location*

Table 5.7   *Computed SAR, eIPD, and absorbed in the phantom power P_{abs} at 60 GHz. The input power is 10 mW. P_a is the accepted by the antenna power*

	SAR [W/kg]			eIPD [mW/cm²]			P_{abs} [mW]	P_a [mW]
	Peak	**1 cm²**	**20 cm²**	**Peak**	**1 cm²**	**20 cm²**		
A_1	9.10	3.64	0.74	0.30	0.12	0.02	0.42	9.95
A_2	498	41.60	2.38	15.91	1.33	0.08	0.98	9.99
A_3	6.34	0.94	0.30	0.20	0.03	0.01	0.05	9.87

allows further decreasing the user exposure as well as distortion of the antenna performances (S_{11}, radiation pattern, directivity, gain, and efficiency) by the body. Note that adding vias might be beneficial to decouple antenna from the body even more. However, this complicates the antenna fabrication process.

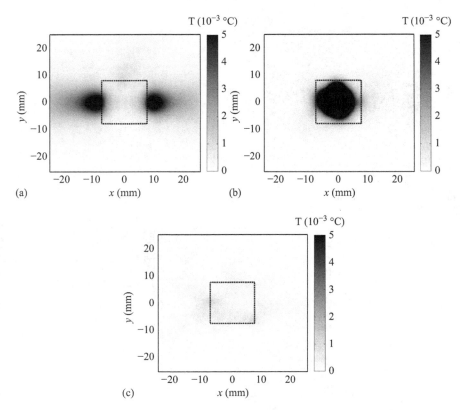

Figure 5.29 *Computed temperature rise distributions on the phantom surface after 30 sec of exposure for (a) A_1, (b) A_2, and (c) A_3*

5.8.3 Heating of tissues

Typical exposure durations in near-field body-centric scenarios are expected to be of the order of seconds or minutes. Here we consider computed temperature rise in a homogeneous skin-equivalent model ($60 \times 60 \times 15\,mm^3$) for the SAR distributions presented in Figure 5.28. The blood flow effect is taken into account (blood flow coefficient is assumed to be equal to $6800\,W/(m^3 °C)$ [50]). First, a short-term exposure is considered (i.e. continuous wave exposure for 30 sec) (Figure 5.29). For completeness, steady-state heating distributions are also analysed (Figure 5.30).

According to the bioheat transfer equation [53], the shorter the exposure is, the closer the temperature distribution to the SAR distribution. As expected, after 30 sec of exposure, heating distribution follows the same trend as SAR deposition (see Figures 5.28 and 5.29). The maximum temperature rise is observed for A_2: the peak temperature and heating averaged over $1\,cm^2$ and $20\,cm^2$ (square area centred around the peak temperature location) equal 0.18°C, 0.04°C, and 0.002°C,

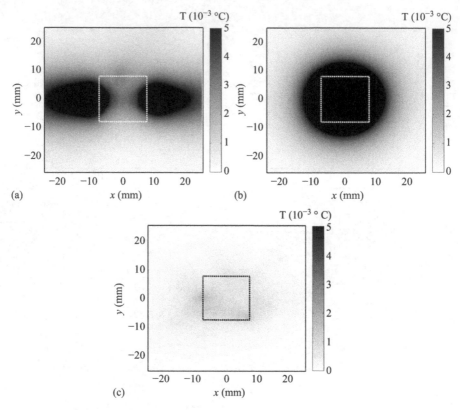

Figure 5.30   Computed steady-state temperature rise distributions on the phantom surface for (a) A_1, (b) A_2, and (c) A_3

respectively. For antennas with a ground plane, i.e. A_1 and A_3, the temperature rise (peak; 1 cm^2 averaged; 20 cm^2 averaged) is (25; 10; 3) and (179; 82; 22) times lower, respectively. Note that the stronger reduction for A_3 is due to the $\lambda_0/2$ spacing between patches resulting in the absence of side lobes.

However, at steady state, reached here after about 20 min of exposure (99% of steady-state temperature rise), the heating spreads under antennas for the structures with a ground plane (A_1 and A_3) due to heat conduction and diffusion (Figure 5.30a,c). For the antenna without a ground plane (A_2), for which the SAR is mainly concentrated under the antenna, steady-state heating spreads beyond the antenna limits (Figure 5.30b). For A_2, the steady-state temperature rise (peak; 1 cm^2 averaged; 20 cm^2 averaged) is of (0.23°C; 0.07°C; 0.008°C), exceeding by (18; 9; 3) and (133; 68; 25) times the peak values for A_1 and A_3, respectively.

Table 5.8 summarizes maximum and averaged temperature rises on the phantom after 30 sec of exposure and at steady state. These data are compared to eIPD in Figure 5.31. This figure also demonstrates the trends related to the exposure duration and averaging for three considered antenna arrays. Similar thermal behaviours are

Table 5.8 *Peak and averaged (over 1 cm² and 20 cm²) simulated temperature rise after 30 sec of exposure and at steady state at the phantom surface for an input power of 10 mW at 60 GHz*

	$T_{30\,sec}$ [°C]			$T_{steady-state}$ [°C]		
	Peak	1 cm²	20 cm²	Peak	1 cm²	20 cm²
A_1	0.0071	0.0042	0.0008	0.0124	0.0084	0.0026
A_2	0.1785	0.0412	0.0024	0.2261	0.0746	0.0078
A_3	0.0010	0.0005	0.00011	0.0017	0.0011	0.00031

noticed for the antennas with a ground plane (A_1 and A_3 – Figure 5.31a,c), while it is different for A_2 (Figure 5.31b).

Note that for the considered input power (10 mW), representing a typical operating power of on-body 60 GHz systems, resulting heating is well below environmental temperature fluctuations. Here we consider only heating induced by the antenna radiation, without taking into account the impact of the antenna presence and heating on the temperature distribution in the phantom. The later together with the presence of textiles can result in an additional temperature rise.

5.8.4 Electrotextiles for the exposure reduction

Exposure reduction by appropriately choosing the antenna feeding type can be efficiently applied to antennas for off-body communications. However, in on-body communications this technique is of a limited interest as along the body radiation of end-fire antennas unavoidably results in the user exposure. Recently, the effect of textiles on propagation along the body at 60 GHz was investigated [54]. It was demonstrated that the presence of a regular textile over a skin-equivalent phantom, as well as an air gap between them, induces a typical decrease of the path gain by 2–5 dB, but it does not significantly affect the path gain exponent. In continuation to this study, it was proposed to use electrotextiles to enhance the propagation along and around the body [55]. This is of importance as at 60 GHz the on-body propagation loss is very high (see Section 5.6.1). As it can be seen in Figure 5.32, the surface current induced on the electrotextile enhances the propagation along the surface in case of V polarization (i.e. E field perpendicular to the phantom surface). The path gain is increased by about 15 dB at 20 cm from the source compared to the scenario without an electrotextile. Besides, electrotextile behaves as a shielding layer for skin, significantly reducing the exposure of the body to almost zero. Absorbed by the body power and peak local SAR with and without electrotextile are compared in Table 5.9 for an input power of 10 mW. Therefore, electrotextiles can be advantageously used at MMW to enhance the on-body propagation while simultaneously reducing the exposure level by more than 95%.

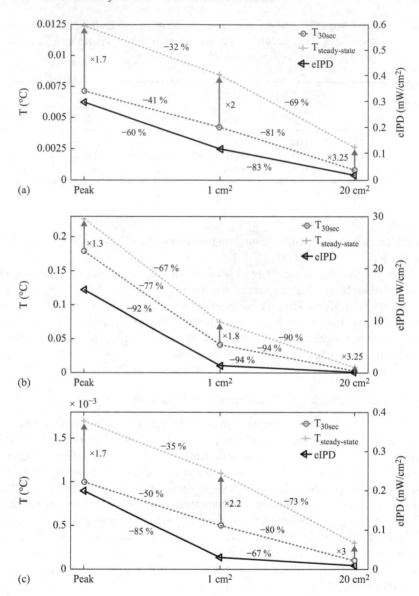

Figure 5.31   Impact of the exposure duration and surface averaging on the heating for (a) A_1, (b) A_2, and (c) A_3. Data are compared for eIPD values for each antenna array

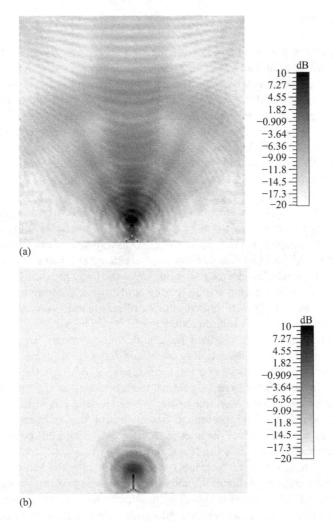

Figure 5.32 *Computed surface current distributions on the electrotextile placed over the skin-equivalent phantom: (a) V polarization and (b) H polarization [55] © 2014 IEEE. Open-ended rectangular waveguide is used as a source*

Table 5.9 *Total absorbed power and maximum SAR averaged over 1 g of skin*

Model	Absorbed power [mW]	Peak SAR_{1g} [W/kg]
Skin	4.14	1.23
Electrotextile/skin	0.068	0.003

5.9   Conclusion

In this chapter, we presented a review of the state-of-the-art, recent advances, and remaining challenges in the field of antenna/human body interactions in the 60 GHz band, with a particular emphasis on the near-field interactions that may occur in emerging body-centric MMW applications. For new near-field exposure scenarios expected to appear in coming years, today there is no any clear regulation in terms of the exposure assessment, and standard methodologies for compliance testing are not available in the 60 GHz band. Most of the exposure guidelines and standards recommend IPD as a dosimetric quantity. It cannot be directly used as a metric for near-field exposures since practically it is very challenging, if not impossible, to determine numerically or experimentally local IPD under near-field conditions. Some reports suggest that temperature rise could be used as an exposure metric at MMW. However, in practice, it is not always possible because of some limitations discussed in this chapter. We suggest to use eIPD as a metric; it can be conveniently retrieved from SAR computations and/or measurements and takes into account the perturbation of the wireless device radiation due to the body proximity.

60 GHz antennas reported so far for on- and off-body communications have been presented as well. Only a limited number of textile antennas have been introduced and improvements in this direction might be of importance, including the issues related to the feeding and their integration into garments. To assess experimentally antenna/human body coupling, semi-solid and solid phantoms have been recently introduced. Further investigations are needed in this direction, in particular on the development of heterogeneous thermal models as well as solid phantoms with tissue-equivalent properties. Example of use of a semi-solid phantom for near-field dosimetry at MMW has been presented, demonstrating that infrared thermometry can be used as a multi-physics dosimetric tool for assessing both heating and EM exposure parameters (SAR, PD, and IPD).

Finally, we reported recent attempts towards minimizing the antenna/body coupling through appropriate choice of the antenna feeding type or using electrotextiles. It was demonstrated that by properly choosing the antenna feeding type, power absorbed in the body can be substantially reduced. Moreover, the use of electrotextiles allows enhancing the on-body propagation and simultaneously reducing the user exposure by more than 95%.

Acknowledgements

The authors acknowledge Denys Nikolayev for his kind assistance in editing this chapter. This work was supported by the European Union Seventh Framework Program under the grant n°619563 (MiWaveS project) and by Labex CominLabs under the French National Research Agency program "Investing for the Future" ANR-10-LABX-07-01 (ResCor/BoWi project).

References

[1] Daniels R.C., Murdock J.N., Rappaport T.S., Heath R.W. "60 GHz wireless: up close and personal". *IEEE Microwave Magazine*. 2010;**11**(7):44–50.

[2] "Cisco visual networking index: global mobile data traffic forecast update 2014–2019 White paper". Available online at www.cisco.com/c/en /us/solutions/collateral/service-provider/visual-networking-index-vni/white _paper_c11-520862.html.

[3] Cotton S.L., Scanlon W.G., Hall P.S. "A simulated study of co-channel inter-BAN interference at 2.45 GHz and 60 GHz". *European Wireless Technology Conference (EuWIT)*; Paris, France, Sept. 2010. pp. 61–4.

[4] Cotton S.L., Scanlon W.G., Madahar B.K. "Millimeter-wave soldier-to-soldier communications for covert battlefield operations". *IEEE Communications Magazine*. 2009;**47**(10):72–81.

[5] Wells J. "Faster than fiber: the future of multi Gb/s wireless". *IEEE Microwave Magazine*. 2009;**10**(3):104–12.

[6] Pellegrini A., Brizzi A., Zhang L., *et al.* "Antennas and propagation for body-centric wireless communications at millimeter-wave frequencies: a review". *IEEE Antennas and Propagation Magazine*. 2013;**55**(4):262–87.

[7] "Beyond 2020 heterogeneous wireless network with millimeter wave small cell access and backhauling (MiWaves)". *EU Project*. Available online at www.miwaves.eu.

[8] Alipour S., Parvaresh F., Ghajari H., Donald F.K. "Propagation characteristics for a 60 GHz wireless body area network (WBAN)". *Military Communications Conference (MILCOM)*, San Jose, CA, USA, Oct.–Nov. 2010. pp. 719–23.

[9] Google ATAP. "Project Soli". Available at www.google.com/atap/project-soli.

[10] Rendevski N., Cassioli D. "UWB and mmWave communication techniques and systems for healthcare". In Yuce M.R. (ed.). *Ultra-Wideband and 60 GHz Communications for Biomedical Applications*. Australia: Springer, 2014. pp. 1–22.

[11] "Glucowise". Available online at www.gluco-wise.com.

[12] Cano-Garcia H., Kosmas P., Kallos E. "Enhancing electromagnetic transmission through biological tissues at millimeter waves using subwavelength metamaterial antireflection coatings". *9th International Congress on Advanced Electromagnetic Materials in Microwaves and Optics – Metamaterials*, Oxford, UK, Sept. 2015. pp. 7–12.

[13] Zhadobov M., Alekseev S.I., Le Dréan Y., Sauleau R., Fesenko E.E. "Millimeter waves as a source of selective heating of skin". *Bioelectromagnetics*. 2015;**36**(6):464–75.

[14] Duck F.A. *Physical Properties of Tissue*. Bath, UK: Academic, 1990.

[15] Chen L.-F., Ong C.K., Neo C.P., Varadan V.K. *Microwave electronics: measurement and materials characterisation*. England: John Wiley & Sons, 2004.

[16]   Ellison W.J. "Permittivity of pure water, at standard atmospheric pressure, over the frequency range 0–25 THz and the temperature range 0–100°C". *Journal of Physical Chemistry.* 2007;**36**:1–18.

[17]   Gandhi O.P., Riazi A. "Absorption of millimeter waves by human beings and its biological implications". *IEEE Transactions on Microwave Theory and Techniques.* 1986;**34**(2):228–35.

[18]   Gabriel S., Lau R.W., Gabriel C. "The dielectric properties of biological tissues: II. Measurements in the frequency range 10 Hz to 20 GHz" *Physics in Medicine and Biology.* 1996;**41**:2251–69.

[19]   Alabaster C.M. "Permittivity of human skin in millimeter wave band". *Electronics Letters.* 2003;**39**(21):1521–22.

[20]   Alekseev S.I., Ziskin M.C. "Human skin permittivity determined by millimeter wave reflection measurements". *Bioelectromagnetics.* 2007;**28**(5):331–9.

[21]   Chahat N., Zhadobov M., Sauleau R. "Broadband tissue-equivalent phantom for BAN applications at millimeter waves". *IEEE Microwave Theory and Techniques.* 2012;**60**(7):2259–66.

[22]   Chahat N., Zhadobov M., Sauleau R., Alekseev S.I. "New method for determining dielectric properties of skin and phantoms at millimeter waves based on heating kinetics". *IEEE Transactions on Microwave Theory and Techniques.* 2012;**60**(3):827–32.

[23]   Zhadobov M., Chahat N., Sauleau R., Le Quément C., Le Dréan Y. "Millimeter-wave interactions with the human body: state of knowledge and recent advances". *International Journal on Microwave and Wireless Technologies.* 2011;**3**(2):237–47.

[24]   Alekseev S.I., Radzievsky A.A., Logani M.K., Ziskin M.C. "Millimeter wave dosimetry of human skin". *Bioelectromagnetics.* 2008;**29**(1):65–70.

[25]   Pavselj N., Miklavcic D. "Resistive heating and electropermeabilization of skin tissue during in vivo electroporation: a coupled nonlinear finite element model". *International Journal of Heat and Mass Transfer.* 2011;**54**: 2294–302.

[26]   Ali K., Brizzi A., Pellegrini A., Hao Y., Alomainy A. "Investigation of the effect of fabric in on-body communication using finite difference time domain technique at 60 GHz". *Antennas and Propagation Conference (LAPC),* Loughborough, UK, Nov. 2012.

[27]   Foster K.R., Kristikos P.T., Schwan H.P. "Effect of surface cooling and blood flow on the microwave heating of tissue". *IEEE Transactions Biomedical Engineering.* 1978;**25**(3):313–16.

[28]   Kanezaki A., Hirata A., Watanabe S., Shirai H. "Parameter variation effects on temperature elevation in a steady-state, one-dimensional thermal model for millimeter wave exposure of one- and three-layer human tissue". *Physics in Medicine and Biology.* 2010;**55**:4647–59.

[29]   Alekseev S.I., Ziskin M.C. "Influence of blood flow and millimeter wave exposure on skin temperature in different thermal models". *Bioelectromagnetics.* 2009;**30**:52–8.

[30] Alekseev S.I., Ziskin M.C. "Millimeter-wave absorption by cutaneous blood vessels: a computational study". *IEEE Transactions on Biomedical Engineering*. 2009;**56**:2380–8.

[31] Nelson D.A., Nelson M.T., Walters T.J., Mason P.A. "Skin heating effects of millimeter-wave irradiation-thermal modelling results". *IEEE Transactions on Microwave Theory Techniques*. 2000;**48**(11);2111–20.

[32] Wu T., Rappaport T.S., Collins C.M. "Safe for generations to come: considerations of safety for millimeter waves in wireless communications". *IEEE Microwave Magazine*. 2015;**16**(2):65–84.

[33] Ryan K.L., D'Andrea J.A., Jauchem J.R., Mason P.A. "Radio frequency radiation of millimeter wave length: potential occupational safety issues relating to surface heating". *Health Physics*. 2000;**78**(2):170–81.

[34] ICNIRP. "Guidelines for limiting exposure to time-varying electric, magnetic, and electromagnetic fields (up to 300 GHz)". *Health Physics*. 1998;**74**(4): 494–522.

[35] *IEEE Standard for safety levels with respect to human exposure to radio frequency electromagnetic fields 3 kHz to 300 GHz*. Apr. 2006.

[36] Hall P.S., Hao Y., Nechayev Y.I., *et al.* "Antennas and propagation for on-body communication systems". *IEEE Antennas and Propagation Magazine*. 2007;**49**(3):41–58.

[37] Nechayev Y.I., Wu X., Constantinou C.C., Hall P.S. "Millimetre-wave path loss variability between two body-mounted monopole antennas". *IET Microwaves Antennas and Propagation*. 2013;**7**(1):1–7.

[38] Guraliuc A.R., Chahat N., Leduc C., Zhadobov M., Sauleau R. "End-fire antenna for BAN at 60 GHz: impact of bending, on-body performances, and study of an on to off-body scenario". *Electronics, Special issue Wearable Electronics*. 2014;**3**:221–33.

[39] Chahat N., Zhadobov M., Sauleau R. "Wearable end-fire textile antenna for on-body communications at 60 GHz". *IEEE Antennas and Wireless Propagation Letters*. 2012;**11**:799–802.

[40] Wu X.Y., Akhoondzadeh-Asl L., Hall P.S. "Printed Yagi-Uda array for on-body communication channels at 60 GHz". *Microwave and Optical Technology Letters*. 2011;**53**(12):2728–30.

[41] Brizzi A., Pellegrini A., Hao Y. "Design of a cylindrical resonant cavity antenna for BAN applications at V band". *IEEE International Workshop on Antenna Technology*, Tucson, AZ, USA, 2012.

[42] Puskely J., Pokorny M., Lacik J., Raida Z. "Wearable disc-like antenna for body centric communications at 61 GHz". *IEEE Antennas and Wireless Propagation Letters*. 2014;**14**:1490–3.

[43] Hall P.S., Hao Y. *Antennas and Propagation for Body Centric Communications Systems*. Norwood, MA, USA: Artech House, 2006.

[44] Chahat N., Zhadobov M., Sauleau R. "Antennas for body centric wireless communications at millimeter wave frequencies". In Huitema L. (ed.). *Progress in Compact Antennas*. InTech, 2014.

[45] Chahat N., Zhadobov M., Le Coq L., Alekseev S.I., Sauleau R. "Characterization of the interactions between a 60-GHz antenna and the human body in an off-body scenario". *IEEE Transactions on Antennas and Propagation.* 2012;**60**(12):5958–65.

[46] Chahat N., Zhadobov M., Sauleau R. "60-GHz textile antenna array for body-centric communications". *IEEE Transactions on Antennas and Propagation.* 2013;**61**(4):1816–24.

[47] Aminzadeh R., Saviz M., Shishegar A.A. "Theoretical and experimental broadband tissue-equivalent phantoms at microwave and millimetre-wave frequencies". *Electronics Letters.* 2014;**50**(8):618–20.

[48] Chahat N., Valerio G., Zhadobov M., Sauleau R. "On-body propagation at 60 GHz". *IEEE Transactions on Antennas and Propagation.* 2013;**61**(4): 1876–88.

[49] Leduc C., Zhadobov M., Sauleau R. "Thermal model of skin-equivalent phantoms at 60 GHz". *European Microwave Week (EuMW)*, Paris, France, Sept. 2015.

[50] Hasgall P.A., Di Gennaro F., Baumgartner C., *et al. IT'IS Database for Thermal and Electromagnetic Parameters of Biological Tissues.* Version 2.5, 2014.

[51] Guraliuc A.R., Zhadobov M., De Sagazan O., Sauleau R. "Solid phantom for body-centric propagation measurements at 60 GHz". *IEEE Transactions on Microwave Theory and Techniques.* 2014;**62**(6):1373–80.

[52] Zhadobov M., Guraliuc A., Chahat N., Sauleau R. "Tissue-equivalent phantoms in the 60-GHz band and their application to the body-centric propagation studies". *International Microwave Workshop on RF and Wireless Technologies for Biomedical and Healthcare Applications (IMWS-Bio 2014),* London, Dec. 2014.

[53] Pennes H.H. "Analysis of tissue and arterial blood temperatures in the resting human forearm". *Journal of Applied Physiology.* 1948;**85**(1):5–34.

[54] Guraliuc A.R., Zhadobov M., Valerio G., Chahat N., Sauleau R. "Effect of textile on the propagation along the body at 60 GHz". *IEEE Transactions on Antennas and Propagation.* 2014;**62**(3):1489–94.

[55] Guraliuc A.R., Zhadobov M., Valerio G., Sauleau R. "Enhancement of on-body propagation at 60 GHz using electro textiles". *IEEE Antennas and Wireless Propagation Letters.* 2014;**13**:603–6.

Chapter 6
Antennas for ingestible capsule telemetry

Denys Nikolayev[*,†,‡,§], *Maxim Zhadobov*[†],
Ronan Sauleau[†] *and Pavel Karban*[‡]

6.1 Introduction

The electronic implants for biomedical applications came out in mid-50s with the invention of the first pacemakers [1]. This healthcare innovation exposed the potential and revealed the vast opportunities in the field of implantable electronic devices, enabling new ways of diagnostics and treatments maintaining the patient mobility. So, the idea of wireless telemetry was born: it permits to efficiently prevent, monitor and treat a disease timely.

The first published statements about the successful development of wireless ingestible capsules for diagnostic and research purposes appeared almost simultaneously in *Nature* magazine in 1957 as a note by Zworykin and Farrar [2] and a letter from Mackay and Jacobson [3]. Zworykin's note reported the capsule that measures pressure variation inside a gastrointestinal (GI) tract. The capsule operated at frequencies around 1 MHz; more technical details were given later in Reference 4. Mackay *et al.* in their further works [5–7] proposed the capsules measuring pressure, temperature and pH. These capsules operated at sub-megahertz frequencies. One of the first successful *in vivo* application of wireless telemetry implant was reported by Sines in 1960 [8]. The first US patent for the 'Pill-type swallowable transmitter' appeared in 1964 [9]. In 1972, Watson *et al.* [10] reported a radiotelemetry capsule for pH profile measurement; the capsule operated in the 270–570 KHz range.

No significant progress was made until the 2000s, when further miniaturisation of electronics, progress in microelectromechanical systems and microfluidics empowered numerous innovations in wireless implants. Apart from telemetering various diagnostic data – endoscopy, temperature, pressure, pH, oxygen and glucose

*Corresponding author. Email: d@deniq.com
†Institute of Electronics and Telecommunications of Rennes (IETR) – UMR CNRS 6164, University of Rennes 1, bat. 11D, 263 Avenue du General Leclerc, 35042 Rennes Cedex, France
‡Department of Theory of Electrical Engineering, University of West Bohemia, Univerzitni 26, 306 14 Pilsen, Czech Republic
§BodyCap, 6 rue de la Girafe, 14000 Caen, France

levels – modern *in-body* biomedical applications include but not limited to: brain–machine interfaces and visual prosthesis, pacemakers and defibrillators, drug delivery and hyperthermia [11–13].

Antenna design and efficiency issues, lossy and dispersive nature of biological tissues, along with the energy and miniaturisation problems, are always among the main challenges to face while developing wireless biotelemetry appliances [14, 15]. Even though the telemetric GI capsules were among the first wireless *in-body* devices, improving the transmission performances still remain one of the main design challenges.

In this chapter, we provide an overview of ingestible capsule wireless telemetry with the main focus on specific challenges and difficulties associated to the design of ingestible GI capsule antenna systems.[1]

6.2   Capsule telemetry in medicine and clinical research

Ingestible capsules are used primarily for wireless endoscopy or telemetry of various physiological parameters: e.g. temperature, pH and pressure [16, 17]. Potential future applications include the drug delivery [18] and surgery [19]. Some GI capsule antennas could be designed for both ingestible and subcutaneous/intraperitoneal implantable applications. It is useful for body temperature telemetry: the same capsule design can be used as ingestible for a human health monitoring or as implantable for a clinical research on animals. Table 6.1 lists the commercial GI systems and working prototypes available as for May 2015.

6.2.1   Wireless endoscopy

The invention of wireless capsule endoscopy not only allowed painless endoscopic imaging, but also facilitated visualisation of whole small bowel and colon. The first published GI images using a wireless transmitter were obtained by Swain *et al.* in 1997 [20]. Comprehensive review of the wireless capsule endoscopy methodology is given from various perspectives in References 21–26. The first part of Table 6.1 lists the up-to-date commercial wireless capsule endoscopes.

Given Imaging provides a family of *PillCam* [27] endoscopes for various purposes: SB series for the visualisation of the small bowel mucosa, COLON series for the colon and ESO series for the oesophageal mucosa (Figure 6.1 right). The first two capsules perform up to 8 h and ESO up to 30 min. All capsules operate at 434.1 MHz. SB 3 and COLON 2 could use either 3.2 MHz bandwidth for 2.7 Mb·s^{-1} data rate or 6.5 MHz for 5.4 Mb·s^{-1}. ESO 3 uses 9.7 MHz bandwidth to transmit 35 fps video. OMOM [28] capsule endoscope transmits in the 2.400–2.4835 GHz range. GFSK modulation enables to communicate 2 fps for QVGA image resolution or 0.5 fps for VGA. Olympus *Endocapsule* [29] (Figure 6.1 left) operates at 433.8 MHz and transmits 2 fps video data. *Sayaka* capsule prototype is batteryless

---

[1]Referred as *capsule antennas* later in the text

Table 6.1   Prototypes and commercial wireless GI capsules

	Company	Reference	Size (mm)	f (MHz)	Features
Endoscopy	Given	PillCam SB 3	26.2 × ø11.4	434	2–6 fps, ⩽8 h
	Imaging	PillCam COLON 2	31.5 × ø11.6	434	4–35 fps, ⩽8 h
		PillCam ESO 3	31.5 × ø11.6	434	35 fps, 30 min
	Chongqing	OMOM	28.3 × ø13	2400	0.5/2 fps, 7–9 h
	Olympus	Endocapsule 10	26 × ø11	434	2 fps, 12 h
	RF	Sayaka	23 × ø9	N/A	*Prot.*, 30 fps, WPT†
	IntroMedic	MiroCam MC1000-W	24.5 × ø10.8	✗	HBC [32], 3 fps, 12 h
		MiroCam MC1000-WM	25.5 × ø10.8	Same + magnetic locomotion, 8 h	
	CapsoVision	CapsoCam SV-1	31 × ø11	✗	20 fps, 15 h
Telemetry	BMedical	VitalSense	23 × ø8.7	40	$T°$, ⩽10 days
	BodyCap	e-Celsius	17.7 × ø8.9	434	$T°$, ≈30 days
	Chongqing	OMOM pH	26 × 6 × 5.5	N/A	pH, ≈4 days
	Given	Bravo pH	25 × 6 × 5.5	434	pH, 2–4 days
	Imaging	SmartPill	26 × ø13	434	$T°$, pH, P, ⩾5 days
	HQ	CorTemp	22 × ø11	0.3	$T°$, 7–10 days
	Philips	IntelliCap	26 × ø11	N/A	*Prot.*, $T°$, pH, *DD*
	Remo	240	—	N/A	*Prot.*, $T°$, pH, P

✗ – the capsule does not transmit data via radio waves; fps – frames per second; *Prot.* – prototype; $T°$ – temperature; P – pressure; *DD* – drug delivery.
† Wireless power transfer.

Figure 6.1   Wireless endoscopy capsules: (left) Olympus Capsule Endoscope –
image courtesy of Olympus America Inc. and (right) Given Imaging
PillCam ESO 2 – image courtesy of Given Imaging Ltd.

and is fed wirelessly; it is claimed to transmit up to 30 fps video data [30]. According to the personal correspondence with RF System Lab, the working frequency has not been defined yet.

Not all capsule endoscopes communicate using radio waves. IntroMedic *MiroCam* [31] employs the patented human body communication technology [32] that uses electrodes in direct contact with a human body to transmit the video data at 3 fps. MC1000-WM model also have magnetic force locomotion control. CapsoView *CapsoCam SV-1* [33] endoscope has four cameras on-board permitting the total 360° viewing angle. The capsule does not have an antenna: it stores all video data on-board to be recovered after the procedure.

Figure 6.2   Wireless telemetry capsules: (left) BodyCap eCelsius capsules and (right) Bravo pH – image courtesy of Given Imaging Ltd.

Wireless capsule endoscopy demands high data rates for transmitting video data [14]. Wideband designs allow transferring high-resolution images without compression, thus reducing the energy consumption and image latency [17]; it requires the antennas with sufficiently large bandwidth [34].

6.2.2   Wireless telemetry of physiological parameters

GI capsules are used in medicine to measure the temperature, pressure and pH level, as well as for the drug delivery.

VitalSense capsule [35] measures temperature in the 25–50°C range and operates at 40.68 MHz; maximal transmission distance reaches 1 m. BodyCap *eCelsius* [36] (Figure 6.2 left) is the smallest temperature telemetry capsule of size (17.7 × ø8.9) mm with the longest operational time (1 month). It transmits at 434 MHz for up to 2 m from human body. Capsule on-board memory permits to recover data if the connection was interrupted. *Bravo pH* [37] (Figure 6.2 right) and *OMOM pH* [38] keep track of the acid reflux; these capsules are fixed in an oesophagus using applicators. Bravo pH operates at 434 MHz with its range up to 1 m. *SmartPill* [39] measures pressure, pH and temperature in the GI tract. The capsule transmits at 434 MHz with the range up to 1 m. HQ *CorTemp* [40] capsule uses 262–300 kHz inductive link to communicate the temperature for up to 40 cm. Philips *IntelliCap* [41] prototype provides the temperature and pH telemetry along with the drug delivery. Remo *240* [42] capsule prototype monitors the temperature, pressure and pH levels with the working range up to a few metres.

6.2.3   Animal-implantable wireless telemetry

Wireless-implant telemetry for animals has similar omnidirectional antenna design requirements as the GI capsules do, since the position of an implant is often undefined (especially for the intraperitoneal implantation). Thereby, some of the capsules could be technically used both as implantable and ingestible. Table 6.2 lists the modern commercial wireless implants for the veterinary and research on animals. Comprehensive overview of these systems can be found in References 43, 44.

Table 6.2 *Commercial animal-implantable telemetry devices*

Company	Product	Size (mm)	Features
BodyCap	Anipill	17.7 × ø8.9	434 MHz, $T°$, range ≈3 m
DSI*	Digital (L) ser.	29 cm³	P, $T°$, BP, *act*. for large animals
	HD series	1.4, 5.9 cm³	P, $T°$, BP, *act*. for small animals
	PA series	1.1–25 cm³	P, *act*. for mice, small and large animals
	TA series	1.1–25 cm³	$T°$, *act*. for the same animals
	EA/CA/ETA/CTA	1.1–20 cm³	BP (EEG, EMG), $T°$, *act*. for the same
	Multiplus ser.	1.9–33 cm³	Respiratory impedance, P, BP, $T°$, *act*.
Millar	Small animal telemeters	7 cm³	2 KHz, wireless battery recharge. O_2, *SNA*, P, BP (ECG, EMG, EEG). Range ⩾5 m
Remo	100	—	$T°$, *act*., ECG, EMG, EEG. Range ⩾3 m
	300	—	Same as '100' + multichannel
	400	—	WPT. $T°$, *act*.
STARR	G2 E-Mitter	15.5 × ø6.5	WPT. $T°$, *act*. for mice
	G2 HR E-Mitter	19.5 × ø6.5	same as above + *HR*
	TA E-Mitter	23.0 × ø8	WPT. $T°$, *act*.
	HR E-Mitter	26.0 × ø8	WPT. *HR*, $T°$, *act*.
Star-Oddi	DST nanoRF-T	17.5 × ø6	$T°$
	DST microRF-HRT	25.4 × ø8.3	*HR*, $T°$
TeleMetronics	TemPlant	34.0 × ø13	30 MHz. Large animals. Range ⩾20 m
	PhysioLinQ	34.0 × ø13	30 MHz. WPT. $T°$, *HR*, *act*. for rats

Abbreviations: P, pressure; $T°$, temperature; *act*., gross motor activity; *SNA*, sympathetic nerve activity; *HR*, heart rate; *BP*, biopotentials; ECG, electrocardiography; EEG, electroencephalography; EMG, electromyography.
*Size varies for different models within the series.

6.2.4 Conclusions

Biotelemetry of physiological data usually does not require high data rates as the endoscopy does. However, the operating range requirements are usually higher, especially when the mobility of a subject – human or animal – should be unrestrained. Reliable long-range communication can facilitate health monitoring to prevent a disease or to diagnose it at an early stage without affecting the life quality of a patient.

6.3 Biological environment

6.3.1 GI capsule passage

Ingestible capsule antennas must efficiently transmit from a GI tract. Antenna performances depend strongly on dielectric parameters of the environment [15]. These properties must be taken into account as a phantom[2] while designing the antennas. Let

---

[2]Environment with tissue equivalent electromagnetic properties

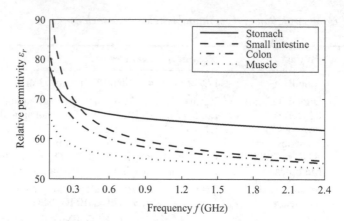

Figure 6.3   Real part of the relative permittivity ε_r of GI tract and muscle tissues (data from Gabriel et al. [45–47])

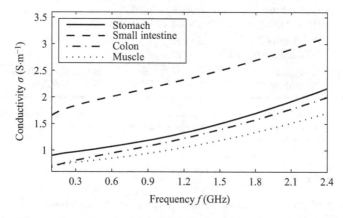

Figure 6.4   The conductivity σ of GI tract and muscle tissues (data from Gabriel et al. [45–47])

us overview the passage of a capsule through GI tract indicating the measured electromagnetic properties at 434 MHz for each section (data from Gabriel *et al.* [45–47], Figures 6.3 and 6.4).

After swallowing, a capsule passes through an oesophagus to stomach; the transit time is a few seconds and the dielectric parameters are $\varepsilon_r = 67.2$, $\sigma = 1.01$ S·m^{-1} for both. In stomach, the capsule was found to spend in average of 33 min for a person without a gastric retention [48]; Worsøe *et al.* state the duration from 35.5 to 57.5 min [49]. Rao *et al.* [50] reported the 3 h gastric empty time for a healthy person. The substantial differences in observed durations are explained by different experimental protocols, i.e. various patient preparation and diet.

Table 6.3 Time-averaged dielectric properties of a human GI tract at 434 MHz

	Time (min)	Time (%)	ε_r (–)	$\sigma(\text{S·m}^{-1})$
Stomach	180	10.53	67.2	1.01
Small intestine	228	13.33	65.3	1.92
Large intestine	1302	76.14	62.0	0.87
Time average			**63.0**	**1.02**

Table 6.4 Tissue dielectric properties for the most common implantation sites

Tissue	403 MHz		434 MHz		2.4 GHz	
	ε_r	σ	ε_r	σ	ε_r	σ
Skin	46.7	0.69	46.1	0.70	38.1	1.44
Muscle	57.1	0.80	56.9	0.81	52.8	1.71
Fat	11.6	0.08	11.6	0.08	10.8	0.26
Cortical bone	13.1	0.09	13.1	0.09	11.4	0.39
Uterus	64.4	1.08	64.0	1.09	57.9	2.21

The conductivity σ is in S·m^{-1}.

These data do not take into account the contents of the GI tract yet to be measured. However, it is an accurate approach for the phantom design when the antenna is robust enough (see Section 6.4.4).

Later, the capsule leaves stomach by duodenum and proceeds to small bowel ($\varepsilon_r = 65.3$, $\sigma = 1.92$ S·m^{-1}); the average transition time is around 212 min in Reference 48 for the patients without the gastric retention and from 260.5 to 275 min in Reference 49. In Reference 50, reported average time is 228 min. After caecum is reached, the capsule begins the passage through colon ($\varepsilon_r = 62.0$, $\sigma = 0.87$ S·m^{-1}), which lasts in average of 21.7 h for a healthy person and 46.7 h for a constipated one [50]. After passing colon, the capsule exits the body by rectum.

Permittivity and conductivity of adjacent tissue (Figures 6.3 and 6.4) affect the antenna impedance and resonance frequency. As the values change during the capsule passage, antenna will detune. By properly averaging the environment parameters and matching the antenna for the obtained values, we can improve the antenna performance over time. One way to do it is to calculate appropriate time-averaged properties of a human GI tract. Table 6.3 shows an example of this derivation for the dielectric properties at 434 MHz using the transition duration data from Reference 50.

6.3.2 Implantable case

As stated in Section 6.2, in some cases, it is useful to design a capsule antenna to operate in both ingestible and implantable scenarios. If so, we must consider the tissue dielectric properties at the implantation site. Table 6.4 lists the parameters (according to Gabriel *et al.* [45–47]) for the most common implantation scenarios at MedRadio and industrial, scientific and medical (ISM) frequencies. If several tissues are adjacent to a capsule, their properties should be properly averaged.

Homogeneous phantom is sufficient in most cases for the ingestible antenna simulation in its environment. The physical phantom for antenna measurement could be fabricated using for instance pure water, salt, sugar and an antibacterial agent (if long-term storage is required) [51, Section 9.2.2.2].

6.4   *In-body* antenna parameters

Before overviewing capsule antennas, let us review its parameters that matter the most for performance evaluations. Comprehensive list of antenna fundamental parameters is given in the *IEEE Standard Definitions of Terms for Antennas* (IEEE Std 145-2013) [52].

6.4.1   *Resonant frequency*

The antenna resonant frequency is defined as 'a frequency at which the input impedance of an antenna is non-reactive' [52], i.e. when the antenna impedance Z_{ANT} is purely real (reactance $X = 0$). In a S_{11} plot, the resonance frequency will appear at the lowest S_{11} value. It is important to note that the resonant frequency f_{res} could differ from the antenna central frequency f_0, which is defined as the middle of a bandwidth (see Section 6.4.2). Dual-band and multi-band antennas have two or more resonant frequencies.

Resonant frequency mainly depends on the electrical size of an antenna. Biological tissues increase the effective dielectric permittivity ε_r^{eff} of an antenna environment and thus the electrical length of an antenna. This results in f_{res} reduction and must be considered for the antenna design. Choosing an optimal frequency depends on multiple factors for a capsule antenna (see Section 6.5).

6.4.2   *Bandwidth*

The bandwidth could be defined as 'range of frequencies within which the performance of the antenna conforms to a specified standard with respect to some characteristic' [52, 53]. For capsule antennas, the bandwidth is characterised mainly with respect to the input impedance of antenna Z_{ANT}. In this way, the bandwidth is the frequency difference between the highest frequency f_{max} and the lowest f_{min} at a specified level of S_{11} (dB) or Voltage standing wave ratio (VSWR):

$$S_{11} = -20 \log_{10} |\Gamma| \quad \text{or} \quad \text{VSWR} = \frac{1 + |\Gamma|}{1 - |\Gamma|}, \tag{6.1}$$

where $\Gamma = \frac{(Z_{ANT} - Z_L)}{(Z_{ANT} + Z_L)}$ is the reflection coefficient and Z_L stands for the load impedance.

The most common S_{11} and VSWR levels for the bandwidth definition of reported capsule antennas are either $S_{11} < -10$ dB or VSWR < 2 (corresponding $S_{11} < -9.54$ dB).

The bandwidth (BW) can be also conveniently reported as *fractional* – in per cents to the antenna central frequency $f_0 = 0.5 \, (f_{max} + f_{min})$:

$$\text{BW}_\% = \frac{f_{max} - f_{min}}{f_0}. \tag{6.2}$$

This metric facilitates comparing the performance of antennas operating at different frequencies.

6.4.3 Radiation efficiency

The radiation efficiency of an antenna relates the power delivered to its input P_{in} to the radiated power P_{rad}:

$$\eta = \frac{P_{rad}}{P_{in}}. \tag{6.3}$$

The delivered power that is not radiated is dissipated due to ohmic and dielectric losses within an antenna [53]. The radiation efficiency does not take into account the mismatch losses. The latter is considered in the *total* efficiency of an antenna.

As discussed later (see Section 6.6), a capsule antenna could not be both efficient and wideband – one must be sacrificed for another.

6.4.4 Detuning immunity

We define a capsule antenna as *robust* (immune to detuning due to the variation of the environment parameters) when the variation between the minimal and maximal frequency within a capsule environment is less than $S_{11} < -10$ dB bandwidth:

$$\left| f_{res}^{max} - f_{res}^{min} \right| < \text{BW}_{-10\,\text{dB}}. \tag{6.4}$$

Alternatively, the robustness can be defined by the matching levels at an operating frequency f_0 as a response to the variation of the environment EM properties. For example, we may characterise an antenna to be robust within the EM properties range from 150% of muscle to 40% of muscle (both permittivity and conductivity) if it stays matched everywhere in this range below $S_{11} < -10$ dB level at $f_0 = 434$ MHz. This method can be preferable for the wideband antennas, especially when multiple resonant frequencies f_{res} exist near the operating frequency f_0.

To maximise the transmission reliability for ingestible capsules, it is important to consider not only the variation of GI tract tissue properties but also its content and probable presence of gas. Thus, the antenna should be robust within the f_{res}^{max} as the antenna frequency in gas and the f_{res}^{min} as operating in the tissue with maximum dielectric properties that the antenna can encounter during GI passage. To achieve such robustness, the obvious choice would be to design an ultrawideband (UWB) antenna covering the detuning range. However, as we will see in Section 6.6, this solution will lead to the cutback of radiation efficiency decreasing the operating range. Per contra, by decoupling the antenna from a body, we can acquire not only enhanced robustness but also increased efficiency and decreased specific absorption rate (SAR) [54]. One way to accomplish it was demonstrated in Reference 54 by designing an efficient narrowband microstrip antenna loaded with high permittivity superstrate.

6.4.5 Directivity and gain

The antenna *directivity D* is defined by the International Electrotechnical Commission (IEC) as the ratio of the maximum radiation intensity U_{max} in a direction from the antenna to the radiation intensity averaged over all directions U_0:

$$D = \frac{U_{max}}{U_0}. \tag{6.5}$$

The antenna *gain G* combines both the directivity and efficiency. Two different definitions of the gain are in use. The commonly named *IEEE gain* considers the radiation efficiency η:

$$G = \eta D, \tag{6.6}$$

and the *realised gain* considers the mismatch losses as well:

$$G_R = \eta \left(1 - |\Gamma|^2\right) D. \tag{6.7}$$

Antenna gain is a good performance metric for the antennas supposed to radiate in a specific direction, as for example subcutaneous and on-body implants. Antenna efficiency does not consider the directivity and thus is useful for omnidirectional devices, as ingestible and some implantable capsules.

Capsule antennas should be as omnidirectional as possible because its position within the human body is undetermined. If the antenna is electrically small, i.e. satisfies the condition (6.12), this requirement will be satisfied naturally as electrically small antennas (ESAs) radiate with an omnidirectional dipole-like 'doughnut' pattern independently of antenna type and design.

6.5 Choice of operating frequency

The systems operating from sub-megahertz to tens of megahertz range use almost exclusively *inductive coupling* links. The first capsules were designed using this type of data transmission [2, 3], and it is still widely used nowadays. The uplink (capsule-to-receiver) can be made either active or passive. Also, the design procedure is somewhat straightforward because of the low Electromagnetic (EM) coupling with the surrounding biological tissue. Main drawbacks of sub-megahertz systems manifest in low data rates – 1–30 kB·s^{-1} [11] – and limited operational range (roughly 10 cm from the body surface) [14]. Also, they are axially directive; omnidirectional transmission is required as capsule position is undefined. Technically speaking, inductive links are not antennas (as they operate in near field), so it is out of scope of this chapter. The comprehensive overview and design guidelines are given in Reference 55, Chapters 5 and 7.

Frequency trade-off for capsule antennas

Given the fixed capsule antenna electrical size:

- *lower frequency* operation deteriorates the capsule antenna radiation efficiency and slightly increases body–air reflection losses;
- *higher frequency* signal attenuates faster within the lossy tissue.

6.5.1 Higher-frequency attenuation losses

Antennas operating in ultra high frequencies (UHF) – about gigahertz – provide good performance in terms of radiation efficiency η but couple more to the surrounding lossy tissue affecting the antenna properties. In particular, the higher frequency f, the higher the attenuation constant α – the real part of a propagation constant γ [56, p. 20]:

$$\alpha = \Re(\gamma) = \Re\left(j\omega\sqrt{\mu\varepsilon'}\sqrt{1 - j\frac{\sigma}{\omega\varepsilon'}}\right),\tag{6.8}$$

where $\omega = 2\pi f$ is the angular frequency, μ denotes the permeability ($\mu \approx \mu_0$ for biological tissues), $\varepsilon' = \varepsilon_0\varepsilon_r$ stands for the real part of the complex permittivity $\hat{\varepsilon}$ and σ is the conductivity.

Another common way to characterise the propagation losses is using the penetration depth:

$$\delta_p = 1/\alpha.\tag{6.9}$$

The physical interpretation of this value is the depth (m) at which the amplitude of the fields inside the material falls by $1/e$ (about 37%) (shown in Figure 6.5).

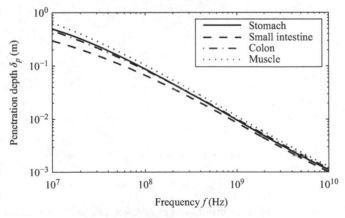

Figure 6.5   *Characteristic depth of penetration within the GI tract and adjacent muscle tissues (equation (6.9))*

Biological tissues are highly dispersive, i.e. their properties depend on the frequency (Figures 6.3 and 6.4). The conductivity increases with frequency that further enhances the attenuation at higher frequencies.

6.5.2   *Lower-frequency efficiency and reflection losses*

At the sub-gigahertz frequencies, we deal mainly with ESAs that have limitations on the maximum achievable efficiency and bandwidth (see Section 6.6).

Furthermore, rapid transitions in dielectric properties can occur at the boundaries between tissues. It deteriorates the through-body transmission due to reflections [14, 57]. Reflection losses are stronger at lower frequencies (Figure 6.6). The sharpest transition arises between a body (skin) and air. The fat layers can create contrast boundaries with skin and muscle too. The transmission efficiency through these boundaries can be characterised using the transmission coefficient T defined as a ratio of transmitted electric field to the incident at an interface between two tissues: $T = \frac{E_t}{E_i}$. For a plane wave under normal incidence on a boundary, it could be estimated from the electromagnetic properties of the media [56, p. 29]:

$$T = \frac{2Z_2}{Z_2 + Z_1},\tag{6.10}$$

where Z_n are the wave impedances of the first and second medium, respectively. The wave impedances can be obtained using the propagation constants γ_n (6.8) of each media [56, p. 20]:

$$Z_n = \frac{j\omega\mu}{\gamma}.\tag{6.11}$$

The plane wave normal incidence is a rough approximation as the skin–air boundary is finite and generally non-planar while the incidence angles are arbitrary; the

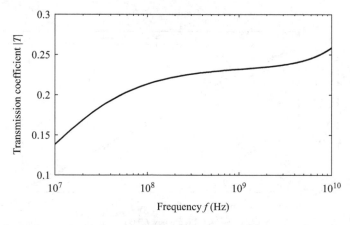

Figure 6.6   Magnitude of the transmission coefficient $|T|$ at the skin–air interface for the normal plane wave incidence (equation (6.10)). Skin properties (dispersive) are from Gabriel et al. [45–47]

spherical wave is more appropriate approximation than planar for *in-body* implants. Moreover, at low frequencies the boundary could be within the near-field region. However, the plane wave approximation gives the general illustration of transmission coefficient frequency behaviour.

So, an optimal operating frequency range must be found as a trade-off between the minimisation of the attenuation and reflection losses and maximisation of the antenna efficiency.

6.5.3 Studies on optimal operating frequency

Chirwa *et al.* [58, 59] reported the propagation study from inside a human intestine in a frequency range from 150 MHz to 1.2 GHz. They found that for the studied phantom (Hugo from Visible Human Project®), the most efficient transmission lies in the frequency range 450–900 MHz and peaks around 650 MHz. Also, the results showed that the radiation characteristics strongly depend on implant position in the body. The tissue dielectric properties in this study have discrepancies with the data commonly used today (e.g. References 45–47). For example, the used small intestine properties are frequency independent within the 150 MHz to 1.2 GHz range with $\varepsilon = 128.9$ and $\sigma = 1.739$. It differs substantially from the data obtained by Gabriel *et al.* [45–47]: $\varepsilon = 61.7$ and $\sigma = 2.037$ at 650 MHz.

Yu *et al.* [60] used the same phantom as in the previous study (adaptive mesh from 140 μm^3 to 4.5 mm^3) to illustrate the body SAR at MICS[3] and ISM[4] bands using the small zigzag dipole antenna placed inside the stomach. The results revealed that the optimal frequency lies around 900 MHz. The authors also state that the optimal frequency choice depends on the antenna size, type and its location in the body.

Sani *et al.* [61] performed the finite-difference time-domain (FDTD) analysis of radiation characteristics of gastric, bladder and cardiac implants operating in MICS band using three distinct computational phantoms. The results also confirmed the strong relationship between the implant location and the radiation characteristics of the antenna.

Xu *et al.* [62] compared the FDTD simulated far fields of four helical antennas at 430, 800, 1200 and 2400 MHz and concluded that the radiation intensity decreases with the increase in operation frequency.

The limitation of all studies mentioned in this subsection is the low-resolution voxel type phantoms. These disregard the reflection losses due to thin tissue layers (when the layer thickness is less than or equal to the phantom voxel side) and will inevitably have the staircase approximation errors. This limitation can be overcome either by increasing the phantom resolution and using multi-grid meshing or by employing new generation boundary representation phantoms [63].

---

[3]Medical Implant Communication Service radio band 402–405 MHz; now part of MedRadio 401–406 MHz sub-band. Regulated by the Federal Communications Commission (FCC)
[4]The industrial, scientific and medical radio bands. Regulated by the International Telecommunication Union (ITU)

Considering the theoretical studies, the most suitable globally available frequency bands for ingestible applications are MedRadio 401–406, 413–419, 426–432, 438–444, 451–457 MHz, and mid-ISM 433–434 MHz. At these frequencies, the main challenge is to properly miniaturise an antenna and to maximise its performance in terms of efficiency or bandwidth. At this point, we must take into account the fundamental limitations of ESAs.

6.6 Fundamental limitations of ESAs

Considering the capsule size and optimal frequency range, in most cases, we will deal with ESAs. By definition, an antenna is considered as electrically small when it satisfies the following condition [64]:

$$ka < 0.5, \tag{6.12}$$

where $k = \frac{2\pi}{\lambda}$ is the wavenumber and a denotes the radius of the minimum size sphere that circumscribes the antenna (also known as the Chu sphere, Figure 6.7).

Bandwidth-efficiency trade-off

In most cases, capsule antennas cannot be both wideband and efficient due to the fundamental limitations of ESAs. Radiation efficiency η must be sacrificed in order to increase the bandwidth, and vice versa.

For the standard ingestible capsule dimensions (see Table 6.1), the typical radius $a \approx 1$ cm. According to (6.12) and considering $\varepsilon_r = 1$, all $a = 1$ cm antennas operating in air at $f_0 \lesssim 2.4$ GHz can be considered as ESAs. However, taking into account adjacent biological tissues and an antenna stack-up, this frequency limit will be lower as the real part of effective relative permittivity around the antenna will be $\varepsilon_r^{\text{eff}} > 1$,

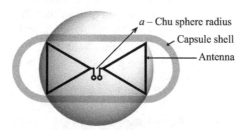

Figure 6.7   Schematic representation of the Chu sphere circumscribing an antenna within a capsule

thus increasing the wavenumber k. Rearranging (6.12), the upper frequency of ESA can be defined as:

$$f_{ESA} < \frac{c}{4\pi a \sqrt{\varepsilon_r^{\text{eff}}}}, \tag{6.13}$$

where $c \approx 3 \times 10^8$ m·s^{-1} is the speed of light. The correct estimation of the $\varepsilon_r^{\text{eff}}$ could be difficult though; it depends on the antenna type, stack-up, capsule materials and geometry, as well as on the adjacent tissue dielectric properties. Figure 6.8 shows how the upper ESA frequency depend on the relative permittivity.

The fundamental ESA performance limits in terms of the quality factor Q were established by Chu [65] and later corrected by McLean [66]. The lower bound on Q_{LB} is inversely proportional to the minimum sphere radius a enclosing an antenna [66]:

$$Q_{LB} = \frac{1}{(ka)^3} + \frac{1}{ka}. \tag{6.14}$$

For a lossy antenna, the right-hand side of (6.14) is multiplied by the antenna radiation efficiency η [67].

The fundamental lower bound of the quality factor Q is inversely proportional to the upper bound of the antenna bandwidth. Yaghjian and Best [68] expressed the antenna Q factor as a function of the bandwidth BW at a given VSWR level as:

$$Q(f_0) = \frac{\text{VSWR} - 1}{\sqrt{\text{VSWR}}} \frac{f_0}{\text{BW}}, \tag{6.15}$$

where f_0 is the antenna central frequency.

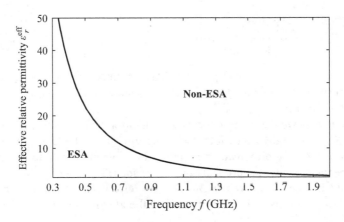

Figure 6.8 *Electrically small antenna ($a = 1$ cm) upper frequency dependence on the effective relative permittivity of the environment according to (6.13)*

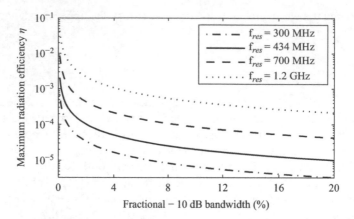

Figure 6.9   Fundamentally achievable antenna radiation efficiency η vs. the
bandwidth for antennas with the circumsphere radius a = 1 cm. The
wavenumber is calculated considering the free space propagation
(ε_r = 1)

Combining (6.14) with (6.15), one obtains the fundamental upper bound limits
of the antenna bandwidth for a given ka:

$$\text{BW}_{\text{UB}} = f_0 \frac{\text{VSWR} - 1}{\sqrt{\text{VSWR}}} \frac{(ka)^3}{\eta\left[(ka)^2 + 1\right]}. \tag{6.16}$$

The above equation gives an important insight: for a given bandwidth, there exists a
maximum achievable radiation efficiency:

$$\eta_{\text{UB}} = \underbrace{\frac{\text{VSWR} - 1}{\sqrt{\text{VSWR}}} \frac{f_0}{\text{BW}}}_{\text{Antenna } Q} \frac{(ka)^3}{(ka)^2 + 1}. \tag{6.17}$$

In other words, while working with ESAs, we must sacrifice the radiation efficiency to decrease the antenna Q factor and thus increase the bandwidth beyond the
fundamental limitations of (6.14).[5]

Figure 6.9 shows the highest achievable radiation efficiencies η for the desired
fractional $S_{11} < -10$ dB bandwidth according to (6.17). The Chu sphere radius is
taken as $a = 1$ cm (approximate maximal radius for an ingestible capsule). Four
wavenumbers k consider the propagation in free space ($\varepsilon_r = 1$).

It could be seen that even 0.1% radiation efficiency is hardly achievable at
434 MHz in air. To raise the upper bounds of the antenna radiation efficiency η_{UB}

---

[5]Theoretically, non-Foster matching circuits can be used to enlarge the bandwidth without sacrificing the
efficiency [69, p. 50]

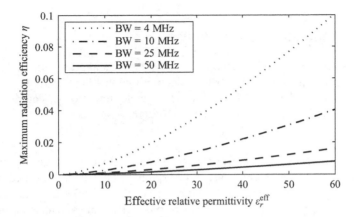

Figure 6.10　Increase in maximum achievable radiation efficiency by dielectric loading of the a = 1 cm antenna operating at $f_0 = 434\,MHz$ with 4, 10, 25 and 50 MHz bandwidths

for a given bandwidth BW at the given frequency f_0, two options exist. The most obvious one consists in increasing the antenna physical size a that leads to a bigger capsule and is often unacceptable.

Other option implies increasing the antenna electrical size without changing its overall physical dimensions. It is achievable by loading the antenna with high permittivity dielectric materials that will increase $\varepsilon_r^{\text{eff}}$. Figure 6.10 shows the effect of increasing $\varepsilon_r^{\text{eff}}$ for $a = 1$ cm antenna operating at $f_0 = 434\,\text{MHz}$. We can notice the distinct exponential effect of dielectric loading for the narrowband antennas.[6]

While choosing an antenna type for a future capsule, we ought to consider the ESA limitations. In particular, the future antenna should make the best use of an available capsule surface to maximise the Chu sphere radius. Also, the antenna type and configuration should allow the dielectric loading – e.g. via substrate, superstrate or filling – to further increase the electrical size. The next section classifies by types and reviews up-to-date capsule antennas.

6.7　Capsule antenna types and overview

At the frequencies lower than very high frequency (VHF) ($f < 30\,\text{MHz}$), wire induction links are used exclusively; as they operate in near field, they are not antennas and so out of scope of this chapter. The comprehensive overview and design guidelines are given in Reference 55, Chapters 5 and 7.

---

[6]Theoretically, magnetic loading could increase the electrical size as well [70]

Capsule antenna classification

- *Wire antennas* – easy to design but cumbersome and poorly integrable.
- *Planar printed antennas* – better integrability but still too bulky.
- *Conformal printed antennas* – well integrable, low profile, could occupy large surface on a capsule but hard to design.

VHF–UHF[7] capsule antennas can be divided into two categories: wire and printed. Reported wire antennas are either spiral or helical, and printed antennas are either planar on a rigid Printed circuit board (PCB) or conformal to the surface of a capsule. Table 6.5 overviews the published VHF–UHF band antennas for ingestible telemetry capsule applications.

6.7.1 Wire antennas

Wire antennas for the ingestible capsules were the first published. They can be either helical or spiral with both circular and rectangular wire cross sections.

Kwak *et al.* [71] reported a helical antenna for a capsule endoscope (Figure 6.11). The size of the antenna is $(5.6 \times \text{ø}8)$ mm. The antenna has a bandwidth of 410–442 MHz (32 MHz or 8%) for VSWR < 2 (or $S_{11} \lesssim -9.5$ dB) when measured in a liquid phantom $(\varepsilon_r = 56, \sigma = 0.8 \, \text{S·m}^{-1})$ in a ø150 mm cylindrical container. The authors used CST Microwave Studio® to simulate the antenna. The gain G and efficiency η were not reported.

Xu *et al.* [62, 72, 73] used a normal-mode helical antenna to study numerically (using FDTD implemented in Xfdtd®) the variations of radiation characteristics and SAR levels at 430, 800, 1200 and 2400 MHz due to the uncertainties of tissue dielectric properties. Four helical antennas were designed to fit the $(15 \times \text{ø}12)$ mm capsule with 1 mm shell thickness. Figure 6.12 shows the generic antenna, and Table 6.5 lists the exact dimensions for each working frequency. The wire is ø1 mm. The male and female phantoms use the 'Visible Human' anatomy data with the dielectric properties reported by Gabriel *et al.* [45–47]. The antenna was meshed with 1 mm³ resolution and the tissue with 3 mm³. The efficiencies η and gains G were not provided.

Rajagopalan and Rahmat-Samii [74] studied inverted conical helical antenna (Figure 6.13) for a $(26 \times \text{ø}11)$ mm capsule designed to operate within the Wireless Medical Telemetry Service (WMTS) 1.4 GHz band. The bandwidth was not reported, but according to the reflection coefficient (Figure 6.13) it is approximately 40 MHz (3%). The wire is ø0.4 mm and the pitch is 1 mm. The first loop is ø1.5 mm and increases by ø2 mm with each winding. The ø9 mm ground plane shields the antenna from the capsule circuitry. Additional ø8 mm metallic surface above the ground plane is used for the antenna tuning. The total antenna

---

[7]Frequency bands as defined by ITU: VHF: 30–300 MHz, UHF: 300–3000 MHz and super high frequency (SHF): 3–30 GHz

	Reference	Antenna type	Footprint	f_0 (MHz)	Bandwidth (MHz)	Bandwidth (%)	Gain (dBi)	η^* (%)	Phantom	Comments
Wire	[62]	Helical	(ø6 to ø10) mm	430, 800, 1200, 2400	—	—	—	—	Visible human	Wire ø1 mm
	[71]	Helical	(5.6 × ø8) mm	426	32	8	—	—	$\varepsilon_r = 56, \sigma = 0.8$	—
	[74]	Inverted conical helical	(5.1 × ø9) mm	1400	40	3	−40.0	<0.1	Muscle [46]	Wire ø0.4 mm
	[76]	Conical spiral	(5 × ø10) mm	450	101	22	—	—	$\varepsilon_r = 56, \sigma = 0.83$	—
	[78]	Large-arm spiral	([3–7] × ø10) mm	450	76–117	17–26	—	—	$\varepsilon_r = 56.91, \sigma = 0.97$	90 mm arm length
	[79]	Dual spiral	(7 × ø10) mm	505	189	38	—	—	$\varepsilon_r = 56, \sigma = 0.83$	—
Planar PCB	[80]	Spiral	(3 × ø10.5) mm	455	110	24	—	—	$\varepsilon_r = 56, \sigma = 0.8$	Sub.: 3 mm, $\varepsilon_r = 2.17$
	[81]	Spiral	(3 × ø8.6) mm	2450	104	4	−0.17	—	Not reported	Sub.: 3 mm, $\varepsilon_r = 3.5$
	[75]	Spiral	—	2450	—	—	—	—	$\varepsilon_r = 53.6, \sigma = 1.8$	—
	[82]	Four-arm spiral	$8.25^2 \times 0.6$ mm³	915	11	1	−46.0	—	Visible Human	Sub.: 0.6 mm, $\varepsilon_r = 6.1$
	[83]	Multilayered spiral	$24.7 \times 8 \times 5.2$ mm³	403	9	2	−28.8	0.058	$\varepsilon_r = 43.5$, tan $\delta = 0.799$	Sub.: $\varepsilon_r = 9.2$, tan $\delta = 0.0022$
				2450	145	6	−18.5	0.530	$\varepsilon_r = 39.2$, tan $\delta = 0.337$	
	[85]	Multilayered helical	$\pi\, 5.5^2 \times 3.81$ mm³	2300	950	41	−32.0	—	$\varepsilon_r = 52.79, \sigma = 1.7$	Sub.: $\varepsilon_r = 10.2$, tan $\delta = 0.0022$
	[86]	Microstrip slot	$23.7 \times 8 \times 1.27$ mm³	3000	UWB	—	—	0.8–20	Various	Sub.: $\varepsilon_r = 9.8$, tan $\delta = 0.002$
Conformal PCB	[89]	Meandered asymmetric dipole	(26 × ø11) mm	1400	136	10	−26.0	0.05	$\varepsilon_r = 58.8, \sigma = 0.84$	Shell: 0.1 mm, $\varepsilon_r = 2.2$
	[92]	Patch with CSRR	10 mil substrate	2400	50	<2	−5.2	0.21	None (air)	Sub.: $\varepsilon_r = 2.2$, tan $\delta = 0.0009$
	[93]	Folded helical	(23 × ø9.5) mm	430	<3	<1	—	—	Not reported	—
	[94]	Meandered loop	(24 × ø11) mm	500	260	52	−33.0	—	$\varepsilon_r = 56.4, \sigma = 0.82$	Shell: 0.5 mm, $\varepsilon_r = 3.15$
	[97]	Microstrip	2 mil substrate	434	50	12	−28.95	—	$\varepsilon_r = 49.6, \sigma = 0.51$	Sub.: $\varepsilon_r = 3.4$, tan $\delta = 0.002$
	[98]	U-shaped loop	(15 × ø10) mm	452	207	46	−28.95	—	$\varepsilon_r = 57.1, \sigma = 0.79$	—
	[99]	Meandered loop	(10 × ø7) mm	429	203	47	−28.4	—	$\varepsilon_r = 57, \sigma = 0.79$	—
	[100]	Meandered loop	On ø10 mm cylinder	404	25	6	−21.0	0.30	Multilayer cylindrical	For bone implants
	[101]	Meandered loop	(40 × ø10) mm	408	92	23	−27.6	0.05	Multilayer cylindrical	—
	[102]	Meandered loop	(10 × ø11) mm	608	785	129	—	—	$\varepsilon_r = 56.4, \sigma = 0.82$	—
	[103]	Loop with CSRR	(15 × ø10) mm	Multi	>2000	—	−29.64	0.1–0.5	Muscle [46]	Shell: 0.2 mm
	[104]	Meandered microstrip	(18 × ø10) mm	402	40	10	—	—	$\varepsilon_r = 57, \sigma = 0.8$	Sub.: $\varepsilon_r = 2.2$; shell: $\varepsilon_r = 2.25$
	[106]	Meander-line	ø10.6 mm cylinder	418	800	145	—	—	$\varepsilon_r = 56, \sigma = 0.8$	—
			ø4.8 mm cylinder	612	564	92	—	—		
	[107]	Meandered asymmetric dipole	(24 × ø11) mm	437	158	36	−37.0	0.02	$\varepsilon_r = 46.76, \sigma = 0.69$	3 mm deep in phantom
	[108]	Trapezoid strip DR	ø8 mm hemisphere	3500	3000	86	−24.2	—	$\varepsilon_r = 51.5, \sigma = 3.2$	Hemisphere: $\varepsilon_r = 3$
	[110]	Helical DR	ø8 mm hemisphere	3500	>3000	—	−32.0	—	$\varepsilon_r = 51.4, \sigma = 3.26$	—
		Loop DR					−30.0	—		

*Antenna radiation efficiency (see Section 6.4.3) DR, Dielectric resonator.

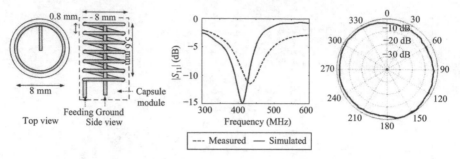

*Figure 6.11   Wire helical antenna by Kwak et al. [71]: dimensions, reflection
coefficient and measured radiation pattern. ©2005 IEEE*

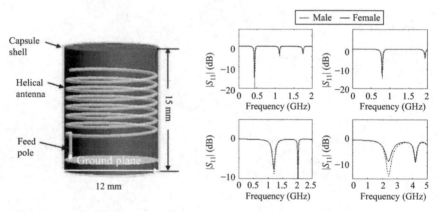

*Figure 6.12   Wire helical antenna by Xu et al. [72]: capsule design and dimensions,
reflection coefficients for four antenna designs (430, 800, 1200 and
2400 MHz) simulated in male and female phantoms. ©2009 IEEE*

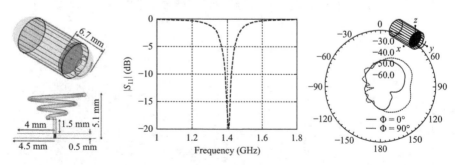

*Figure 6.13   Wire helical antenna by Rajagopalan and Rahmat-Samii [74]:
dimensions, reflection coefficient and radiation pattern. ©2012 IEEE*

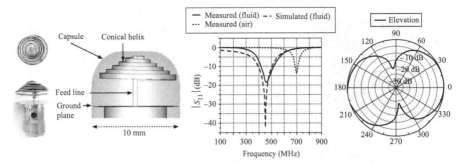

Figure 6.14   *Wire conical spiral antenna by Lee et al. [76]: prototype, design,*
reflection coefficient and simulated elevation plane radiation pattern
(azimuth plane is isotropic). ©2008 IEEE

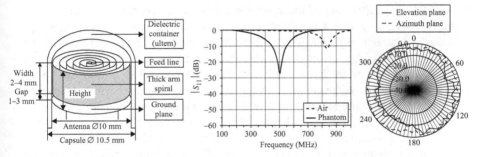

Figure 6.15   *Large-arm spiral wire antenna by Lee et al. [78]: design, reflection*
coefficient and radiation pattern measured at 500 MHz. ©2011 IEEE

size is thus $(5.1 \times ø9)$ mm. The authors report the radiation efficiency $\eta < 0.1\%$ and
the simulated peak gain $G = -40$ dBi at 1.4 GHz. The antenna is linearly polarised.
The equivalent tissue parameters were set as frequency dependent with the values
$\varepsilon_r \in [53.55, 54.81]$ and $\sigma \in [0.98, 1.34]$ S·m^{-1} for 1–1.8 GHz. The permittivity of
the capsule shell is $\varepsilon_r = 3.2$. For the antenna measurements, a tissue equivalent liquid
phantom contained deionised water, sugar, salt and cellulose.

Lee *et al.* [76] proposed a conical spiral antenna (Figure 6.14) at 450 MHz.
Antenna has 101 MHz (or 22%) bandwidth for VSWR < 2 when measured in the liq-
uid phantom with $\varepsilon_r = 56, \sigma = 0.83$ S·m^{-1} in a ø300 mm cylindrical container. Due
to the conical structure, this antenna exhibits more broadband characteristics compared to the conventional narrowband spiral antennas and helical antennas mentioned
earlier. The antenna size is $(5 \times ø10)$ mm. The authors used a 50 Ω matched coaxial
cable to connect the antenna to a vector network analyser for measurements and CST
Microwave Studio® for simulations. The efficiency and gain were not reported.

Large-arm spiral antenna by Lee *et al.* [77, 78] uses flat wire (Figure 6.15) and
operates around 450 MHz with the $S_{11} < -10$ dB bandwidth from 76 MHz (17%)
to 117 MHz (26%) depending on the antenna dimensions. The liquid phantom has

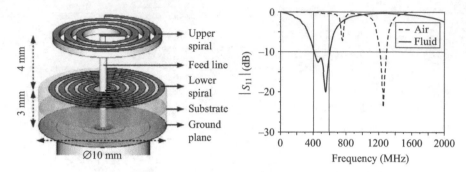

Figure 6.16   Dual spiral wire antenna by Lee and Yoon [79]: design and measured reflection coefficient. ©2008 IEEE

$\varepsilon_r = 56.91$, $\sigma = 0.97$ S·m^{-1} and occupies a ø300 mm cylindrical container. The phantom consists of pure water, methanol and sodium chloride. The 0.25 mm capsule shell has a relative permittivity $\varepsilon_r = 3.15$. The total antenna height h (including the ground plane) varies from 3 to 7 mm affecting the bandwidth; the diameter ø is 10 mm. The 0.5 mm thick antenna arm width varies from 2 to 4 mm: wider arms give more broadband characteristics. The total length of the arm – which defines the f_{res} – was experimentally found to be 90 mm ($\lambda/4$) for the antenna with the parameters $w = 4$ mm and $g = 1$ mm. The radiation efficiency η and the gain G were not given.

Dual spiral antenna for wireless endoscope is another development by Lee and Yoon [79]. The total antenna size is (7 × ø10) mm. The antenna consists of two spirals: wire and printed on a PCB (Figure 6.16). The PCB is 1.574 mm thick with $\varepsilon_r = 2.5$. One feed line powers both spirals that are mounted above a ground plane. Two spirals have different overall length for enhancing the bandwidth (the printed one is longer than the wire one, the exact lengths were not reported). The authors reported a 411–600 MHz bandwidth for VSWR < 2 (189 MHz or 37.8%), when measured in a liquid phantom with $\varepsilon_r = 56$, $\sigma = 0.83$ S·m^{-1} in a ø300 mm cylindrical container. As for the previously mentioned antenna, the phantom consists of pure water, methanol and sodium chloride. The radiation efficiency η and gain G were not reported.

Despite the high achievable bandwidths of wire antennas, they are voluminous, poorly integrable in a capsule design and ill-suited for mass production. Better integration and more technological manufacturing could be achieved with printed antennas.

6.7.2   Planar printed antennas

The first reported printed antenna for ingestible capsules was the spiral antenna on a rigid PCB reported by Kwak *et al.* [80] (Figure 6.17). The antenna bandwidth is comparable with wire antennas: 110 MHz or 24% (400–510 MHz for VSWR < 2) when measured in a liquid phantom with $\varepsilon_r = 56$, $\sigma = 0.8$ S·m^{-1} in a ø300 mm cylindrical container. The substrate is 3 mm thick (ø10.5 mm) and has $\varepsilon_r = 2.17$. The $\lambda/4$ spiral trace width is 0.5 mm. The radiation efficiency η and gain G were not reported.

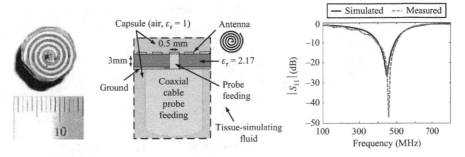

Figure 6.17 Printed spiral antenna by Kwak et al. [80]: prototype, design and reflection coefficient. ©2005 IEEE

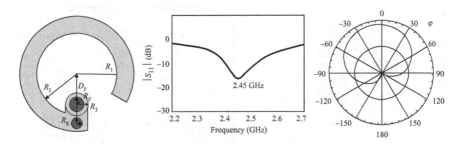

Figure 6.18 Printed spiral antenna by Xie et al. [81]: design, reflection coefficient and φ radiation pattern. ©2006 IEEE

Xie *et al.* [81] reported a 2.45 GHz printed spiral antenna on a 3 mm thick F4BK335 substrate ($\varepsilon_r = 3.5, \tan\delta = 0.001$). The dimensions are (Figure 6.18): $R_1 = 4.3$ mm, $R_2 = 3.3$ mm, $R_3 = 1$ mm, $R_F = 0.65$ mm, $D_F = 1$ mm and $R_S = 0.5$ mm. Antenna uses the shortening pin miniaturisation technique. The reported bandwidth is 104 MHz (4%). The maximum gain reaches $G = -0.17$ dBi. The antenna lacks the omnidirectionality for ingestible capsule applications. The phantom characteristics and radiation efficiency η were not reported.

Another 2.45 GHz spiral antenna on a rigid PCB for capsule applications was reported by Zhao *et al.* [75] (Figure 6.19). The authors used the following tissue parameters: $\varepsilon_r = 53.6, \sigma = 1.8$ S·m^{-1}. No dimensions were given along with the bandwidth, neither other performance indications as radiation pattern, gain and efficiency, nor the measurement data.

Huang *et al.* [82] proposed a microstrip four-arm spiral antenna (Figure 6.20) for 915 MHz ISM band printed on a rigid 0.6 mm thick PCB with $\varepsilon_r = 6.1$. The antenna is narrowband (1.2% or 11 MHz bandwidth at $S_{11} < -10$ dB). The gain is approximately $G = -46$ dBi. A shortening strip contributes to the miniaturisation; the total antenna size is $8.25 \times 8.25 \times 0.6$ mm^3. The capsule casing and circuitry was not taken into account in simulations. The antenna was analysed using FDTD method inside a heterogeneous phantom (Visible Human, 25 distinct tissues). Adaptive mesh

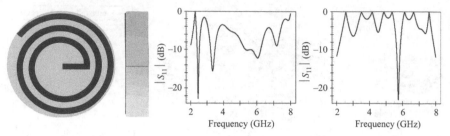

*Figure 6.19   Printed spiral antenna by Zhao et al. [75]: design, reflection
coefficient within a phantom (middle) and free space (right). ©2010
IEEE*

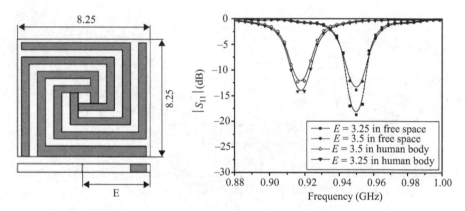

*Figure 6.20   Printed four-arm spiral antenna by Huang et al. [82]: design and
dimensions (mm), and reflection coefficient. ©2007 IET*

was used with the $0.25 \times 0.25 \times 0.3\,\text{mm}^3$ resolution for the antenna and $5\,\text{mm}^3$ for
the rest of the computational domain. The radiation pattern in Figure 6.20 depicts the
case when the antenna is inside a simulated human small intestine.

Merli *et al.* [83] reported the dual-band multilayered spiral antenna (Figure 6.21)
that operates in the 401–406 MHz MedRadio band and 2.4–2.5 GHz ISM band;
the *in vitro* measured $S_{11} < -10$ dB bandwidths are 9.3 MHz (2.3%) and 145 MHz
(6%), respectively, for MedRadio and ISM. The gains $G_{\text{MedRadio}} = -29.4$ dBi and
$G_{\text{ISM}} = -17.7$ dBi, the directivities $D_{\text{MedRadio}} = 3.5$ dBi and $D_{\text{ISM}} = 4.4$ dBi, and the
radiation efficiencies $\eta_{\text{MedRadio}} = 0.051\%$ and $\eta_{\text{ISM}} = 0.605\%$ within a liquid phantom in a ø80 mm cylindrical container. Unlike the previously mentioned phantoms,
the capsule is displaced from the central axe (Figure 6.21 middle right). The measured
parameters of the liquid phantoms are $\varepsilon_r = 57.36$, $\tan \delta = 0.580$ for the MedRadio
band and $\varepsilon_r = 53.76$, $\tan \delta = 0.241$ for the ISM. The MedRadio phantom consists of
(per cent by weight): distilled water (51.3%), sugar (47.3%) and salt (1.4%). The ISM
phantom is made of distilled water (73.2%), salt (1.4%) and diethylene glycol butyl
ether (26.76%). Further analysis of the antenna was performed in SEMCAD X using

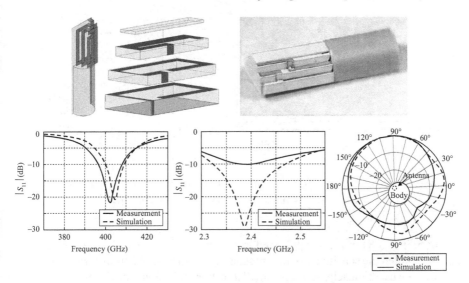

Figure 6.21   *Multilayer antenna by Merli et al. [83]: design, prototype, reflection*
coefficients (MedRadio and ISM) and ISM radiation patterns (dB).
© 2010 IEEE

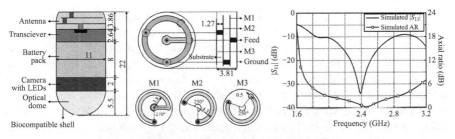

Figure 6.22   *Circularly polarised multilayer antenna by Liu et al. [85]: capsule*
and antenna dimensions (mm), reflection coefficient and axial ratio.
©2014 IEEE

the Duke phantom from the Virtual Family [84]. The antenna size without the casing is
$24.7 \times 8 \times 5.2$ mm^3. It fits inside the capsule of $(32.1 \times \varnothing 10)$ mm. The 0.8 mm shell
is made of Polyetheretherketone (PEEK) $(\varepsilon_r \approx 3.5)$. Antenna 35 μm copper layers are
printed on Rogers TMM® 10 substrate $(\varepsilon_r = 9.2, \tan \delta = 0.0022)$ and united using
30 μm thick epoxy adhesive (MedRadio: $\varepsilon_r = 3$; ISM: $\varepsilon_r = 4$; $\tan \delta = 0.001$).

Liu *et al.* [85] proposed a circularly polarised multilayer helical antenna for the
ISM 2.4–2.48 GHz band (Figure 6.22). The bandwidth is 1.85–2.8 GHz (950 MHz
or 41%) for $S_{11} < -10$ dB when simulated in a 100 mm^3 cubic phantom with
$\varepsilon_r = 52.79, \sigma = 1.7$ S·m^{-1}. The peak gain is $G \approx -32$ dBi. The antenna consists of
three layers interconnected by vias, each layer containing an open loop. The substrate
is 50 mil Rogers RO3010 with $\varepsilon_r = 10.2, \tan \delta = 0.0022$. The total antenna footprint

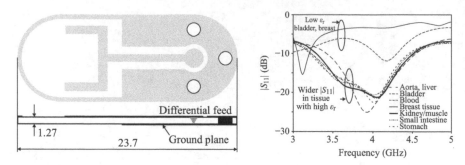

Figure 6.23   Printed slot UWB antenna by Dissanayake et al. [86]: design and reflection coefficients for various tissues. ©2009 IEEE

is $\pi \times 5.5^2 \times 3.81$ mm^3; it is designed to fit within the (26 × ø11) mm capsule. The authors studied the sensitivity of the antenna by varying the phantom parameters; four tissues were used: muscle, stomach, small intestine and colon with the properties from References 45–47. Ansys® HFSS™ was used to design and study the antenna within a homogeneous phantom and CST Microwave Studio® to study the antenna in a voxel phantom – Gustav from the CST Voxel Family – the antenna radiation, sensitivity and SAR.

Dissanayake *et al.* [86] reported a slot microstrip UWB antenna that operates at 3.5–4.5 GHz (Figure 6.23). The slot design is based on a magnetic dipole principle that is less affected by the near field perturbations caused by the varying dielectric parameters of the surrounding biological tissue. The antenna is printed on the capsule-shaped 1.27 mm Rogers TMM® 10i substrate ($\varepsilon_r = 9.8$, tan$\delta = 0.002$). The total antenna size is $23.7 \times 8 \times 1.27$ mm^3, the capsule shell is 0.5 mm thick. The slot occupies half of the substrate, which is loaded with the glycerine ($\varepsilon_r = 50$). This contributes to the miniaturisation as well as the impedance matching with the biological tissue. For the simulation in CST Microwave Studio®, various tissue parameters were used (Figure 6.23, middle); the cylindrical computational phantom size is (40 × ø29) mm. The reported radiation efficiencies vary from $\eta = 0.8\%$ for the small intestine to $\eta = 20\%$ for the breast tissue. The proof-of-concept measurement used pork midloin chops as an equivalent tissue. The authors claim that the high permittivity dielectric loading of an antenna (ideally with the ε_r equal to the surrounding tissue) contributes to lower reflection coefficients, so that the wider band matching of the antenna could be achieved. The same antenna was used in the power absorption study by Thotahewa *et al.* [87]; Dissanayake *et al.* also reported the similar antenna design earlier [88].

Planar printed antennas perform well enough for the *in-body* capsule applications. They occupy less volume than wire antennas, they integrate easily with a capsule circuitry, and they are more technological and inexpensive. However, their footprint is the critical drawback for capsule applications. Conformal printed antennas occupy negligibly small volume and they easily integrate within a capsule, yet they possess all the advantages of planar antennas, and could outperform them.

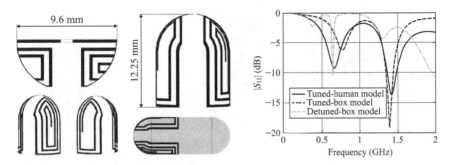

Figure 6.24   Conformal chandelier meandered dipole antenna by Izdebski et al. [89]: design and reflection coefficient. ©2009 IEEE

6.7.3   Conformal printed antennas

Izdebski *et al.* [89] reported a 1.4 GHz conformal chandelier meandered dipole antenna (Figure 6.24). The $S_{11} < -10$ dB bandwidth is about 136 MHz ($\approx 10\%$). The antenna is based on the offset meandered dipole design that permits better matching to 50 Ω than the balanced meandered dipole. The antenna conforms the capsule extremity. To validate the design and numerical models, the antenna was printed on a planar PCB (RT/duroid® 5880, $\varepsilon_r = 2.2$). The homogeneous phantom has cuboid shape with the sides $350 \times 350 \times 200$ mm^3 and the dielectric properties $\varepsilon_r = 58.8$, $\sigma = 0.84$ S·m^{-1} at $f_{res} = 1.4$ GHz. The capsule shell is 0.1 mm thick with $\varepsilon_r = 2.2$; the antenna is printed on the interior surface. The final lengths of the antenna arms (i.e. tuned inside the phantom) are 25.2 mm for the shorter arm and 42 mm for the longer one. Numerical analysis of the influence of the batteries inside the capsule showed no critical effect on reflection coefficient. The simulation results for the cuboid phantom were verified by the simulation inside the Ansoft human body model. The radiation efficiency and gain in free space are $\eta = 11\%$ and $G = -7.86$ dBi and inside the anatomically realistic Ansoft phantom are $\eta = 0.05\%$ and $G = -26$ dBi. The authors used the Finite element method (FEM) implemented in Ansys® HFSS™ to analyse the antenna. Similar antennas were also reported [74, 90, 91].

Cheng *et al.* [92] designed a 2.4 GHz (ISM band) conformal patch antenna loaded with the complementary split-ring resonator (CSRR; Figure 6.25). The reported gain is $G = -5.2$ dBi and radiation efficiency is $\eta = 21\%$ (both in air). No phantom was used for the simulation. For measurements, in addition to the air, a $75 \times 75 \times 50$ mm^3 sponge was used damped with 30 ml of water. The bandwidth was not reported, but according to the reflection coefficient it does not exceed 50 MHz (<2% for $S_{11} < -10$ dB, Figure 6.25). The bending diameter is ⌀10 mm. The patch length is 10.5 mm (roughly $\lambda/10$) and the width could be varied for the impedance matching (consequently, the gap between the folded patch opposite edges will change). The 17 μm thick patch top layer is printed on the 0.254 mm (10 mil) RT/duroid® 5880 substrate with $\varepsilon_r = 2.2$, $\tan \delta = 0.0009$. Ground plane reduces the effect of a capsule circuitry on the antenna. The authors used the lumped element approximation along with Ansys® HFSS™ to analyse the antenna.

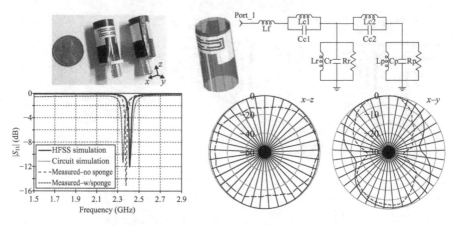

Figure 6.25   Conformal CSRR-loaded patch antenna by Cheng et al. [92]: fabricated antenna, design and equivalent circuit, reflection coefficients and radiation pattern (solid: simulated, dashed: measured). ©2011 IEEE

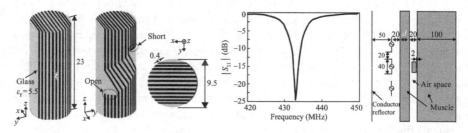

Figure 6.26   Conformal modified helical antenna by Kumagai et al. [93]: design (mm), reflection coefficient and WPT calculation model. ©2011 IEEE

Kumagai *et al.* [93] reported a 430 MHz conformal helical antenna. The antenna was designed for wireless power transfer to *in-body* capsules. The $S_{11} < -10$ dB bandwidth is about 2.5 MHz (0.6%) according to the simulated reflection coefficient (Figure 6.26). The environment parameters (phantom) were not reported; conceivably it was air or vacuum according to the narrow bandwidth characteristic. Antenna conforms to a (23 × ø9.5) mm cylindrical volume (Figure 6.26). The trace is 0.4 mm wide and is fed symmetrically. One end of the antenna is open and the opposite end is shortened, thus the impedance match could be achieved. The power transfer to the antenna was analysed using the antenna model in 20 mm air layer between two phantom layers: 15–35 mm the outer layer and 100 mm the inner layer (Figure 6.26 right). Both phantom layers are muscle equivalent with $\varepsilon_r = 56.9$, $\sigma = 0.81$ S·m^{-1}. The antenna was analysed using the FDTD method.

Yun *et al.* [94] reported an UWB conformal meandered-loop antenna (Figure 6.27). The operation frequencies are 370–630 MHz (260 MHz or 52%

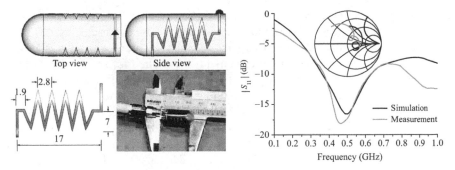

*Figure 6.27   Conformal meandered-loop antenna by Yun et al. [94]: design,
dimensions (mm), manufactured antenna and reflection coefficients.
©2010 IEEE*

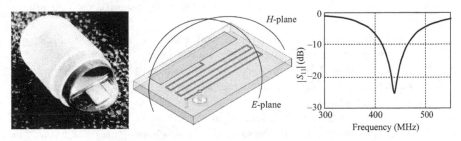

*Figure 6.28   Conformal microstrip antenna by Mahe et al. [97]: manufactured
prototype, design and simulated reflection coefficient. ©2012 Yann
Mahe et al.*

VSWR < 2 bandwidth) when simulated in the phantom ($\varepsilon_r = 56.4$, $\sigma = 0.82$ S·m^{-1} at 500 MHz). ø300 mm cylindrical liquid phantom was used for the measurement; the ingredients were not reported. The full effective wavelength antenna is printed on the outer surface of the 0.5 mm capsule shell ($\varepsilon_r = 3.15$), maximising the achievable Chu sphere radius, thus increasing the efficiency. The approach to increase the radiation efficiency η is opposite to the one proposed by Merli *et al.* [95]. The height of meanders and the gap between each could be used to tune the antenna. The batteries affect the antenna characteristics as it is unshielded. The authors claimed 43.7% measured antenna efficiency (calculated using the Frii's formula). The total capsule size is (24 × ø11) mm. The metallisation is 18 μm thick copper. The prototype was printed on a 25.4 μm polyimide film and fixed on the outer surface of the capsule shell. The authors used CST Microwave Studio® to analyse the antenna. Kim *et al.* reported an *in vivo* measurement of the endoscope prototype using this antenna [96].

Mahe *et al.* [97] reported a conformal microstrip antenna for the 434 MHz ISM band (Figure 6.28). The antenna on a polyimide substrate ($\varepsilon_r = 3.4$) has a bandwidth about 50 MHz (or 12%) for $S_{11} < -10$ dB. The microstrip antenna design includes a ground plane, which implies antenna shielding and minimisation of capsule circuitry effect on the radiation performance. Three miniaturisation techniques were used: short circuiting, stepped impedance and meandering. The authors achieved 90% size reduction compared to the standard $\lambda/2$ patch antenna. The proposed analytical model permits to calculate the antenna dimensions for the desired f_{res} and miniaturisation constraints to fit various capsules. Also, the model permits to shrink the numerical antenna tuning to one dimension (the length of meandered element). The proof-of-concept prototype was built on $18 \times 9 \times 1.58\,\text{mm}^3$ FR4 substrate ($\varepsilon_r = 4.4$) with the $3 \times 18\,\text{mm}^2$ first element size and $0.2 \times 75\,\text{mm}^2$ the second. The maximum gain is $G = -33\,\text{dBi}$, the radiation efficiency η was not reported. As for the quarter wave patch, the antenna input impedance Z_{ANT} depends on a feeding point with the maximum on the radiating edge and minimum near the short circuit. A flexible prototype was built on a $50\,\mu\text{m}$ Pyralux® AP8525 substrate (polyimide, $\varepsilon_r = 3.4$), which conforms to a $(17 \times \varnothing 7)\,\text{mm}$ capsule. The metallisation is $18\,\mu\text{m}$ thick. The antenna was measured in a liquid phantom within a $\varnothing 200\,\text{mm}$ cylindrical container with $\varepsilon_r = 49.6$, $\sigma = 0.51\,\text{S·m}^{-1}$. The effect of bending is insignificant. The authors used Ansys® HFSS™ to analyse the antenna.

Alrawashdeh *et al.* reported a range of conformal loop antennas. In Reference 98, the authors reported a conformal U-shaped loop antenna covering both MedRadio 403 MHz and ISM 433 MHz bands. The antenna is UWB: 348–555 MHz or 207 MHz (46%) for VSWR < 2 when simulated inside a $(110 \times \varnothing 120)\,\text{mm}$ homogeneous cylindrical phantom ($\varepsilon_r = 57.1$, $\sigma = 0.79\,\text{S·m}^{-1}$ at 404.5 MHz). In air, the antenna resonates at 4 GHz (left reflection coefficient plot). The authors also compared a three-layer body model of the same size with a homogeneous phantom. The gain values are $G = -28.38\,\text{dBi}$ and $G = -28.95\,\text{dBi}$, respectively, within the multilayer and homogeneous phantoms. The total antenna size conforms to the surface of the $(15 \times \varnothing 10)\,\text{mm}$ cylinder. The $18\,\mu\text{m}$ metallisation is approximated by Perfect electric conductor (PEC) for the model. The unfolded antenna electrical size is about $\lambda/3 \times \lambda/6$, including the $25.4\,\mu\text{m}$ thick polyimide substrate. Another $17\,\mu\text{m}$ polyimide layer was added above the antenna to insulate it from the tissue.

Further development with more meanders of the previous antenna was reported in Reference 99 (Figure 6.29). The bandwidth is 327–530 MHz (203 MHz or 47%)

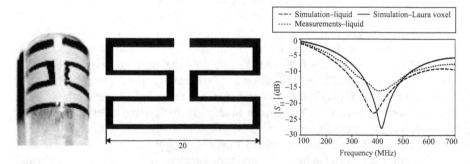

Figure 6.29   *Conformal meandered loop antenna by Alrawashdeh et al. [99]: prototype, planar design (mm) and reflection coefficients. © 2013 IET*

for $S_{11} < -10$ dB when simulated in $180 \times 100 \times 50$ mm^3 elliptical cylinder phantom (human arm model) with $\varepsilon_r = 57$, $\sigma = 0.79$ S·m^{-1}. The reported gain is $G = -28.4$ dBi. The radiation efficiency η was not reported. Antenna conforms to the $(10 \times \varnothing 7)$ mm cylinder.

Another MedRadio conformal loop antenna by Alrawashdeh *et al.* [100] was designed for cylindrical knee implants (Figure 6.30). The $\varnothing 10$ mm antenna bending diameter is in the same range as for the GI capsules. The antenna resonates at 404 MHz with a bandwidth about 25 MHz (6%) when simulated in a $(444 \times \varnothing 130)$ mm cylindrical multilayer phantom (Figure 6.30 right). The simulated gain is $G = -21$ dBi and the radiation efficiency is $\eta = 0.3\%$. It must be noted that the gain and the efficiency values are higher because of lower loss for this kind of implant application. The antenna is designed to be printed on a 20 µm polyimide substrate. The trace width is 1 mm and the separation distance varies from 1 to 3 mm. The radiation pattern is omnidirectional; however, the maximum gain is directed along the phantom. The transverse directivity would be preferable for the reported application.

Similar design of a conformal meandered loop antenna for the same application was proposed by Alrawashdeh *et al.* in Reference 101 (Figure 6.31). The bandwidth is 362–454 MHz (92 MHz or 23%) for $S_{11} < -10$ dB when simulated in a multilayer cylindrical phantom: $\varnothing 30$ mm bone $\rightarrow$ 70 mm muscle $\rightarrow$ 4 mm fat $\rightarrow$ 2 mm skin

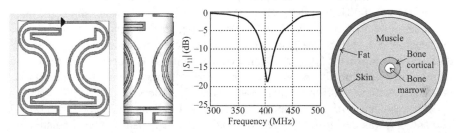

Figure 6.30   Conformal loop antenna by Alrawashdeh et al. [100]: planar and conformal designs, reflection coefficient and multilayer phantom. © 2013 IEEE

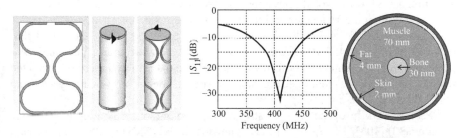

Figure 6.31   Conformal loop antenna by Alrawashdeh et al. [101]: planar and conformal designs, reflection coefficient and multilayer phantom. © 2014 IEEE

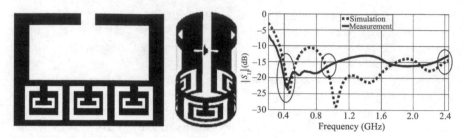

Figure 6.32   Conformal CSRR-loaded loop antenna by Alrawashdeh et al. [103]:
planar and conformal designs, and reflection coefficient.
© 2015 IEEE

(Figure 6.31 right). The simulated radiation efficiency is $\eta = 0.05\%$ and the gain is $G = -27.6$ dBi; as for the previous case, the values are higher because the environment is less lossy. The antenna conforms to the $(40 \times \varnothing 10)$ mm cylinder. The total radiator length is 137 mm (0.67λ) and the trace is 1 mm wide. In this study, the authors reported also the analysis of antenna detuning due to the variation of dielectric parameters of the surrounding tissue.

One more meandered loop antenna by Alrawashdeh *et al.* was reported in Reference 102. The antenna has one of the largest bandwidth among all reported capsule antennas: 215–1000 MHz (785 MHz or 129%) for VSWR < 2 when simulated in the phantom $(\varepsilon_r = 56.4, \sigma = 0.82$ S·m^{-1} at 500 MHz). In free space, the antenna resonates around 3.1 GHz. The gain and efficiency were not reported. The antenna is designed to conform the $(10 \times \varnothing 11)$ mm cylinder while printed on a 25.4 μm substrate with $\varepsilon_r = 3.5$ (presumably polyimide). The metallisation is copper; the antenna trace is 0.5 mm width with 0.5 mm separation between meanders.

Alrawashdeh *et al.* also proposed a conformal loop antenna loaded with CSRR. The antenna was designed to cover MedRadio and ISM bands. The simulated and measured bandwidths are extremely large (Figure 6.32 right). The authors motivated the use of CSRR to improve the impedance matching as well as to increase the antenna radiation efficiency η and gain G. The antenna was simulated using a homogeneous elliptic cylindrical muscle equivalent phantom $(180 \times 100 \times 50$ mm$^3)$ as well as within a CST Katja voxel body model (43-year-old female, 62 kg). The claimed simulated radiation efficiencies within the homogeneous phantom are 0.12%, 0.2%, 0.3%, 0.35%, 0.53% and realised gains G_R are $-26, -25.1, -24, -21, -15$ dBi at 403, 433, 868, 915, 2450 MHz, respectively. A prototype was manufactured and measured within pork. Antenna conforms to the $(15 \times \varnothing 10)$ mm cylinder and insulated with 0.2 mm shell. All mentioned antennas by Alrawashdeh *et al.* were analysed using CST Microwave Studio®.

Psathas *et al.* proposed a microstrip meandered antenna operating at 402 MHz (Figure 6.33) [104]. The $S_{11} < -10$ dB bandwidth is 39.95 MHz (9.9%) and the gain is $G = -29.64$ dBi when simulated in a homogeneous 100 mm^3 cubic phantom with $\varepsilon_r = 57, \sigma = 0.8$ S·m^{-1}. The antenna was printed on a $18 \times 18 \times 0.127$ mm^3 Rogers RT/duroid® 5880 substrate with $\varepsilon_r = 2.2, \tan \delta < 0.0009$ and conforms to $\varnothing 10$ mm

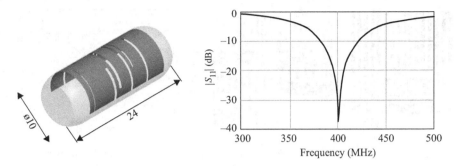

Figure 6.33   *Conformal microstrip meandered antenna by Psathas et al. [105]: design (mm) and reflection coefficient. ©2013 IEEE*

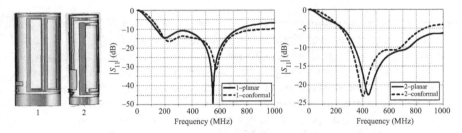

Figure 6.34   *Two conformal meander-line antennas by J. C. Wang et al. [106]: designs and reflection coefficients (solid: planar, dashed: conformal). ©2013 IEEE*

cylinder. Superstrate layer is 0.1 mm polyethylene ($\varepsilon_r = 2.25$, $\tan \delta = 0.001$). Antenna miniaturisation is achieved using shorting pin and meandering techniques, which permits to obtain dual-band characteristics as well. The proposed feed is an L-shape coaxial probe. The antenna sensitivity to the environment variation was analysed for four tissues [105]: oesophagus, stomach, small and large intestines (dielectric properties from References 45–47). The authors used FEM implemented in Ansys® HFSS™ to design the antenna; the mesh adaptivity was set-up to achieve 2% S_{11} error.

J. C. Wang *et al.* [106] reported two conformal antennas based on a planar meander-line design. The first conformal antenna (Figure 6.34) has a bandwidth of 306–529 MHz (223 MHz or 53%) for VSWR < 2 when simulated in the phantom having $\varepsilon_r = 56$, $\sigma = 0.8$ S·m^{-1}. However, there is a discrepancy between the reported values and reflection coefficient plot (roughly 150–950 MHz on the plot). The phantom has an elliptical cylinder shape with the dimensions $100 \times 150 \times 200$ mm^3. The conformal antenna was simulated with bending diameters ø10.6 mm (reported antenna results), ø14 mm, ø20 mm and ø30 mm. The planar antenna size is 33×23 mm^2. The capsule shell has $\varepsilon_r = 4.4$. The authors analysed also the impact of a capsule shell with variable thickness: 0.5, 1.6 and 2.4 mm. The second antenna (Figure 6.34) has a reported bandwidth of 330–894 MHz (564 MHz or 92%) for VSWR < 2 when simulated in the same phantom as for the first antenna. The same discrepancy as for the first

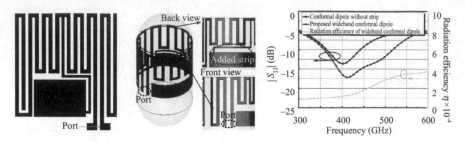

*Figure 6.35   Enhanced bandwidth dipole antenna by Xu et al. [107]: design
(planar and conformal), reflection coefficient and radiation efficiency.
©2015 IEEE*

antenna exists between written and plotted values: about 310–690 MHz on the plot. The conformal antenna was simulated with bending diameters ø4.8 mm (reported antenna results), ø10 mm, ø14 mm, ø20 mm and ø30 mm. The planar antenna size is $18.2 \times 15\,\text{mm}^2$. Efficiencies and gain were not reported for both antennas; however, authors claim the transmission for at least 15 cm from the body surface. The authors used CST Microwave Studio® to analyse the antenna.

Xu *et al.* [107] proposed an enhanced bandwidth dipole antenna for 401–406 MHz MedRadio band (Figure 6.35). The bandwidth is 358–516 or 158 MHz (36.1%) when simulated in a $180\,\text{mm}^3$ cubic phantom with $\varepsilon_r = 46.76$, $\sigma = 0.69\,\text{S·m}^{-1}$. The capsule is located at the centre of x–y-plane and 3 mm deep along the z-axis of the phantom. This phantom set-up is apparently improper for the suggested capsule application – endoscopy. The reported gain is $G = -37$ dBi and efficiency is about $\eta = 0.02\%$. The planar prototype was printed on 0.635 mm thick Rogers 3010 substrate ($\varepsilon_r = 10.2, \tan \delta = 0.0035$) and covered with the same material superstrate. The planar antenna was measured in the phantom containing sugar (56.18%), salt (2.33%) and deionised water. The measured dielectric characteristics of the phantom were not reported. The final antenna is optimised to be printed on the polyimide substrate ($\varepsilon_r = 3.5, \tan \delta = 0.008$) to conform the $(24 \times \text{ø}11)$ mm capsule. As the antenna is based on a dipole radiation principle, it does not have a ground plane (hence the mirror currents will affect the radiation efficiency).

Q. Wang *et al.* [108] proposed an UWB conformal trapezoid strip dielectric resonator antenna (Figure 6.36). The antenna bandwidth is 2–5 GHz (3 GHz or 86%) at $S_{11} < -10$ dB when simulated in a $200\,\text{mm}^3$ cubic homogeneous phantom with $\varepsilon_r = 51.5, \sigma = 3.2\,\text{S·m}^{-1}$ at 4 GHz. The maximum gain is $G = -24.2$ dBi. The permittivity of ø8 mm hemisphere is $\varepsilon_r = 3$ (presumably Polyvinylchloride (PVC)); it is designed to fit a $(26 \times \text{ø}11)$ mm capsule. The phantom for the measurement was prepared using the recipe from Reference 109 containing $C_{12}H_{22}O_{11}$.

Morimoto *et al.* [110] proposed two UWB (3.4–4.8 GHz band) hemispherical capsule-conformal antennas: helical and loop (Figure 6.37). The −10 dB bandwidth begins at 3 GHz for the loop antenna and at 3.5 GHz for the helical. Antennas were simulated using the FDTD method inside a $50\,\text{mm}^3$ cubic phantom with

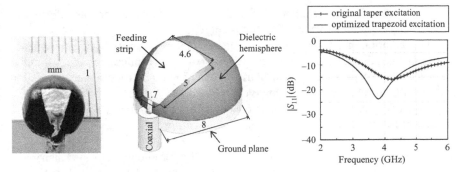

Figure 6.36 *Trapezoid strip dielectric resonator antenna by Q. Wang et al. [108]: prototype, design (mm) and reflection coefficients.* © *2010 IEEE*

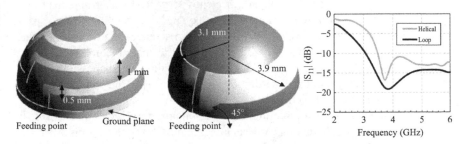

Figure 6.37 *Conformal hemispherical antennas by Morimoto et al. [110]: helical and loop design and dimensions, reflection coefficients.* © *2013 IEEE*

$\varepsilon_r = 51.4, \sigma = 3.26$ S·m^{-1} at 4.1 GHz. The permittivity of ø8 mm hemisphere is $\varepsilon_r = 3$. The antenna is covered with a 0.1 mm polyethylene shell ($\varepsilon_r = 2.2$). The maximum gains are $G_{loop} = -30$ dBi and $G_{helical} = -32$ dBi. The loop antenna has zeros in y–z-plane radiation pattern; both antennas are omnidirectional in x–y-plane. The authors manufactured a planar loop antenna to validate the model.

6.8 Conclusions

Emerging biomedical wireless telemetry permits to continuously monitor physiological parameters unaffecting the mobility of a patient or an animal. It relies on an efficient and reliable communication with an external monitor via a radio frequency link. Although alternatives to radio frequency transmission exist for wireless GI capsules (inductive links and human body communication (HBC) [32]), it is limited both in operational range and throughput. Wireless transmission in VHF–SHF bands yields the longest range and highest data transfer rates for ingestible wireless devices.

Biological media start to notably influence the antenna impedance and performances for the frequencies beyond a few tenth of megahertz. Tissue dielectric properties must be taken into account while designing antennas as well as when comparing the existent: the reported efficiencies and bandwidths are not informative without considering the media dielectric properties.

The radiation efficiency – instead of gain – is a more appropriate performance metric for capsule antennas since they are required to have an omnidirectional radiation pattern, as a capsule position in GI tract is undefined. This requirement is satisfied naturally for all capsule-sized antennas operating below $f \lesssim 2$ GHz[8] according to the theory of ESAs [64]. Several studies concluded that frequencies above 2 GHz are inefficient for ingestible capsule antennas (see Section 6.5), so the directivity could be neglected in most cases.

Modern VHF–SHF capsule antennas can be divided into wire, printed planar and printed conformal. Wire antennas are easy to design and tune, they provide sufficient bandwidth and, in helical implementation, they occupy a sufficiently large surface augmenting the fundamental bandwidth versus efficiency limits (see Section 6.6). Spiral antennas can have a ground plane sufficiently large to shield an antenna from a capsule circuitry. However, wire antennas are bulky, poorly integrable into a capsule design and non-technological (i.e. expensive and poorly repeatable for mass production). Printed planar antennas are more technological to manufacture and its spiral designs are less voluminous, but still lack the integrability. The bandwidth is comparable to wire antennas but it is harder to achieve the same Chu sphere radii, thus planar antennas are less efficient for a given bandwidth.

Flexible substrates permit to minimise an antenna footprint within a capsule by conforming to the shell. In this way, we can increase the Chu sphere radius by making use of all available capsule surface, thus fending off the performance limits of ESAs. So, theoretically achievable radiation efficiency η for a given bandwidth could be greater than for wire and planar antennas. Moreover, printed conformal antennas are technological and could be fully integrated into a circuit design; for instance, it could be printed on the same flexible PCB with electronics and rolled to fit into the capsule as in Reference 97. Unshielded designs (e.g. meandered dipole and loop antennas) interact strongly with capsule circuitry. Conformal microstrip antennas are less coupled to the circuitry because of a ground plane. It is convenient from a perspective of system design: once tuned, the antenna will maintain its characteristics when capsule circuitry or sensing units are revised. So, microstrip designs contribute to the faster implementation of new biotelemetry sensing technologies since the antenna coupling to capsule electronics could be intricate and time consuming to analyse.

While designing capsule antennas of whatsoever type, a frequency of operation must be properly chosen. Optimal frequency maximises the antenna radiation efficiency without deteriorating too much the through-body transmission due to the attenuation. MedRadio, UWB and some of ISM bands are globally available, WMTS can be used only in the USA and Canada. When the frequency is chosen, another compromise should be considered between the bandwidth and radiation efficiency. Ideally, a capsule antenna should have a bandwidth as narrow as possible to reach

---

[8]This frequency limit depends strongly on antenna size and effective permittivity of the medium. The details are given in Section 6.6

the target data rate and to cover the frequency fluctuation due to the uncertainty of tissue properties. In this way, the maximum achievable radiation efficiency will be maximised for a given antenna electrical size.

A capsule antenna must be robust enough to remain tuned while the adjacent tissue parameters vary during the GI passage. This is achievable either by sufficiently wide-band designs to cover the resonant frequency detuning (efficiency must be sacrificed) or using the microstrip designs – as in References 92, 97, 104 – with an insulation layer (superstrate). These narrowband antennas are preferable as their maximum achievable radiation efficiency is higher. Microstrip antennas require miniaturisation to operate within the optimal frequency range for GI capsules. Several miniaturisation techniques exist and comprehensive guidelines are given in References 64, 111.

A properly designed antenna will reliably transmit during the GI tract passage within a specified operational range, provide sufficient bandwidth for desired data transmission rates, occupy minimal volume within a capsule and integrate well with control and sensing electronics.

Capsule antenna safety and dosimetry issues are out of the scope of this chapter; relevant information can be found in References 11, 12, 55. Remaining research challenges on capsule antennas involve maximising the radiation efficiency, reducing the coupling with biological tissue and thus minimising the SAR levels as well as studying new biocompatible materials. Future implantable and ingestible wireless telemetry systems have a great potential to improve the diagnosis, treatment and prevention of disease, illness or injury.

Acknowledgements

We would like to express our gratitude and appreciation to the BodyCap company without which this research would have been impossible. Special thanks to Professor Ala Sharaiha for giving a deep insight into electrically small antennas. Support for this work is provided by the Eiffel scholarship (the French Ministry of Foreign Affairs and International Development) and by the Grant project GACR P102/11/0498 (the Grant Agency of the Czech Republic).

References

[1] W. Greatbatch and C. F. Holmes, 'History of implantable devices,' *IEEE Engineering in Medicine and Biology Magazine*, vol. 10, no. 3, pp. 38–41, 1991.

[2] V. K. Zworykin and J. T. Farrar, 'A "radio pill",' *Nature*, vol. 179, no. 4566, p. 898, 1957,

[3] R. S. Mackay and B. Jacobson, 'Endoradiosonde,' *Nature*, vol. 179, no. 4572, pp. 1239–1240, 1957.

[4] J. T. Farrar, V. K. Zworykin, and J. Baum, 'Pressure-sensitive telemetering capsule for study of gastrointestinal motility,' *Science*, vol. 126, no. 3280, pp. 975–976, 1957.

[5] R. S. Mackay and B. Jacobson, 'Pill telemeters from digestive tract,' *Electronics*, vol. 31, pp. 51–53, 1958.

[6] R. S. Mackay, 'Radio telemetering from within the human body,' *IRE Transactions on Medical Electronics*, vol. ME-6, no. 2, pp. 100–105, 1959.

[7] R. S. Mackay, 'Endoradiosondes: further notes,' *IRE Transactions on Medical Electronics*, vol. ME-7, no. 2, pp. 67–73, 1960.

[8] J. O. Sines, 'Permanent implants for heart rate and body temperature recording in the rat,' *AMA Archives of General Psychiatry*, vol. 2, no. 2, pp. 182–183, 1960.

[9] M. Herbert, *Pill-type swallowable transmitter*, US3144017 A, Aug. 1964.

[10] B. W. Watson, S. J. Meldrum, H. C. Riddle, R. L. Brown, and G. E. Sladen, 'pH profile of gut as measured by radiotelemetry capsule,' *British Medical Journal*, vol. 2, no. 5805, pp. 104–106, 1972.

[11] A. Kiourti, K. A. Psathas, and K. S. Nikita, 'Implantable and ingestible medical devices with wireless telemetry functionalities: a review of current status and challenges,' *Bioelectromagnetics*, vol. 35, no. 1, pp. 1–15, 2014.

[12] R. Bashirullah, 'Wireless implants,' *IEEE Microwave Magazine*, vol. 11, no. 7, pp. S14–S23, 2010.

[13] T. F. Budinger, 'Biomonitoring with wireless communications,' *Annual Review of Biomedical Engineering*, vol. 5, pp. 383–412, 2003.

[14] E. Y. Chow, M. M. Morris, and P. P. Irazoqui, 'Implantable RF medical devices: the benefits of high-speed communication and much greater communication distances in biomedical applications,' *IEEE Microwave Magazine*, vol. 14, no. 4, pp. 64–73, 2013.

[15] M. R. Yuce and T. Dissanayake, 'Easy-to-swallow antenna and propagation,' *IEEE Microwave Magazine*, vol. 14, no. 4, pp. 74–82, 2013.

[16] C. McCaffrey, O. Chevalerias, C. O'Mathuna, and K. Twomey, 'Swallowable-capsule technology,' *IEEE Pervasive Computing*, vol. 7, no. 1, pp. 23–29, 2008.

[17] M. R. Yuce and T. Dissanayake, 'Easy-to-swallow wireless telemetry,' *IEEE Microwave Magazine*, vol. 13, no. 6, pp. 90–101, 2012.

[18] F. Munoz, G. Alici, and W. Li, 'A review of drug delivery systems for capsule endoscopy,' *Advanced Drug Delivery Reviews*, vol. 71, pp. 77–85, 2013.

[19] J. L. Toennies, G. Tortora, M. Simi, P. Valdastri, and R. J. Webster, 'Swallowable medical devices for diagnosis and surgery: the state of the art,' *Proceedings of the Institution of Mechanical Engineers, Part C: Journal of Mechanical Engineering Science*, vol. 224, no. 7, pp. 1397–1414, Jul. 2010.

[20] C. P. Swain, F. Gong, and T. N. Mills, 'Wireless transmission of a colour television moving image from the stomach using a miniature CCD camera, light source and microwave transmitter,' *Gastrointestinal Endoscopy*, vol. 45, no. 4, p. AB40, 1997.

[21] G. Iddan, G. Meron, A. Glukhovsky, and P. Swain, 'Wireless capsule endoscopy,' *Nature*, vol. 405, no. 6785, pp. 417–417, 2000.

[22] P. Swain, 'Wireless capsule endoscopy,' *Gut*, vol. 52, no. Suppl. 4, pp. iv48–iv50, 2003.

[23]   A. Karargyris and N. Bourbakis, 'Wireless capsule endoscopy and endoscopic imaging: a survey on various methodologies presented,' *IEEE Engineering in Medicine and Biology Magazine*, vol. 29, no. 1, pp. 72–83, 2010.

[24]   K. Twomey and J. Marchesi, 'Swallowable capsule technology: current perspectives and future directions,' *Endoscopy*, vol. 41, no. 4, pp. 357–362, 2009.

[25]   M. R. Basar, F. Malek, K. M. Juni, M. S. Idris, and M. I. M. Saleh, 'Ingestible wireless capsule technology: a review of development and future indication,' *International Journal of Antennas and Propagation*, vol. 2012, pp. 1–14, 2012.

[26]   G. Pan and L. Wang, 'Swallowable wireless capsule endoscopy: progress and technical challenges,' *Gastroenterology Research and Practice*, vol. 2012, pp. 1–9, Dec. 2011.

[27]   Given Imaging Ltd. *PillCam Capsule Endoscopy* [online]. Available from http://www.givenimaging.com/en-int/innovative-solutions/capsule-endoscopy [Accessed 2 Feb 2015].

[28]   Chongqing Jinshan Science and Technology (Group) Co. Ltd. *OMOM Capsule Endoscopy* [online]. Available from http://english.jinshangroup .com/index.php/Index/product/id/2 [Accessed 2 Feb 2015].

[29]   Olympus Inc. *Capsule Endoscopy* [online]. Available from http://medical .olympusamerica.com/procedure/capsule-endoscopy [Accessed 2 Feb 2015].

[30]   RF Co. Ltd. *Sayaka* [online]. Available from http://rfsystemlab.com/en /sayaka [Accessed 2 Feb 2015].

[31]   IntroMedic Co. Ltd. *Mirocam* [online]. Available from http://www .intromedic.com/eng/sub_products_2.html [Accessed 2 Feb 2015].

[32]   C. H. Hyoung, S. W. Kang, J. H. Hwang, *et al.*, *Human body communication system and method*, US20100274083 A1, Oct 2010.

[33]   CapsoVision. *CapsoCam SV-1* [online]. Available from http://www.capso vision.com/index.php/capsocam [Accessed 2 Feb 2015].

[34]   K. M. S. Thotahewa, J.-M. Redoute, and M. R. Yuce, 'A UWB wireless capsule endoscopy device,' in *2014 36th Annual International Conference of the IEEE Engineering in Medicine and Biology Society (EMBC)*, Chicago, IL, 2014, pp. 6977–6980.

[35]   BMedical. *VitalSense* [online]. Available from http://www.bmedical.com.au /shop/core-body-temperature-capsule-ingestable-jonah.html [Accessed 2 Feb 2015].

[36]   BodyCap Medical. *eCelsius* [online]. Available from http://www.bodycap -medical.com/index.php?view=article&id=101 [Accessed 2 Feb 2015].

[37]   Given Imaging Ltd. *Bravo pH* [online]. Available from http://www.given imaging.com/en-int/Innovative-Solutions/Reflux-Monitoring/Bravo-pH/ [Accessed 2 Feb 2015].

[38]   Chongqing Jinshan Science and Technology (Group) Co. Ltd. *Wireless pH Monitoring Capsule* [online]. Available from http://english.jinshangroup .com/index.php/Index/product/id/6 [Accessed 2 Feb 2015].

[39]   Given Imaging Ltd. *SmartPill* [online]. Available from http://www.givenimag ing.com/en-int/Innovative-Solutions/Motility/SmartPill [Accessed 2 Feb 2015].

[40]   HQ Inc. *CorTemp Sensor* [online]. Available from http://www.hqinc.net /cortemp-sensor-2/ [Accessed 2 Feb 2015].

[41]   Philips. *IntelliCap* [online]. Available from http://www.research.philips .com/initiatives/intellicap/ [Accessed 2 Feb 2015].

[42]   Remo Technologies Ltd. *Biotelemetry PProducts* [online]. Available from http://www.remotechnologies.com/products.html [Accessed 2 Feb 2015].

[43]   K. Kramer, L. Kinter, B. P. Brockway, H.-P. Voss, R. Remie, and B. L. M. Van Zutphen, 'The use of radiotelemetry in small laboratory animals: recent advances,' *Journal of the American Association for Laboratory Animal Science*, vol. 40, no. 1, pp. 8–16, 2001.

[44]   K. Kramer and L. B. Kinter, 'Evaluation and applications of radioteleme-try in small laboratory animals,' *Physiological Genomics*, vol. 13, no. 3, pp. 197–205, 2003.

[45]   C. Gabriel, S. Gabriel, and E. Corthout, 'The dielectric properties of biolog-ical tissues: I. Literature survey,' *Physics in Medicine and Biology*, vol. 41, pp. 2231–2249, 1996.

[46]   S. Gabriel, R. W. Lau, and C. Gabriel, 'The dielectric properties of biological tissues: II. Measurements in the frequency range 10 Hz to 20 GHz,' *Physics in Medicine and Biology*, vol. 41, pp. 2251–2269, 1996.

[47]   S. Gabriel, R. W. Lau, and C. Gabriel, 'The dielectric properties of biological tissues: III. Parametric models for the dielectric spectrum of tissues,' *Physics in Medicine and Biology*, vol. 41, pp. 2271–2293, 1996.

[48]   E. Ben-Soussan, G. Savoye, M. Antonietti, S. Ramirez, E. Lerebours, and P. Ducrotté, 'Factors that affect gastric passage of video capsule,' *Gastrointestinal Endoscopy*, vol. 62, no. 5, pp. 785–790, 2005.

[49]   J. Worsøe, L. Fynne, T. Gregersen, *et al.*, 'Gastric transit and small intesti-nal transit time and motility assessed by a magnet tracking system,' *BMC Gastroenterology*, vol. 11, no. 145, pp. 1–10, 2011.

[50]   S. S. C. Rao, B. Kuo, R. W. McCallum, *et al.*, 'Investigation of colonic and whole-gut transit with wireless motility capsule and radiopaque markers in constipation,' *Clinical Gastroenterology and Hepatology*, vol. 7, no. 5, pp. 537–544, 2009.

[51]   P. S. Hall and Y. Hao, *Antennas and Propagation for Body-Centric Wireless Communications*, 2nd ed. Artech House, Norwood, MA, 2012.

[52]   *IEEE Standard for Definitions of Terms for Antennas*, IEEE Std 145-2013, Mar. 2014.

[53]   C. A. Balanis, *Antenna Theory: Analysis and Design*, 3rd ed. John Wiley & Sons, New York, 2005.

[54]   D. Nikolayev, M. Zhadobov, R. Sauleau, P.-A. Chapon, E. Blond, and P. Karban, 'Decreasing SAR and enhancing transmission of an in-body biotelemetry capsule by reducing the near-field coupling with surrounding tissues,' presented at *The BioEM 2015*, Monterey, CA, 2015.

[55] K. S. Nikita, *Handbook of Biomedical Telemetry*. John Wiley & Sons, New York, 2014.

[56] D. M. Pozar, *Microwave Engineering*, 4th ed. John Wiley & Sons, New York, 2012.

[57] F. S. Barnes and B. Greenebaum, *Handbook of Biological Effects of Electromagnetic Fields. Bioengineering and Biophysical Aspects of Electromagnetic Fields*, 3rd ed. CRC/Taylor & Francis, Boca Raton, FL, 2007.

[58] L. C. Chirwa, P. A. Hammond, S. Roy, and D. R. S. Cumming, 'Electromagnetic radiation from ingested sources in the human intestine between 150 MHz and 1.2 GHz,' *IEEE Transactions on Biomedical Engineering*, vol. 50, no. 4, pp. 484–492, 2003.

[59] L. C. Chirwa, P. A. Hammond, S. Roy, and D. R. S. Cumming, 'Radiation from ingested wireless devices in biomedical telemetry bands,' *Electronics Letters*, vol. 39, no. 2, pp. 178–179, 2003.

[60] H. Yu, G. S. Irby, D. M. Peterson, *et al.*, 'Printed capsule antenna for medication compliance monitoring,' *Electronics Letters*, vol. 43, no. 22, pp. 1179–1181, 2007.

[61] A. Sani, A. Alomainy, and H. Yang, 'Numerical characterization and link budget evaluation of wireless implants considering different digital human phantoms,' *IEEE Transactions on Microwave Theory and Techniques*, vol. 57, pp. 2605–2613, 2009.

[62] L. Xu, M. Q.-H. Meng, H. Ren, and Y. Chan, 'Radiation characteristics of ingestible wireless devices in human intestine following radio frequency exposure at 430, 800, 1200, and 2400 MHz,' *IEEE Transactions on Antennas and Propagation*, vol. 57, no. 8, pp. 2418–2428, 2009.

[63] D. Nikolayev, 'An anatomically realistic boundary representation phantom for studying VHF–UHF radiation effects,' *Computational Problems in Electrical Engineering*, vol. 3, no. 2, pp. 77–82, 2013.

[64] J. Volakis, C.-C. Chen, and K. Fujimoto, *Small Antennas: Miniaturization Techniques and Applications*. McGraw-Hill Professional, New York, 2010.

[65] L. J. Chu, 'Physical limitations of omni-directional antennas,' *Journal of Applied Physics*, vol. 19, no. 12, pp. 1163–1175, 1948.

[66] J. S. McLean, 'A re-examination of the fundamental limits on the radiation Q of electrically small antennas,' *IEEE Transactions on Antennas and Propagation*, vol. 44, no. 5, pp. 672–676, 1996.

[67] S. R. Best and A. D. Yaghjian, 'The lower bounds on Q for lossy electric and magnetic dipole antennas,' *IEEE Antennas and Wireless Propagation Letters*, San Diego, CA, vol. 3, no. 1, pp. 314–316, 2004.

[68] A. D. Yaghjian and S. R. Best, 'Impedance, bandwidth, and Q of antennas,' *IEEE Transactions on Antennas and Propagation*, Washington, DC, vol. 53, no. 4, pp. 1298–1324, 2005.

[69] R. C. Hansen, *Electrically Small, Superdirective, and Superconducting Antennas*, 1st ed. Wiley-Interscience, New York, 2006.

[70] N. Altunyurt, M. Swaminathan, P. M. Raj, and V. Nair, 'Antenna miniaturization using magneto-dielectric substrates,' in *59th Electronic Components and Technology Conference (ECTC)*, 2009, pp. 801–808.

[71]   S.-I. Kwak, K. Chang, and Y.-J. Yoon, 'The helical antenna for the capsule endoscope,' in *2005 IEEE Antennas and Propagation Society International Symposium*, 2005, vol. 2B, pp. 804–807.

[72]   L. Xu, M. Q.-H. Meng, and Y. Chan, 'Effects of dielectric parameters of human body on radiation characteristics of ingestible wireless device at operating frequency of 430 MHz,' *IEEE Transactions on Biomedical Engineering*, vol. 56, no. 8, pp. 2083–2094, 2009.

[73]   L. Xu, M. Q.-H. Meng, and C. Hu, 'Effects of dielectric values of human body on specific absorption rate following 430, 800, and 1200 MHz RF exposure to ingestible wireless device,' *IEEE Transactions on Information Technology in Biomedicine*, vol. 14, no. 1, pp. 52–59, 2010.

[74]   H. Rajagopalan and Y. Rahmat-Samii, 'Wireless medical telemetry characterization for ingestible capsule antenna designs,' *IEEE Antennas and Wireless Propagation Letters*, vol. 11, pp. 1679–1682, 2012.

[75]   D. Zhao, X. Hou, X. Wang, and C. Peng, 'Miniaturization design of the antenna for wireless capsule endoscope,' presented at *Fourth International Conference on Bioinformatics and Biomedical Engineering (iCBBE)*, San Diego, CA, 2010, pp. 1–4.

[76]   S.-H. Lee, K. Chang, K. J. Kim, and Y.-J. Yoon, 'A conical spiral antenna for wideband capsule endoscope system,' in *IEEE Antennas and Propagation Society International Symposium*, New Orleans, LA, 2008, pp. 1–4.

[77]   S.-H. Lee and Y.-J. Yoon, 'Fat arm spiral antenna for wideband capsule endoscope systems,' in *IEEE Radio and Wireless Symposium (RWS)*, 2010, pp. 579–582.

[78]   S.-H. Lee, J. Lee, Y. J. Yoon, *et al.*, 'A wideband spiral antenna for ingestible capsule endoscope systems: experimental results in a human phantom and a pig,' *IEEE Transactions on Biomedical Engineering*, Chiba, Japan, vol. 58, no. 6, pp. 1734–1741, 2011.

[79]   S.-H. Lee and Y.-J. Yoon, 'A dual spiral antenna for ultra-wideband capsule endoscope system,' in *International Workshop on Antenna Technology: Small Antennas and Novel Metamaterials*, Suzhou, China, 2008, pp. 227–230.

[80]   S.-I. Kwak, K. Chang, and Y.-J. Yoon, 'Ultra-wide band spiral shaped small antenna for the biomedical telemetry,' in *Microwave Conference Proceedings, APMC*, Shanghai, China, 2005, vol. 1, pp. 1–4.

[81]   X. Xie, G. Li, B. Chi, X. Yu, C. Zhang, and Z. Wang, 'Micro-system design for wireless endoscopy system,' presented at *The 27th Annual International Conference IEEE-EMBS*, Chengdu, China, 2005, pp. 7135–7138.

[82]   B. Huang, G. Z. Yan, and P. Zan, 'Design of ingested small microstrip antenna for radiotelemetry capsules,' *Electronics Letters*, vol. 43, no. 22, pp. 1–2, 2007.

[83]   F. Merli, L. Bolomey, J. Zurcher, G. Corradini, E. Meurville, and A. K. Skrivervik, 'Design, realization and measurements of a miniature antenna for implantable wireless communication systems,' *IEEE Transactions on Antennas and Propagation*, vol. 59, no. 10, pp. 3544–3555, 2011.

[84]   A. Christ, W. Kainz, E. G. Hahn, *et al.*, 'The Virtual Family – development of surface-based anatomical models of two adults and two children for

dosimetric simulations,' *Physics in Medicine and Biology*, vol. 55, pp. N23–N38, 2010.

[85] C. Liu, Y.-X. Guo, and S. Xiao, 'Circularly polarized helical antenna for ISM-band ingestible capsule endoscope systems,' *IEEE Transactions on Antennas and Propagation*, vol. 62, no. 12, pp. 6027–6039, 2014.

[86] T. Dissanayake, K. P. Esselle, and M. R. Yuce, 'Dielectric loaded impedance matching for wideband implanted antennas,' *IEEE Transactions on Microwave Theory and Techniques*, vol. 57, no. 10, pp. 2480–2487, 2009.

[87] K. M. S. Thotahewa, J.-M. Redoute, and M. R. Yuce, 'Electromagnetic power absorption of the human abdomen from IR-UWB based wireless capsule endoscopy devices,' presented at *The 2013 IEEE International Conference on Ultra-Wideband (ICUWB)*, Sydney, NSW, 2013, pp. 79–84.

[88] T. Dissanayake, M. R. Yuce, and C. Ho, 'Design and evaluation of a compact antenna for implant-to-air UWB communication,' *IEEE Antennas and Wireless Propagation Letters*, vol. 8, pp. 153–156, 2009.

[89] P. M. Izdebski, H. Rajagopalan, and Y. Rahmat-Samii, 'Conformal ingestible capsule antenna: a novel chandelier meandered design,' *IEEE Transactions on Antennas and Propagation*, vol. 57, no. 4, pp. 900–909, 2009.

[90] H. Rajagopalan, P. Izdebski, and Y. Rahmat-Samii, 'A miniaturized ingestible antenna for capsule medical imaging system,' presented at *The IEEE Antennas and Propagation Society International Symposium*, Charleston, SC, 2009, pp. 1–4.

[91] H. Rajagopalan and Y. Rahmat-Samii, 'Novel ingestible capsule antenna designs for medical monitoring and diagnostics,' in *Proceedings of the Fourth European Conference on Antennas and Propagation (EuCAP)*, Barcelona, Spain, 2010, pp. 1–5.

[92] X. Cheng, D. E. Senior, C. Kim, and Y.-K. Yoon, 'A compact omnidirectional self-packaged patch antenna with complementary split-ring resonator loading for wireless endoscope applications,' *IEEE Antennas and Wireless Propagation Letters*, vol. 10, pp. 1532–1535, 2011.

[93] T. Kumagai, K. Saito, M. Takahashi, and K. Ito, 'A 430 MHz band receiving antenna for microwave power transmission to capsular endoscope,' presented at *The General Assembly and Scientific Symposium, 2011 XXXth URSI*, Istanbul, Turkey, 2011, pp. 1–4.

[94] S. Yun, K. Kim, and S. Nam, 'Outer-wall loop antenna for ultrawideband capsule endoscope system,' *IEEE Antennas and Wireless Propagation Letters*, vol. 9, pp. 1135–1138, 2010.

[95] F. Merli, B. Fuchs, J. R. Mosig, and A. K. Skrivervik, 'The effect of insulating layers on the performance of implanted antennas,' *IEEE Transactions on Antennas and Propagation*, vol. 59, no. 1, pp. 21–31, Jan. 2011.

[96] K. Kim, S. Yun, S. Lee, S. Nam, Y. J. Yoon, and C. Cheon, 'A design of a high-speed and high-efficiency capsule endoscopy system,' *IEEE Transactions on Biomedical Engineering*, vol. 59, no. 4, pp. 1005–1011, 2012.

[97] Y. Mahe, A. Chousseaud, M. Brunet, and B. Froppier, 'New flexible medical compact antenna: design and analysis,' *International Journal of Antennas and Propagation*, vol. 2012, Article ID 837230, 6 pages, 2012. doi:10.1155/2012/837230.

[98]   R. Alrawashdeh, Y. Huang, and P. Cao, 'A conformal U-shaped loop antenna for biomedical applications,' in *2013 Seventh European Conference on Antennas and Propagation (EuCAP)*, 2013, pp. 157–160.

[99]   R. Alrawashdeh, Y. Huang, and P. Cao, 'Flexible meandered loop antenna for implants in MedRadio and ISM bands,' *Electronics Letters*, vol. 49, no. 24, pp. 1515–1517, 2013.

[100]   R. Alrawashdeh, Y. Huang, and P. Cao, 'A flexible loop antenna for total knee replacement implants in the MedRadio band,' in *Antennas and Propagation Conference (LAPC)*, Loughborough, UK, 2013, pp. 225–228.

[101]   R. Alrawashdeh, Y. Huang, and A. A. B. Sajak, 'A flexible loop antenna for biomedical bone implants,' presented at *The Eighth European Conference on Antennas and Propagation (EuCAP 2014)*, The Hague, the Netherlands, 2014, pp. 999–1002.

[102]   R. Alrawashdeh, Y. Huang, P. Cao, and E. Lim, 'A new small conformal antenna for capsule endoscopy,' in *2013 Seventh European Conference on Antennas and Propagation (EuCAP)*, 2013, pp. 220–223.

[103]   R. S. Alrawashdeh, Y. Huang, M. Kod, and A. A. B. Sajak, 'A broadband flexible implantable loop antenna with complementary split ring resonators,' *IEEE Antennas and Wireless Propagation Letters*, vol. 14, pp. 1322–1325, 2015.

[104]   K. A. Psathas, A. Kiourti, and K. S. Nikita, 'A novel conformal antenna for ingestible capsule endoscopy in the MedRadio band,' in *34th Progress in Electromagnetics Research Symposium (PIERS 2013)*, Stockholm, Sweden, 2013.

[105]   K. A. Psathas, A. P. Keliris, A. Kiourti, and K. S. Nikita, 'Operation of ingestible antennas along the gastrointestinal tract: detuning and performance,' in *IEEE 13th International Conference on Bioinformatics and Bioengineering (BIBE)*, 2013, pp. 1–4.

[106]   J. C. Wang, E. G. Lim, Z. Wang, *et al.*, 'UWB planar antennas for wireless capsule endoscopy,' in *2013 International Workshop on Antenna Technology (iWAT)*, 2013, pp. 340–343.

[107]   L.-J. Xu, Y.-X. Guo, and W. Wu, 'Bandwidth enhancement of an implantable antenna,' *IEEE Antennas and Wireless Propagation Letters*, vol. 14, 2015, pp. 1510–1513.

[108]   Q. Wang, K. Wolf, and D. Plettemeier, 'An UWB capsule endoscope antenna design for biomedical communications,' in *2010 Third International Symposium on Applied Sciences in Biomedical and Communication Technologies (ISABEL)*, 2010, pp. 1–6.

[109]   J. Zhou, D. Hara, and T. Kobayashi, 'Development of ultra wideband electromagnetic phantoms for antennas and propagation studies,' presented at *The First European Conference on Antennas and Propagation (EuCAP)*, 2006, pp. 1–6.

[110]   Y. Morimoto, D. Anzai, and J. Wang, 'Design of ultra wide-band low-band implant antennas for capsule endoscope application,' in *2013 Seventh International Symposium on Medical Information and Communication Technology (ISMICT)*, 2013, pp. 61–65.

[111]   K.-L. Wong, *Compact and Broadband Microstrip Antennas*. John Wiley & Sons, New York, 2002.

Chapter 7
In vivo wireless channel modeling

A. Fatih Demir*, Z. Esat Ankarali*, Yang Liu*,
Qammer H. Abbasi†, Khalid Qaraqe‡, Erchin Serpedin†,
Huseyin Arslan*,**, and Richard D. Gitlin*

7.1   Introduction

Technological advances in biomedical engineering have significantly improved the quality of life and increased the life expectancy of many people. In recent years, there has been increased interest in wireless body area networks (WBANs) research with the goal of satisfying the demand for innovative biomedical technologies and improved healthcare quality [1, 2]. One component of such advanced technologies is represented by the devices such as wireless *in vivo* sensors and actuators, e.g., pacemakers, internal drug delivery devices, nerve stimulators, wireless capsule endoscopes (WCEs), etc. *In vivo* wireless medical devices and their associated technologies represent the next stage of this evolution and offer a cost efficient and scalable solution along with the integration of wearable devices. *In vivo*-WBAN devices (Figure 7.1) are capable of providing continuous health monitoring and reducing the invasiveness of surgeries. Furthermore, patient information can be collected over a larger period of time, and physicians are able to perform more reliable analysis by exploiting *big data* [3] rather than relying on the data recorded in short hospital visits [4–6].

In order to fully exploit and increase further the potential of WBANs for practical applications, it is necessary to accurately assess the propagation of electromagnetic (EM) waveforms in an *in vivo* communication environment (implant-to-implant and implant-to-external device) and obtain accurate channel models that are necessary to optimize the system parameters and build reliable, high-performance communication systems. In particular, creating and accessing such a model is necessary for achieving high data rates, target link budgets, determining optimal operating frequencies, and designing efficient antennas and transceivers including digital baseband transmitter/receiver algorithms [7, 8]. Therefore, investigation of the *in vivo* wireless communication channel is crucial to obtain better performance for *in vivo*-WBAN devices. However, research on the *in vivo* wireless communication is still in the early

*Department of Electrical Engineering, University of South Florida, Tampa, FL, USA
**Department of Electrical Engineering, Istanbul Medipol University, Istanbul, Turkey
†Department of ECEN Texas A&M University, College Station, TX, USA
‡Department of ECEN Texas A&M University, Doha, Qatar

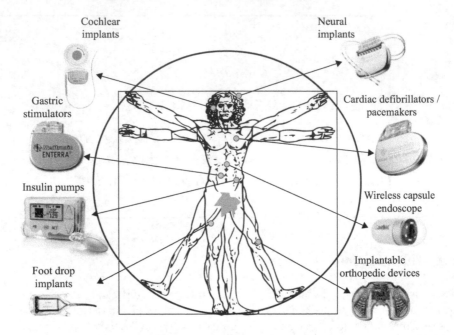

Figure 7.1   In vivo-WBAN

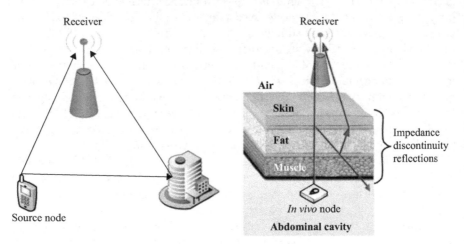

Figure 7.2   The classical communication channel compared with the in vivo channel. ©2016 IEEE. Reprinted with permission from Reference 8

stages, and heretofore there have been relatively few studies compared to the on-body wireless communication channel [2, 9–11].

The *in vivo* channel exhibit different characteristics than those of the more familiar wireless cellular and Wi-Fi environments since the EM wave propagates through a very lossy environment inside the body, and the predominant scatterers are present in the vicinity of the antenna (Figure 7.2). In this chapter, the state of the art of *in vivo* channel characterization is presented, and several research challenges are discussed by considering various communication methods, operational frequencies, and antenna designs. Furthermore, a numerical and experimental characterization of the *in vivo* wireless communication channel is described in detail. This chapter aims to provide a more complete picture of this fascinating communications medium and stimulate more research in this important area.

7.2   EM modeling of the human body

In order to investigate the *in vivo* wireless communication channel, accurate body models, and knowledge of the EM properties of the tissues are crucial [2]. Human autopsy materials and animal tissues have been measured over the frequency range 10 Hz to 20 GHz [12] and the frequency-dependent dielectric properties of the tissues are modeled using the four-pole Cole–Cole equation, which is expressed as:

$$\varepsilon(\omega) = \varepsilon_\infty + \sum_{m=1}^{4} \frac{\Delta\varepsilon_m}{1 + (j\omega\tau_m)^{(1-a_m)}} + \frac{\sigma}{j\omega\varepsilon_0} \tag{7.1}$$

where ε_∞ stands for the body material permittivity at terahertz frequency, ε_0 denotes the free-space permittivity, σ represents the ionic conductivity and ε_m, τ_m, a_m are the body material parameters for each anatomical region. The parameters for anatomical regions are provided in Reference 13, and the EM properties such as conductivity, relative permittivity, loss tangent, and penetration depth can be derived using these parameters in (7.1).

Various physical and numerical phantoms have been designed in order to simulate the dielectric properties of the tissues for experimental and numerical investigation [14]. These can be classified as homogeneous, multilayered, and heterogeneous phantom models. Although heterogeneous models provide a more realistic approximation to the human body, design of physical heterogeneous phantoms is quite difficult and performing numerical experiments on these models is very complex and resource intensive. On the other hand, homogeneous or multilayer models cannot differentiate EM wave radiation characteristics for different anatomical regions. Figure 7.3 shows examples of heterogeneous physical and numerical phantoms.

Analytical methods are generally viewed as infeasible and require extreme simplifications. Therefore, numerical methods are used for characterizing the *in vivo* wireless communication channel. Numerical methods provide less complex and appropriate approximations to Maxwell's equations via various techniques, such as uniform theory of diffraction (UTD), method of moments (MoM), finite element method (FEM), and finite-difference time-domain method (FDTD). Each method has its own pros and cons and should be selected based on the simulation model and size, operational frequency, available computational resources, and interested characteristics, such as power delay profile (PDP), specific absorption rate (SAR), etc. A detailed comparison of these methods is available in References 2 and 15.

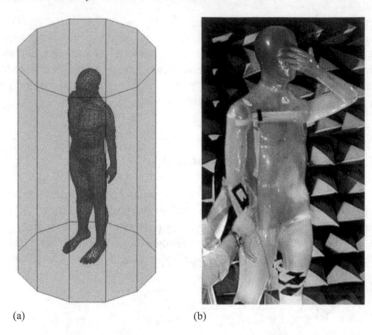

(a)   (b)

Figure 7.3   Heterogeneous human body models: (a) HFSS® model [19] and
(b) physical phantom [14]. ©2016 IEEE. Reprinted with permission
from Reference 55

It may be preferable that numerical experiments should be confirmed with real measurements. However, performing experiments on a living human is carefully regulated. Therefore, anesthetized animals [16, 17] or physical phantoms, allowing repeatability of measurement results [14, 18] are often used for experimental investigation. In addition, the first such study conducted on a human cadaver was reported in Reference 20.

7.3   EM wave propagation through human tissues

Propagation in a lossy medium, such as human tissues, results in a high absorption of EM energy [21]. The absorption effect varies with the frequency-dependent electrical characteristics of the tissues, which mostly consist of water and ionic content [22]. The SAR provides a metric for the amount of absorbed power in the tissue and is expressed as follows [23]:

$$\text{SAR} = \frac{\sigma |E|^2}{\rho} \tag{7.2}$$

where σ, E, and ρ represent the conductivity of the material, the RMS magnitude of the electric field, and the mass density of the material, respectively. The Federal

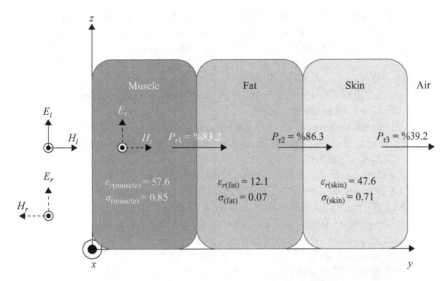

Figure 7.4   *Multi-layer human tissue model at 403 MHz (ε_r: permittivity,*
σ: conductivity, P_τ: power transmission factor). ©2016 IEEE.
Reprinted with permission from Reference 55

Communications Commission (FCC) recommends the SAR to be less than 1.6 W/kg taken over a volume having 1 g of tissue [24].

When a plane EM wave propagates through the interface of two media having different electrical properties, its energy is partially reflected and the remaining portion is transmitted through the boundary of this media. Superposition of the incident and the reflected wave can cause a standing wave effect that may increase the SAR values [22]. A multilayer tissue model at 403 MHz, where each layer extends to infinity (much larger than the wavelength of EM waves) and the dielectric values are calculated using Reference 25, is illustrated in Figure 7.4. If there is a high contrast in the dielectric values of media/tissues, wave reflection at the boundary increases and transmitted power decreases.

In addition to the absorption and reflection losses, EM waves also suffer from expansion of the wave fronts (which assume an ever-increasing sphere shape from an isotropic source in free space), diffraction and scattering (which depend on the wavelength of EM wave). Section 7.6 discusses *in vivo* propagation models by considering these effects in detail.

7.4 Frequency of operation

Since EM waves propagate through the frequency-dependent materials inside the body, the operating frequency has an important effect on the communication channel. Accordingly, the allocated and recommended frequencies are summarized including their effects for the *in vivo* wireless communications channel in this section. The

IEEE 802.15.6 standard [1] was released in 2012 to regulate short-range wireless communications inside or in the vicinity of the human body, and are classified as narrow-band (NB) communications, ultra-wide band communications (UWB), and human body communications (HBC) [26, 27]. The frequency bands and channel bandwidths (BW) allocated for these communication methods are summarized in Table 7.1. An IEEE 802.15.6 compliant *in vivo*-WBAN device must operate in at least one of these frequency bands.

NB communications operate at lower frequencies compared to UWB communications and hence suffers less from absorption. This can be appreciated by considering (7.1) and (7.2) that describe the absorption as a function of frequency. The medical device radio communications service (MedRadio uses discrete bands within the 401–457 MHz spectrum including the international medical implant communication service (MICS) band) and medical body area network (MBAN, 2360–2400 MHz) are allocated by the FCC for medical devices usage. Therefore, co-user interference problems are less severe in these frequency bands. However, NB communications are only allocated small bandwidths (1 MHz at most) in the standard as shown in Table 7.1. The IEEE 802.15.6 standard does not define a maximum transmit power and the local regulatory bodies limit it. The maximum power is restricted to 25 W EIRP (equivalent radiated isotropic power) by FCC, whereas it is set to 25 W ERP (equivalent radiated power) by ETSI (European Telecommunication Standards Institute) for the 402–405 MHz band.

UWB communications is a promising technology to deploy inside the body due to its inherent features including high data rate capability, low power, improved penetration (propagation) abilities through tissues, and low probability of intercept. The large bandwidths for UWB (499 MHz) enable high data rate communications and applications. Also, UWB signals are inherently robust against detection and smart

Table 7.1　Frequency bands and bandwidths for the three different propagation methods in IEEE 802.15.6. ©2016 IEEE. Reprinted with permission from Reference 55

Propagation method	IEEE 802.15.6 operating freq. bands		Selected references
	Frequency band	**BW**	
Narrow band communications	402–405 MHz	300 kHz	[7, 14, 22, 31, 32, 36, 47]
	420–450 MHz	300 kHz	
	863–870 MHz	400 kHz	[7, 14, 31, 36, 45, 47]
	902–928 MHz	500 kHz	
	950–956 MHz	400 kHz	
	2360–2400 MHz	1 MHz	[7, 14, 36, 50]
	2400–2438.5 MHz	1 MHz	
UWB communications	3.2–4.7 GHz	499 MHz	[17, 28, 36, 50]
	6.2–10.3 GHz	499 MHz	
Human-body communications	16 MHz	4 MHz	[26, 27]
	27 MHz	4 MHz	

jamming attacks because of their extremely low maximum effective isotropic radiated power (EIRP) spectral density, which is −41.3 dBm/MHz [28, 29]. On the other hand, UWB communications inside the body suffers from pulse distortion caused by frequency-dependent tissue absorption and compact antenna design. Recently, the terahertz frequency band has also been a subject of interest for *in vivo* propagation, and it is regarded as one of the most promising bands for the EM paradigm of nano-communications [30].

7.5   *In vivo* antenna design considerations

Unlike free space communications, *in vivo* antennas are often considered to be an integral part of the channel, and they generally require different specifications than *ex vivo* antennas [2, 31–33]. In this section, we will describe their salient differences as compared to the *ex vivo* antennas.

In vivo antennas are subject to strict size constraints and in addition need to be biocompatible. Although copper antennas have better performance, only specific types of materials such as titanium or platinum should be used for *in vivo* communications due to their noncorrosive chemistry [6]. The standard definition of the gain is not valid for *in vivo* antennas since it includes body effects [34, 35]. As noted above, the gain of the *in vivo* antennas cannot be separated from the body effects as the antennas are considered to be an integral part of the channel. Hence, the *in vivo* antennas should be designed and placed carefully in order not to harm the biological tissues and to provide power efficiency. When the antennas are placed inside the human body, their electrical dimensions and gains decrease due to the high dielectric permittivity and high conductivity of the tissues, respectively [36]. For instance, fat has a lower conductivity than skin and muscle. Therefore, *in vivo* antennas are usually placed in a fat (usually subcutaneous fat –SAT–) layer to increase the antenna gain. This placement also provides less absorption losses due to shorter propagation path. However, the antenna size becomes larger in this case. In order to reduce high losses inside the tissues, a high permittivity, low loss coating layer can be used. As the coating thickness increases, the antenna becomes less sensitive to the surrounding material [36, 37].

Lossy materials covering the *in vivo* antenna change the electrical current distribution in the antenna and radiation pattern [18]. It is reported in Reference 31 that directivity of *in vivo* antennas increases due to the curvature of body surface, losses, and dielectric loading from the human body. Therefore, this increased directivity should be taken into account as well in order not to harm the tissues in the vicinity of the antenna [23].

In vivo antennas can be classified into two main groups as electrical and magnetic antennas. Electrical antennas, e.g., dipole antennas, generate electric fields (E-field) normal to the tissues, while magnetic antennas, e.g., loop antennas produce E-fields tangential to the human tissues [38]. Normal E-field components at the medium interfaces overheat the tissues due to the boundary condition requirements [39] as illustrated in Figure 7.5. The muscle layer has a larger permittivity value than the fat layer, and hence, the E-field increases in the fat layer. Therefore, magnetic antennas allow higher transmission power for *in vivo*-WBAN devices as can be appreciated

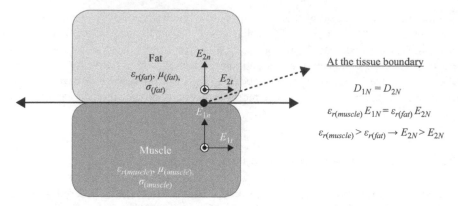

Figure 7.5   *EM propagation through tissue interface. ©2016 IEEE. Reprinted with permission from Reference 55*

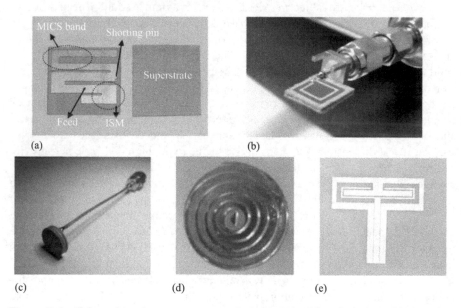

Figure 7.6   *Selected in vivo antenna samples: (a) A dual-band implantable antenna [41], (b) a miniaturized implantable broadband stacked planar inverted-F antenna (PIFA) [42], (c) a miniature scalp-implantable [43], (d) a wideband spiral antenna for WCE [16], and (e) an implantable folded slot dipole antenna [44]. ©2016 IEEE. Reprinted with permission from Reference 55*

by (7.2). In practice, magnetic loop antennas require large sizes, which is a challenge to fit inside the body. Accordingly, smaller size spiral antennas having a similar current distribution as loop antennas can be used for *in vivo* devices [40]. Several selected sample antennas designed for *in vivo* communications are shown in Figure 7.6.

7.6 *In vivo* EM wave propagation models

The important factors for *in vivo* wireless communication channel characterization, such as EM modeling of the human body, propagation through the tissues, and selection of the operational frequency, have been discussed in detail in the preceding sections. Further, the main differences between *in vivo* and *ex vivo* antenna designs were discussed – principally that the antenna must be considered as an integral part of the *in vivo* channel. In this section, the focus is on EM wave propagation inside the human body considering the anatomical features of organs and tissues. Then, the analytical and statistical path loss models will be presented. Since the EM wave propagates through a very lossy environment inside the body and predominant scatterers are present in the near-field region of the antenna, the *in vivo* channel exhibits different characteristics than those of the more familiar wireless cellular and Wi-Fi environments.

EM wave propagation inside the body is subject-specific and strongly related to the location of the antenna as demonstrated in References 7, 18, 20, 31, and 45. Therefore, channel characterization is generally investigated for a specific part of the human body. Figure 7.7 shows several investigated anatomical regions for various

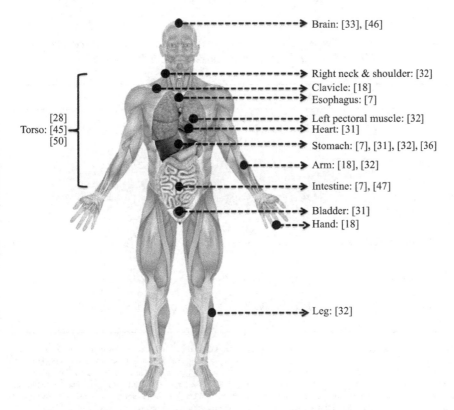

Figure 7.7 Investigated anatomical human body regions. ©2016 IEEE. Reprinted with permission from Reference 55

in vivo-WBAN applications. For example, the heart area has been studied for implantable cardioverter defibrillators and pacemakers, while the gastrointestinal (GI) tract including esophagus, stomach, and intestine has been investigated for WCE applications. The bladder region is studied for wirelessly controlled valves in the urinary tract, and the brain is investigated for neural implants [33, 46]. Also, clavicle, arm, and hands are specifically studied as they are affected less by the *in vivo* medium.

When the *in vivo* antenna is placed in an anatomically complex region, path loss, a measure of average signal power attenuation, increases [7]. This is the case with the intestine which presents a complex structure with repetitive, curvy-shaped, dissimilar tissue layers, while the stomach has a smoother structure. As a result, the path loss is greater in the intestine than in the stomach even at equal *in vivo* antenna depths [7].

Various analytical and statistical path loss formulas have been proposed for the *in vivo* channel in the literature as listed in Table 7.2. These formulas have been derived considering different shadowing phenomena for the *in vivo* medium. The initial three models are functions of the Friis transmission equation [51], return loss, and absorption in the tissues. These models are valid, when either the far field conditions are fulfilled or few scattering objects exist between the transmitter and receiver antennas. In the first model, the free space path loss (FSPL) is expressed by the Friis transmission

Table 7.2   *A review of selected studied path loss models for various scenarios. ©2016 IEEE. Reprinted with permission from Reference 55*

Model	Formulation
FSPL [31]	$P_r = P_t G_t G_r \left(\dfrac{\lambda}{4\pi R} \right)^2$
FSPL with RL [31], [36]	$P_r = P_t G_t (1 - |S_{11}|^2) G_r (1 - |S_{22}|^2) \left(\dfrac{\lambda}{4\pi R} \right)^2$
FSPL with RL and absorption [40]	$P_r = P_t G_t (1 - |S_{11}|^2) G_r (1 - |S_{22}|^2) \left(\dfrac{\lambda}{4\pi R} \right)^2 (e^{-aR})^2$
PMBA for near and far fields [48]	$P_{rn} = \dfrac{16\delta(P_t - P_{NF})}{\pi L^2} A_e, \quad P_{rf} = \left(\dfrac{(P_t - P_{NF} - P_{FF})\lambda^2}{4\pi R^2} \right) G_t G_r$
Statistical model – A [45], [50]	$PL(d) = PL_0 + n(d/d_0) + S \quad (d_0 \leq d)$
Statistical model – B [14], [31], [32]	$PL(d) = PL(d_0) + 10n \log_{10}(d/d_0) + S \quad (d_0 \leq d)$

P_r/P_t, stands for the received/transmitted power; G_r/G_t denotes the gain of the receiver/transmitter antenna; λ represents the free space wavelength; R is the distance between transmitter and receiver antennas, $|S_{11}|$ and $|S_{22}|$ stand for the reflection coefficient of receiver/transmitter antennas; a is the attenuation constant, P_{NF}/P_{FF} is the loss in the near/far fields; δ is A_e/A where A_e is the effective aperture and A is the physical aperture of the antenna; L is the largest dimension of the antenna; d is the depth distance from the body surface, d_0 is the reference depth distance, n is the path loss exponent; PL_0 is the intersection term in dB; S denotes the random shadowing term. Abbreviations: FSPL represents the free space path loss in the far field, RL is the return loss, and PMBA denotes the propagation loss model.

equation. FSPL mainly depends on the gain of antennas, distance, and operating frequency. Its dependency on distance is a result of expansion of the wave fronts as explained in Section 7.3. Additionally, FSPL is frequency dependent due to the relationship between the effective area of the receiver antenna and wavelength. The two equations of the FSPL model in Table 7.2 are derived including the antenna return loss and absorption in the tissues. Another analytical model, PMBA [48], calculates the SAR over the entire human body for the far and near fields and gives the received power using the calculated absorption. Although these analytical expressions provide intuition about each component of the propagation models, they are not practical for link budget design as is the case with the wireless cellular communication environment.

The channel modeling subgroup (Task Group 15.6), which worked on developing the IEEE 802.15.6 standard, submitted their final report on body area network (BAN) channel models in November 2010. In this report, it is determined that the Friis transmission equation can be used for *in vivo* scenarios by adding a random variation term, and the path loss is modeled statistically with a log-normal distributed random shadowing S and path loss exponent n [29, 49]. The path loss exponent (n) heavily depends on environment and is obtained by performing extensive simulations and measurements. In addition, the shadowing term (S) depends on the different body materials (e.g., bone, muscle, fat, etc.) and the antenna gain in different directions [32]. The research efforts on assessing the statistical properties of the *in vivo* propagation channel are not finalized. There are still many open research efforts dedicated to building analytical models for different body parts and operational frequencies [14, 20, 31, 32, 45, 50].

7.7 *In vivo* channel characterization

The numerical *in vivo* channel characterization was performed in [45] using ANSYS HFSS® 15.0, which is a full-wave EM field simulator based on the FEM. ANSYS also provides a detailed human body model of adult male. The numerical investigation was validated by conducting experiments on a human cadaver in a laboratory environment [20]. Istanbul Medipol University provided the ethical approval for the study and medical assistance for this study.

7.7.1 Simulation setup

The simulations [45] have been designed based on an implant-to-external device (in-body to on-body) communications scenario. The human male torso area was divided into four sub regions considering the major internal organs: heart, stomach, kidneys, and intestine as shown in Figure 7.8. The measurements were performed in each sub region by rotating both the receiver (*ex vivo*) and transmitter antennas (*in vivo*) around the body on a plane at 22.5° angle increments as shown in Figure 7.9. For each location of the *ex vivo* antenna (5 cm away from the body surface), the *in vivo* antenna was placed at 10 different depths (10–100 mm). Moreover, the antennas were placed in the same direction in order to prevent polarization losses.

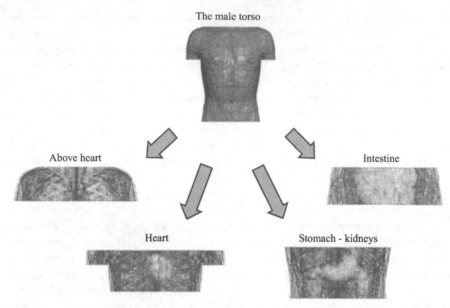

Figure 7.8   Investigated anatomical human body regions. ©2014 IEEE. Reprinted with permission from Reference 45

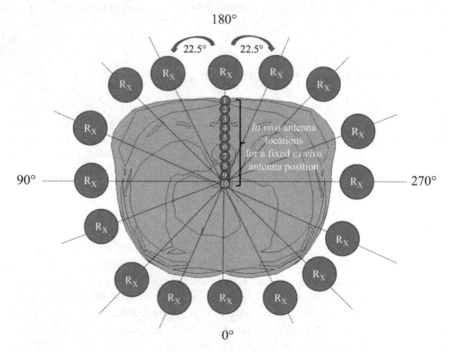

Figure 7.9   In vivo and ex vivo antenna locations in the simulation. 16 (angles) × 10 (depth) × 4 (regions) = 640 simulations were performed for path loss in total. ©2014 IEEE. Reprinted with permission from Reference 45

Omni-directional dipole antennas at 915 MHz were deployed in simulations for simplicity. The dipole antenna size is proportional to the wavelength, which changes with respect to both frequency and permittivity. Although the frequency of operation was fixed in this study, the permittivity of the environment was variable. Therefore, the antennas were optimized inside the body with respect to the average torso permittivity in order to obtain maximum power delivery. In addition, a few antenna locations with high return loss (i.e., > -7 dB) discarded from the data.

7.7.2 Experimental Setup

In order to validate our simulation results in [45], we conducted experiments on a human cadaver with a similar setup [20]. The human male torso area is investigated at 915 MHz by measuring the channel response through a vector network analyzer (VNA), while using two antennas, one (*in vivo*) [52], and other a dipole antenna (*ex vivo*) as illustrated in Figure 7.10. The *in vivo* antenna was placed at six different locations (Figure 7.11) inside the body around heart, stomach, and intestine by a physician. The antennas were located in the same orientation, and all return loss values were less than -7 dB in the experiment dataset.

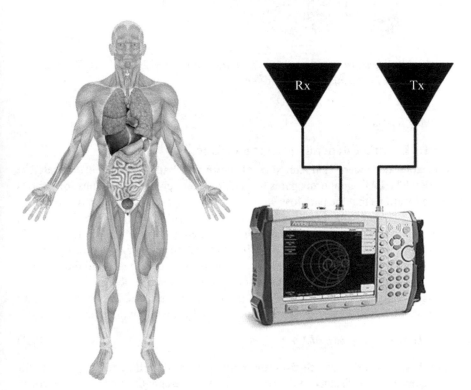

Figure 7.10 Experiment setup for in vivo channel. ©2015 IEEE. Reprinted with permission from Reference 20

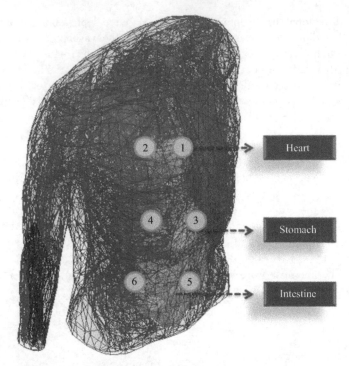

*Figure 7.11   Measurement locations on human cadaver. ©2015 IEEE. Reprinted
with permission from Reference 20*

7.7.3   Results

7.7.3.1   Location-dependent characteristics

The location-dependent characteristics of the *in vivo* path loss have investigated at
915 MHz. The EM signal propagates through different organs and tissues for various
antenna locations that the path loss varies significantly even for equal *in vivo* depths.
Figure 7.12 presents the mean path loss for each angular position (see Figure 7.9).
It is observed that 0° has the highest path loss, whereas symmetric locations, 112.5°
and 247.5°, have the lowest attenuation. In addition, the number of scattering objects
(random variables) increases as the *in vivo* antenna is placed deeper and the variance
of path loss increases as well due to summation of random variables.

Figure 7.13 shows the scatter plot of path loss versus depth, and the *in vivo* path
loss is modeled statistically as a function of depth by the following equation in dB:

$$PL(d) = PL_0 + m(d/d_0) + S(d_0 \leq d) \tag{7.3}$$

where d is the depth distance from the body surface in millimeters, d_0 stands for
the reference depth distance (i.e., 10 mm), PL_0 represents the intersection term in
dB, m denotes the decay rate of received power, and S is the shadowing term in dB,
which is a normally distributed random variable with zero mean and variance σ. The

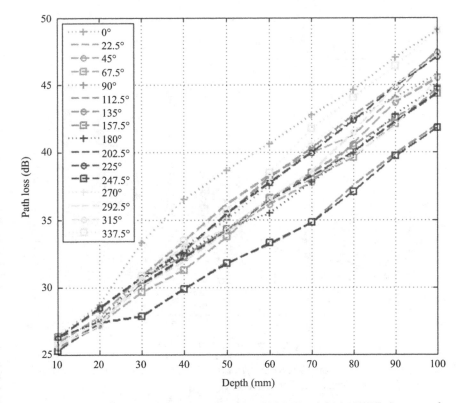

Figure 7.12   Path loss versus in vivo depth at 915 MHz. ©2014 IEEE. Reprinted with permission from Reference 45

parameters for the statistical *in vivo* path loss model are provided in Tables 7.3 and 7.4. There exists a 30% difference between the received power decay rates (m) of heart and stomach areas. In addition, the path loss at heart and intestine areas exhibits more deviation around the mean than other two regions. It could be concluded that the path loss increases significantly, when the *in vivo* antenna is placed in an anatomically complex region as also reported in Reference 7.

The numerical studies were validated with experiments on human cadaver at 915 MHz. The *in vivo* antennas were placed at six different locations as shown in Figure 7.11, and the *ex vivo* antenna was placed 2 cm away from the body surface. Table 7.5 presents the path loss values for the selected *in vivo* locations, and a comparison of experimental results with numerical studies is provided in Figure 7.14. The discrepancies should have occurred due to additional losses which are not considered in simulations.

The angular-dependent characteristics of the *in vivo* channel were investigated by performing further simulations at 0.4 GHz, 1.4 GHz, and 2.4 GHz. The *in vivo* antenna was fixed inside the abdomen (78 mm in depth from body surface), and the *ex vivo* antenna was rotated on the body surface with the azimuth angle of 0°–355° with

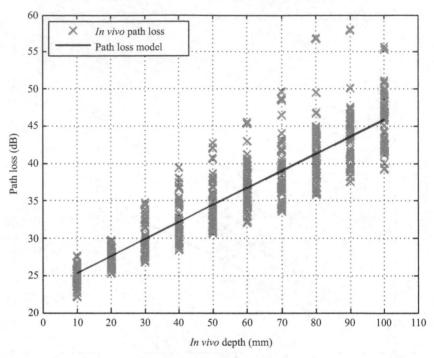

Figure 7.13   Path loss versus in vivo depth at 915 MHz. ©2014 IEEE. Reprinted with permission from Reference 45

Table 7.3   Parameters for the statistical path loss model (body region). ©2014 IEEE. Reprinted with permission from Reference 45

Parameters/Body area	PL_0[dB]	m	σ
Above heart	24.75	2.30	3.73
Heart	22.70	1.96	2.38
Stomach–kidneys	22.56	2.55	1.79
Intestine	24.23	2.31	3.47
Overall torso area	23.56	2.28	3.38

Table 7.4   Parameters for the statistical path loss model (body side). ©2014 IEEE. Reprinted with permission from Reference 45

Parameters/Body area	PL_0[dB]	m	σ
Anterior	23.83	2.46	3.51
Posterior	23.76	2.21	1.92
Left lateral	23.34	2.28	3.67
Right lateral	23.22	2.27	3.51
Overall torso area	23.56	2.28	3.38

*Table 7.5   Path loss values for selected in vivo locations. ©2015
IEEE. Reprinted with permission from Reference 20*

Location	*In vivo* depth (cm)	Path loss (dB)
1) Above heart	3	45.32
2) Below heart	8	55.61
3) Above stomach	5	48.19
4) Inside stomach	9	50.80
5) Above intestine	2	29.95
6) Below intestine	10	50.47

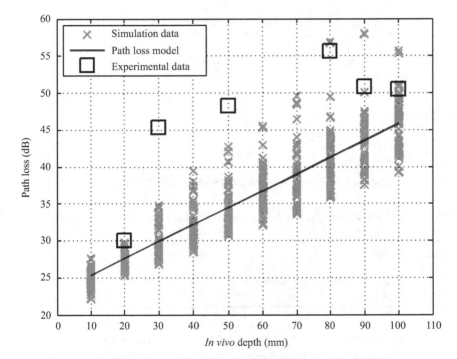

*Figure 7.14   Path loss versus in vivo depth from body surface. ©2015 IEEE.
Reprinted with permission from Reference 20*

5° increment. The results are presented in Figure 7.15 and Table 7.6. It could be
observed that the angular dependency (i.e., the variation of the path loss versus
azimuth angle) in terms of peak to average ratio is similar for different frequencies.

7.7.3.2   Frequency-dependent characteristics

Since the EM waves propagate through the frequency-dependent materials inside the
body, the operating frequency has an important effect on the path loss model as well.
The frequency-dependent characteristics of the *in vivo* channel were investigated by

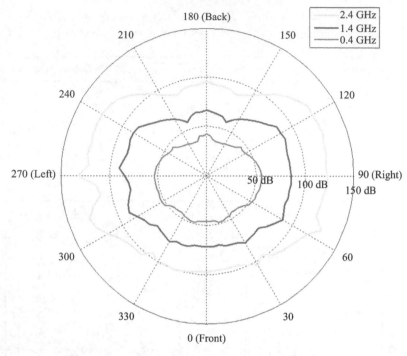

Figure 7.15   Angular-dependent path loss for on body receiver

Table 7.6   Comparison of angular-dependent path loss at different frequencies

Frequencies (GHz)	0.4	1.4	2.4
Average (dB)	46.316	76.74442	108.8819
Maximum difference (dB)	20.3373	33.04337	45.38211
Peak to average ratio	1.197665	1.171730	1.208047

performing simulations from 0.4 GHz to 6 GHz at 0.1 GHz increment. The *in vivo* antenna was implanted in the abdomen (78 mm in depth from the body surface), and the *ex vivo* antenna was placed at three different locations: $d = 50$ mm (in body), $d = 78$ mm (on body), $d = 200$ mm (far external node), where d denotes the distance between the transmitter and receiver. The path loss was measured for these three scenarios (implant to implant, implant to on-body, and implant to far external node), and the results are plotted in Figure 7.16 [53]. It is observed that the frequency-dependent path loss [in dB] increases linearly. Therefore, the frequency-dependent *in vivo* path loss [in ratio] increases exponentially, which is faster than that in free space.

7.7.3.3   Time dispersion characteristics

In addition to the path loss, the time dispersion characteristics of the *in vivo* channel were investigated for different body regions using a PDP in the simulation environment as shown in Figure 7.17. It is observed that greater dispersion is

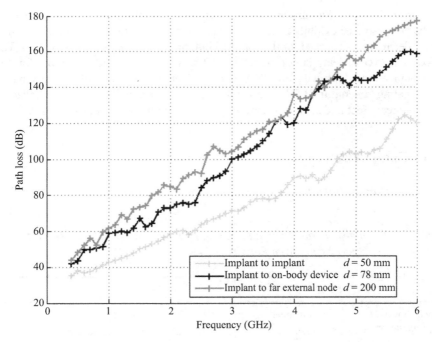

Figure 7.16   *Frequency-dependent path loss at different locations. ©2015 IEEE.*
Reprinted with permission from Reference 53

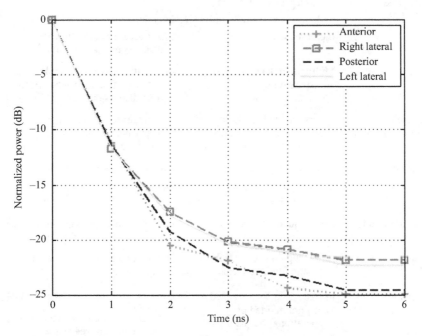

Figure 7.17   *Power delay profile for each body side. ©2014 IEEE. Reprinted with*
permission from Reference 45

present in the sides than anterior or posterior body locations at 915 MHz. Interestingly, the torso area exhibits an exponential decaying behavior on the dB scale while linear decaying is observed for the classical indoor/outdoor channel models [28]. The maximum excess delay is not more than 10 ns which might be negligible for NB communications. However, for UWB communications, which is also a very popular signaling scheme in WBAN research, this dispersion may lead to a significant interference effect and should be carefully considered in the waveform design.

7.8   Comparison of *in vivo* and *ex vivo* channels

We summarize the differences between the *in vivo* and *ex vivo* channels in Table 7.7.

Table 7.7   Comparison of in vivo and ex vivo channels. ©2015 IEEE. Reprinted with permission from Reference 54

Feature	*Ex vivo*	*In vivo*
Physical wave propagation	Constant speed Multipath	Variable speed Multipath – plus penetration in biological tissues
Attenuation and path loss	Lossless medium (losses are negligible) Path loss is essentially uniform Increases with distance	Very lossy medium Location dependent Increases exponentially with distance inside the body
Dispersion	Multipath delays → time dispersion	Multipath delays of variable speed → frequency-dependent time dispersion
Near-field communications	Deterministic near-field region around the antenna	Inhomogeneous medium → near-field region changes with angles and position inside body
Power limitations	Average and peak	Average and peak – plus SAR
Shadowing	Follows a *log-normal* distribution	Follows a *log-normal* distribution
Multipath fading	Flat/frequency selective fading	Lower speed of propagation causes longer dispersion than in free space
Antenna	Antenna gain is essentially location independent	"Implant location"-dependent antenna gain
Wavelength	In free space → the speed of light divided by operational frequency	$\lambda = \dfrac{c}{\sqrt{\varepsilon_{rf}}}$ (e.g., $\varepsilon_r = 35$ at 2.4 GHz → roughly six times shorter than the wavelength in free space)

7.9 Summary

In this chapter, the state of the art of *in vivo* wireless channel characterization has been presented. Various studies described in the literature are dedicated to the *in vivo* communication channel, and they consider different parameters in studying various anatomical regions. Furthermore, the location-dependent characteristics of *in vivo* wireless communication at 915 MHz are analyzed in detail via numerical and experimental investigations. A complete model for the *in vivo* channel is not available and remains an open research problem. However, considering the expected future growth of implanted technologies and their potential use for the detection and diagnosis of various health-related issues in the human body, the channel modeling studies should be further extended to develop better and more efficient communications systems for future *in vivo* systems.

Acknowledgements

This publication was made possible by NPRP grant # 6-415-3-111 from the Qatar National Research Fund (a member of Qatar Foundation). The statements made herein are solely the responsibility of the authors.

References

[1] "IEEE standard for local and metropolitan area networks: Part 15.6: Wireless body area networks," IEEE submission, February 2012, IEEE Std.

[2] P. S. Hall and Y. Hao, Antennas and Propagation for Body-Centric Wireless Communications, 2nd Edition. Norwood, MA: Artech House, 2012.

[3] S. Yu, X. Lin, and J. Misic, "Networking for big data [guest editorial]," Network, IEEE, vol. 28, no. 4, pp. 4–4, July–August 2014.

[4] A. Taparugssanagorn, A. Rabbachin, M. Hamalainen, J. Saloranta, and J. Iinatti, "A review of channel modelling for wireless body area network in wireless medical communications," in Proceeding of the 11th International Symposium on Wireless Personal Multimedia Communications (WPMC), 2008.

[5] A. Kiourti, K. A. Psathas, and K. S. Nikita, "Implantable and ingestible medical devices with wireless telemetry functionalities: A review of current status and challenges," Bioelectromagnetics, vol. 15, pp. 1–15, August 2013.

[6] S. Movassaghi, M. Abolhasan, J. Lipman, D. Smith, and A. Jamalipour, "Wireless body area networks: A survey," Communication Surveys and Tutorials, IEEE, vol. 16, pp. 1–29, 2014.

[7] M. R. Basar, F. Malek, K. M. Juni, *et al.*, "The use of a human body model to determine the variation of path losses in the human body channel in wireless capsule endoscopy," Progress in Electromagnetics Research, vol. 133, pp. 495–513, 2013.

[8] T. P. Ketterl, G. E. Arrobo, A. Sahin, T. J. Tillman, H. Arslan, and R. D. Gitlin, "*In vivo* wireless communication channels," in IEEE 13th Annual Wireless and Microwave Technology Conference (WAMICON), 2012.

[9] D. Smith, D. Miniutti, T. Lamahewa, and L. Hanlen, "Propagation models for body-area networks: A survey and new outlook," Antennas and Propagation Magazine, IEEE, vol. 55, pp. 97–117, Oct. 2013.

[10] Q. H. Abbasi, A. Sani, A. Alomainy, and Y. Hao, "On-body radio channel characterisation and system-level modelling for multiband OFDM ultra wideband body-centric wireless network," IEEE Transactions on Microwave Theory and Techniques, vol. 58, no. 12, pp. 3485–3492, December 2010.

[11] Q. H. Abbasi, A. Sani, A. Alomainy, and Y. Hao, "Numerical characterisation and modelling of subject-specific ultra wideband body-centric radio channels and systems for healthcare applications," IEEE Transaction on Information and Technology in Biomedicine, vol. 16, no. 2, pp. 221–227, March 2012.

[12] S. Gabriel, R. Lau, and C. Gabriel, "The dielectric properties of biological tissues: II. Measurements in the frequency range 10 Hz to 20 GHz," Physics in Medicine and Biology, vol. 40, no. 11, p. 2251, 1996.

[13] S. Gabriel, R. W. Lau, and C. Gabriel, "The dielectric properties of biological tissues: III. Parametric models for the dielectric spectrum of tissues," Physics in Medicine and Biology, vol. 41, pp. 2271–2293, 1996.

[14] A. Alomainy and Y. Hao, "Modeling and characterization of biotelemetric radio channel from ingested implants considering organ contents," IEEE Transactions on Antennas and Propagation, vol. 57, pp. 999–1005, April 2009.

[15] A. Pellegrini, A. Brizzi, L. Zhang, *et al.*, "Antennas and propagation for body-centric wireless communications at millimeter-wave frequencies: A review [wireless corner]," Antennas and Propagation Magazine, IEEE, vol. 55, no. 4, pp. 262–287, 2013.

[16] S. H. Lee, J. Lee, Y. J. Yoon, *et al.*, "A wideband spiral antenna for ingestible capsule endoscope systems: Experimental results in a human phantom and a pig," IEEE Transactions on Biomedical Engineering, vol. 58, no. 6, pp. 1734–1741, June 2011.

[17] R. Chavez-Santiago, I. Balasingham, J. Bergsland, *et al.*, "Experimental implant communication of high data rate video using an ultra wideband radio link," in Engineering in Medicine and Biology Society (EMBC), 2013 35th Annual International Conference of the IEEE. IEEE, 2013, pp. 5175–5178.

[18] H.-Y. Lin, M. Takahashi, K. Saito, and K. Ito, "Characteristics of electric field and radiation pattern on different locations of the human body for in-body wireless communication," IEEE Transactions on Antennas and Propagation, vol. 61, pp. 5350–5354, October 2013.

[19] http://www.ansys.com/Products.

[20] A. F. Demir, Q. H. Abbasi, Z. E. Ankarali, M. Qaraqe, E. Serpedin and H. Arslan, "Experimental Characterization of *In Vivo* Wireless Communication Channels," Vehicular Technology Conference (VTC Fall), 2015 IEEE 82nd, Boston, MA, 2015, pp. 1–2.

[21] B. Latre, B. Braem, I. Moerman, C. Blondia, and P. Demeester, "A survey on wireless body area networks," Wireless Networks, vol. 17, pp. 1–18, November 2010.

[22] K. Y. Yazdandoost, Wireless Mobile Communication and Healthcare. Springer Berlin Heidelberg, 2012, ch. A Radio Channel Model for In-body Wireless Communications, pp. 88–95.

[23] "C95.1-200S: IEEE Standard for Safety Levels With Respect to Human Exposure to Radio Frequency Electromagnetic Fields, 3 kHz to 300 GHz," 2006, IEEE Std.

[24] T. P. Ketterl, G. E. Arrobo, and R. D. Gitlin, "SAR and BER evaluation using a simulation test bench for *in vivo* communication at 2.4 GHz," in Wireless and Microwave Technology Conference (Wamicon), IEEE, 2013.

[25] W. G. Scanlon, "Analysis of tissue-coupled antennas for UHF intra-body communications," *Antennas and Propagation, (ICAP 2003)*. Twelfth International Conference on (Conf. Publ. No. 491), vol. 2, 2003, pp. 747–750.

[26] M. S. Wegmueller, A. Kuhn, J. Froehlich, *et al.*, "An attempt to model the human body as a communication channel," IEEE Transactions on Biomedical Engineering, vol. 54, no. 10, pp. 1851–1857, 2007.

[27] A. T. Barth, M. A. Hanson, H. C. Powell Jr, D. Unluer, S. G. Wilson, and J. Lach, "Body-coupled communication for body sensor networks," in Proceedings of the ICST 3rd international conference on Body area networks. ICST (Institute for Computer Sciences, Social-Informatics and Telecommunications Engineering), 2008, p. 12.

[28] R. C.-S. Khaleghi and I. Balasingham, "Ultra-wideband statistical propagation channel model for implant sensors in the human chest," IET Microwaves, Antennas & Propagation, vol. 5, p. 1805, 2011.

[29] R. Chavez-Santiago, K. Sayafian Pour, A. Khaleghi, *et al.*, "Propagation models for IEEE 802.15.6 standardization of implant communication in body area networks," in IEEE Communications Magazine, vol. 51, no. 8, pp. 80–87, August 2013.

[30] K. Yang, Q. Abbasi, K. Qaraqe, A. Alomainy, and Y. Hao, "Bodycentric nanonetworks, EM channel characterisation in water at the terahertz band," in Asian Pacific Microwave Conference (APMC), Japan, November 2–5 2014, pp. 1–5.

[31] A. Sani, A. Alomainy, and Y. Hao, "Numerical characterization and link budget evaluation of wireless implants considering different digital human phantoms," IEEE Transactions on Microwave Theory and Techniques, vol. 57, pp. 2605–2613, October 2009.

[32] K. Sayrafian-Pour, W.-B. Yang, J. Hagedorn, *et al.*, "Channel models for medical implant communication," International Journal of Wireless Information Networks, vol. 17, pp. 105–112, December 2010.

[33] H. Bahrami, B. Gosselin, and L. A. Rusch, "Realistic modeling of the biological channel for the design of implantable wireless UWB communication systems," in Engineering in Medicine and Biology Society (EMBC) Annual International Conference, IEEE, 2012.

[34]   A. Johansson, "Wireless communication with medical implants: Antennas and propagation," Ph.D. dissertation, Lund University, 2004.

[35]   J. Kim and Y. Rahmat-Samii, "Implanted antennas inside a human body: Simulations, designs, and characterizations," IEEE Transactions on Microwave Theory and Techniques, vol. 52, no. 8, pp. 1934–1943, 2004.

[36]   J. Gemio, J. Parron, and J. Soler, "Human body effects on implantable antennas for ISM bands applications: Models comparison and propagation losses study," Progress in Electromagnetics Research, vol. 110, pp. 437–452, October 2010.

[37]   F. Merli, B. Fuchs, J. R. Mosig, and A. K. Skrivervik, "The effect of insulating layers on the performance of implanted antennas," IEEE Transactions on Antennas and Propagation, vol. 59, no. 1, pp. 21–31, 2011.

[38]   K. Y. Yazdandoost and R. Kohno, "Wireless communications for body implanted medical device," in Asia-Pacific Microwave Conference, 2007.

[39]   J. R. Reitz, J. M. Frederick, and R. W. Christy, Foundations of Electromagnetic Theory (4th ed.). Addison-Wesley, Reading. ISBN 0-201-52624-7, 1993.

[40]   S. H. Lee, J. Lee, Y. J. Yoon, et al., "A wideband spiral antenna for ingestible capsule endoscope systems: Experimental results in a human phantom and a pig," IEEE Transactions on Biomedical Engineering, vol. 58, pp. 1734–1741, June 2011.

[41]   T. Karacolak, A. Hood, and E. Topsakal, "Design of a dual-band implantable antenna and development of skin mimicking gels for continuous glucose monitoring," IEEE Transactions on Microwave Theory and Techniques, vol. 56, pp. 1001–1008, April 2008.

[42]   A. Laskovski and M. Yuce, "A MICS telemetry implant powered by a 27 MHz ISM inductive link," in Engineering in Medicine and Biology Society, EMBC, 2011 Annual International Conference of the IEEE, 2011.

[43]   A. Kiourti and K. Nikita, "Miniature scalp-implantable antennas for telemetry in the MICS and ISM bands: Design, safety considerations and link budget analysis," IEEE Transactions on Antennas and Propagation, vol. 60, no. 8, pp. 3568–3575, August 2012.

[44]   M. L. Scarpello, D. Kurup, H. Rogier, et al., "Design of an implantable slot dipole conformal flexible antenna for biomedical applications," IEEE Transactions on Antennas and Propagation, vol. 59, no. 10, pp. 3556–3564, 2011.

[45]   A. F. Demir, Q. H. Abbasi, Z. E. Ankarali, E. Serpedin and H. Arslan, "Numerical characterization of in vivo wireless communication channels," RF and Wireless Technologies for Biomedical and Healthcare Applications (IMWS-Bio), IEEE MTT-S International Microwave Workshop Series, London, pp. 1–3, 2014.

[46]   Z. N. Chen, G. C. Liu, and T. S. See, "Transmission of RF signals between MICS loop antennas in free space and implanted in the human head," IEEE Transactions on Antennas and Propagation, vol. 57.6, pp. 1850–1854, 2009.

[47]   L. C. Chirwa, P. Hammond, S. Roy, and D. R. S. Cumming, "Electromagnetic radiation from ingested sources in the human intestine between 150 MHz and 1.2 GHz," IEEE Transactions on Biomedical Engineering, vol. 50, pp. 484–492, April 2003.

[48] S. K. S. Gupta, S. Lalwani, Y. Prakash, E. Elsharawy, and L. Schwiebert, "Towards a propagation model for wireless biomedical applications," in IEEE International Conference on Communications (ICC), 2003.

[49] K. Y. Yazdandoost and K. Sayrafian-Pour, "Channel model for body area network (BAN)," IEEE P802, vol. 15, 2009.

[50] S. Stoa, C. S. Raul, and I. Balasingham, "An ultra wideband communication channel model for the human abdominal region," in GLOBECOM Workshops (GC Workshops), IEEE, 2010.

[51] "ANSYS HFSS®." [Online]. Available: http://www.ansys.com/Products /Electronics/ANSYS+HFSS®. [Accessed: 12-Nov-2015].

[52] A. Rahman and Y. Hao, "A novel tapered slot CPW-fed antenna for ultra-wideband applications and its on/off-body performance," 2007 International workshop on Antenna Technology: Small and Smart Antennas Metamaterials and Applications, Cambridge, 2007, pp. 503–506.

[53] Y. Liu and R. D. Gitlin, "A phenomenological path loss model of the *in vivo* wireless channel," in IEEE 16th Wireless and Microwave Technology Conference (WAMICON), April 2015.

[54] C. He, Y. Liu, G. E. Arrobo, T. P. Ketterl, and R. D. Gitlin, "*In Vivo* wireless communications and networking," in Information Theory and Applications Workshop (ITA), 2015, San Diego, CA, February 2015.

[55] A. F. Demir, Z. E. Ankarali, Q. H. Abbasi, *et al.*, "*In Vivo* communications: Steps toward the next generation of implantable devices," in IEEE Vehicular Technology Magazine, to be published on June 2016.

Chapter 8

Diversity and MIMO for efficient front-end design of body-centric wireless communications devices

Imdad Khan*, Khalida Ghanem† and Peter S. Hall‡

Abstract

The advancement of intelligent, small sensors, microelectronics, integrated circuit, and low power wireless communication has led us close to the deployment of body area networks (BANs). There is considerable ongoing research on antennas and propagation for BANs. Diversity and Multiple Input Multiple Output (MIMO) are the two well-known multiple antenna techniques to overcome fading and provide channel capacity improvement. In this chapter, we discuss the use of antenna diversity and MIMO for on-body channels to support reliable and high data rate communication. Diversity is also used to cancel the co-channel interference from a nearby BAN device. Probabilistic channel models of diversity and MIMO on-body channels are proposed and validated by measurements.

8.1 Introduction

Human body is a hostile environment for the propagation of electromagnetic waves in the wireless body-centric communication systems. Fading and interference are the two major concerns that affect the reliability and quality of service of wireless links. Signals transmitted from a body-worn transmitting antenna come across various impairments while propagating to a receiving antenna mounted on the same, or another human body in the close vicinity. Apart from the large attenuation due to the lossy nature of the human body, the propagating signal encounters reflection and scattering due to the human body parts, which cause significant variation in the received signal. Moreover, in the on-body channels, both the transmitter and receiver can move in a random manner relative to each other, causing Doppler shift and pathloss variations.

*Islamic University, Madinah Munawwarah, KSA
†Centre for Development of Advanced Technologies, Algiers, Algeria
‡University of Birmingham, Birmingham, UK

Apart from the mentioned sources of signal fading, the body itself causes fading due to the shadowing of links and multiple signal paths on the body, such as the two possible paths from front to back around the body from both sides. The surrounding environment is also responsible to cause signal fading. Antenna diversity is a powerful tool to combat fading by using multiple antennas either on the receiver side (receive diversity) or at the transmitter side (transmit diversity). The diversity gain is limited by the correlation among the diversity branch signals and the power imbalance [1]. Diversity exploits the multipath phenomena and works better in non-line of sight (NLOS) channels.

Another important issue related to body-worn sensors and devices is the transmitted power. It is desirable to reduce the power level to increase the battery life and reduce the specific absorption rate value. Antenna diversity can help in keeping the transmitted power low by providing better signal reception. Body area network (BAN) devices can also be prone to co-channel interference from unwanted devices mounted on the same human body or worn by other individuals using the same facilities. Rejection of the interference from a nearby BAN becomes more significant when the BANs are operating very close to each other and the level of the desired, and interference signals are almost the same. Antenna diversity can provide good interference cancellation capability [2].

The ever increasing demand of wireless devices in personal healthcare, entertainment, security and personal identification, fashion, personalized communications, etc. requires high data rates and reliable links. BAN is believed to be an integral part of the fourth-generation wireless communication standards and has many applications that require data transmission at high data rates. This triggers the use of multiple-input multiple-output (MIMO) techniques for the BAN devices. MIMO provides excellent channel capacity improvement as well as diversity gain if used effectively [3]. Due to the presence of line of sight (LOS) and the short-range nature of body-centric communications, it may be a common perception that diversity and MIMO may not be effective for body-worn devices. Diversity and MIMO both exploit multipath fading. This chapter provides the significance of multiple antennas for front-end design of body-worn devices for the two ISM bands of 2.45 GHz and 5.8 GHz. It shows that in the presence of significant amount of fading in on-body channels from various sources, diversity and MIMO provide very significant improvement in terms of reliability and channel capacity. The analysis is based on real-time measurements with antennas mounted on human body. In order to complete the analysis and provide a better understanding of the multiple antenna links, the stochastic channel models are also presented for the diversity and MIMO channels. These models can prove to be very helpful in the design of effective front end with multiple antennas.

The chapter is organized as follows. Section 8.2 provides an overview of diversity systems initially and then gives the detailed diversity performance analysis of various on-body channels along with the measurement setup and channel models. The later subsections provide the details of the stochastic diversity channel model with its first-order and second-order statistics and spectral analysis. It also signifies the use of diversity for interference cancellation. The use of MIMO in on-body channels for capacity improvement and for diversity gains is presented in Section 8.3. The MIMO stochastic model is presented in this section as well. Section 8.4 summarizes the chapter with some significant conclusions.

8.2 Receive diversity for body-worn devices

Receive diversity is achieved by using multiple diversity branches at the receiver side and a single antenna at the transmitter side. These multiple diversity branches can be achieved by different ways including space diversity, pattern or angle diversity, and polarization diversity. In space diversity, the diversity branch signals are achieved by more than one antenna elements placed at some distance in space from each other. The distance between the elements is usually more than half wavelength to decrease the correlation among the branches to a minimum [1]. The pattern or angle diversity makes use of the directional radiation patterns of antenna elements. The patterns of the diversity branch elements are directed in different directions with minimum overlap to achieve the diversity branches. The elements producing the directional patterns can be placed as close as possible as long as they are in the far-field region of each other, and the overlap in the main radiation lobe is kept to a minimum. Pattern diversity can also be achieved by using beam switching antenna arrays and keeping the switching time between the main lobes less than the channel coherence time. In the third type of antenna diversity technique, called polarization diversity, the diversity branches are achieved by using antenna elements having different polarization, preferably orthogonal. Due to limited changes in the antenna orientations for most of the on-body channels, the power imbalance can be larger if the patterns or the polarizations of the diversity elements are significantly different. So care must be taken in designing pattern and polarization diversity antennas for on-body channels to prevent one branch signal being dominant at all times. Apart from the antenna diversity, frequency, and time diversity are the other types of diversity techniques in which the diversity branches differ in frequency and time slots using frequency and time multiplexing, respectively [4].

The signals received at the multiple diversity branches can be combined at the receiver in various ways to effectively increase the receive Signal-to-Noise Ratio (SNR) [5]. The diversity performance is quantified as the Diversity Gain (DG), which is defined as the SNR increase due to diversity combining at a given probability level compared to single antenna system [5]. Diversity gain mainly depends upon the power imbalance (difference between the average signal levels) and the correlation among the various diversity branch signals. In general, the smaller the values of these two parameters are, the more effective diversity combining is and higher diversity gain can be achieved. The correlation among the branch signals depends upon the propagation channel and the separation of the branch signals. The complex signal correlation coefficient (ρ_s) and envelope correlation coefficient (ρ_e) between the two-branch signals can be calculated as [1, 5]:

$$\rho_s = \frac{\sum\limits_{i=1}^{N} V_1(i)V_2^*(i)}{\sqrt{\sum\limits_{i=1}^{N} V_1(i)V_1^*(i)}\sqrt{\sum\limits_{i=1}^{N} V_2(i)V_2^*(i)}} \tag{8.1}$$

$$\rho_e = \frac{\sum\limits_{i=1}^{N} \sqrt{S_1(i)S_2(i)}}{\sqrt{\sum\limits_{i=1}^{N} \sqrt{S_1(i)}\sqrt{S_1(i)}}\sqrt{\sum\limits_{i=1}^{N} \sqrt{S_2(i)}\sqrt{S_2(i)}}} \qquad (8.2)$$

where N is the total number of samples, S_1 and S_2, V_1 and V_2 represent the zero-meaned received power signals and complex voltage signals, respectively. With LOS, the correlation and power imbalance are usually high. A correlation coefficient below 0.7 is considered adequate for acceptable diversity performance [1].

The simplest of the diversity combining techniques is the selection combining (SC) [5] in which the signal with the higher instantaneous SNR is connected to the receiver at a particular instant. This technique is easy to implement but gives the lowest diversity gains. In equal gain combining (EGC), the diversity branch signals are added together with equal gain (usually 1) [5]. This method gives slightly better diversity gain than SC. The maximum ratio combining (MRC) method produces the best theoretically possible SNR improvement [5]. In this case, the branch signals are weighted according to their respective SNR and then added thus resulting in more dominance for the branch signal with better SNR. However, the better performance is achieved at a price of increased receiver complexity and cost. The expressions to achieve the diversity combined signal for these three combining techniques with two-branch diversity system are given in (8.3)–(8.5) [5]:

$$SC(t) = \max(r_1(t), r_2(t)) \qquad (8.3)$$

$$EGC(t) = \frac{(r_1(t) + r_2(t))}{\sqrt{2}} \qquad (8.4)$$

$$MRC(t) = \sqrt{(r_1(t)^2 + r_2(t)^2} \qquad (8.5)$$

where $r_1(t)$ and $r_2(t)$ are the fast fading envelopes of the two received branch signals.

As stated above, diversity can play a significant role in improving the receive SNR and in making the channels more robust and reliable. However, in the on-body channels, the transmitter and receiver are supposed to be in LOS for majority of cases. Consequently, the diversity performance can be affected by the resulting high correlation and high power imbalance. In order to investigate the significance of diversity in on-body channels, the diversity performance was quantified for a variety of on-body channels in different frequency bands in the studies [6–10]. Measurements in these studies were carried out in an indoor office environment with random movements of the body by mounting transmitting and receiving diversity antennas on the same human body at various locations, as shown in Figure 8.1. The received signal envelopes were cophased, demeaned, and combined using (8.3)–(8.5) to achieve the diversity combined signals. Various types of diversity antennas were designed and used to quantify the diversity performance analysis of on-body channels. These include monopole, printed inverted-F antenna (printed-IFA) [7, 8], planar inverted-F antenna (PIFA) [7, 8], dual-polarized monopole-loop polarization diversity

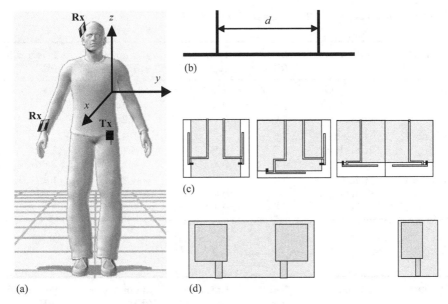

Figure 8.1   Antennas used in the measurement and the on-body channel. (a) The on-body channel, (b) space diversity monopole antenna with variable spacing d, (c) printed-IFA diversity antennas, and (d) single Tx and diversity PIFA

antenna [9], and annular ring-slot antenna (ARSA) [10]. Details of the measurements setup and antennas used in these studies can be found in References 7–10. Figure 8.1 also shows the layout and structure of few of the diversity antennas used.

8.2.1 Diversity performance analysis

The empirical diversity analysis is done for three on-body channels, namely belt-chest, belt-head, and belt-wrist channels [7, 8]. Measurements were conducted by mounting the various types of antennas given and described above. The three on-body channels were selected to investigate the static, quasi-static, and dynamic environments. The belt-chest channel is a representative of a static channel since the transmitter and receiving antennas are mounted on the belt and chest positions, respectively, which involve very little and slow movement of the antennas and hence results in almost constant path length. The belt-head channel represents a quasi-static channel due to limited movement of the head. The belt-wrist channel, in contrast to the above two channels, is highly dynamic and involves large variation in path length due to the random motion of hand.

   Antennas, such as monopole, generate polarization which is perpendicular to the body surface when mounted perpendicular to the body surface and hence the propagating wave is less attenuated compared to the wave which is polarized tangential to the body [7]. The presence of a strong dominant ray results in high correlation as

Table 8.1   Results for the three channels at 2.45 GHz

Channel	Antenna	MRC (dB)	Env. corr (ρ_e)	Mean power (dB)	Power diff (dB)
Belt-chest	Monopole	2.1	0.3	−29.1	3.8
	Printed-IFA	6.5	0.5	−44.64	2.8
	PIFA	4.9	0.4	−39.2	3.1
Belt-head	Monopole	7.6	0.1	−46.3	2.6
	Printed-IFA	7.7	0.5	−54.1	2.2
	PIFA	9.2	0.2	−48.2	2.3
Belt-wrist	Monopole	9.4	0.1	−39.7	1.4
	Printed-IFA	8.1	0.4	−59.7	1.2
	PIFA	7.8	0.3	−49.5	2.5

Table 8.2   Results for the three channels at 5.8 GHz

Channel	Antenna	MRC (dB)	Env. corr (ρ_e)	Mean power (dB)	Power diff (dB)
Belt-chest	Monopole	5.4	0.3	−44.4	2.5
	Printed-IFA	9.0	0.1	−55.7	2.5
	PIFA	5.3	0.3	−50.8	3.1
Belt-head	Monopole	9.8	0.2	−57.2	0.2
	Printed-IFA	8.8	0.1	−67.0	2.6
	PIFA	8.9	0.2	−62.8	1.7
Belt-wrist	Monopole	8.5	0.2	−51.2	1.6
	Printed-IFA	8.5	0.1	−63.9	1.7
	PIFA	8.6	0.1	−59.8	2.6

well as high power imbalance and hence very low diversity gain values are expected for this on-body channel [7, 8], as reflected in Tables 8.1 and 8.2. On the other hand, antennas with polarization parallel to the surface of the body, like printed-IFA, result in high pathloss due to much more attenuation of wave while propagating along the surface of the body. For this reason, the direct ray may not be as strong as in the case of monopole antenna, resulting in dominant or comparable multipath components. This results in low correlation and low power imbalance and diversity is proved to be useful in this case with slightly higher values of diversity gain, as given in Tables 8.1 and 8.2. PIFA has reasonably well link budget and significant diversity performance, as shown in Tables 8.1 and 8.2. The presence of strong LOS or direct ray in case of belt-chest channel also depends upon the physical structure and texture of the human body on which the body-worn devices are mounted. The direct ray may be obstructed and attenuated more in case of a fat person, and diversity can be useful in this scenario. Despite the presence of a dominant ray or LOS component in case of the belt-chest channel, diversity offers some improvement and the decision to use diversity for this

Table 8.3   Results for the three channels at 2.45 GHz with pattern and polarization diversity antennas

Channel	Antenna	MRC (dB)	Env. corr (ρ_e)	Mean power (dB)	Power diff (dB)
Belt-chest	ARSA in polarization diversity mode	6.29	0.22	−49.61	1.80
	ARSA in pattern diversity mode	8.68	0.13	−50.32	0.61
	Monopole-loop polarization diversity antenna	0.70	0.20	−39.60	14.80
Belt-head	ARSA in polarization diversity mode	8.13	0.07	−60.98	3.81
	ARSA in pattern diversity mode	7.79	0.15	−62.91	2.94
	Monopole-loop polarization diversity antenna	9.20	0.10	−49.40	0.40
Belt-wrist	ARSA in polarization diversity mode	9.15	0.20	−59.10	0.79
	ARSA in pattern diversity mode	9.26	0.24	−57.51	0.53
	Monopole-loop polarization diversity antenna	4.20	0.30	−44.10	13.90

on-body channel depends upon the improvement offered and the cost involved in implementing diversity.

The belt-head on-body channel involves LOS, partial LOS, or complete NLOS conditions while the head is moved randomly. Receive diversity offers good improvement for this channel compared to the static channel. The results are presented for different antennas and frequencies in Tables 8.1 and 8.2. It can be noticed that the correlation coefficient values are relatively low compared to the belt-chest channel.

The belt-wrist channel undergoes more severe fading and diversity proves to be more effective in this case, as clear from the low correlation and high diversity gain values on average given in Tables 8.1 and 8.2. Due to strong depolarization of the transmitted wave and the random orientation of the receiver antenna, the antenna type has no significant effect on the diversity performance of this channel.

Table 8.3 summarizes the DG, envelope correlation and power imbalance values with the pattern diversity ARSA, polarization diversity ARSA and monopole-loop antennas. Since low depolarization is expected in case of belt-chest link, polarization diversity offers no significant diversity gains. On the other hand, belt-head and belt-wrist links offer reasonable improvement in diversity gains by virtue of depolarization of the propagating wave. High DG values of pattern diversity mode in ARSA suggest the presence of multiple angle of arrivals of the propagating wave along the surface of the body.

8.2.2   *Diversity channel characterization and spectral analysis*

8.2.2.1   Channel characterization and modelling

In order to design efficient front end for body-worn devices and systems, the channel characteristics need to be known. The channel variation resulting from random movements may at times become very unpredictable. The channel behaviour is also dependent upon the type of antenna used. So far, the antenna de-embedding from the on-body transmission channels has not been successful. Thus the on-body propagation channel characterization includes the antenna embedded as part of the channel. The channel can behave differently for different body parts and even for different body shape and gender. Hence characterization of on-body channels is a complex scenario. On-body channel models are significantly different from the mobile radio channels which are mainly characterized by Rayleigh fading that assumes a random variation in the amplitude of the fading envelope and uniform distribution of the phase in NLOS scenario. The majority of on-body channels have a dominant component due to LOS or due to the wave propagating along the surface of the body [11] and hence behave differently than the mobile channel. Moreover, the scattering is of different nature as well. Various efforts have been made to model the on-body channels [12–15]. The empirical models of short-term and long-term fading for on-body diversity channels are presented in Reference 14. Statistical analysis and Doppler spread is also given. Reference [16] provides the second-order statistics of the same channels. The data recorded in measurement campaign described above was used for characterizing the fading envelopes with the same antenna types and same frequencies of 2.45 GHz and 5.8 GHz. For the purpose of channel characterization, the movements were divided into static and dynamic postures. Static postures involve very less movements of the antennas and path length remains almost constant, whereas, dynamic postures involve rapid and random antenna movements and have varying path lengths and more scattering. Details can be found in Reference 14.

The short-term and long-term fading envelopes of the received signals can be separated using (8.6) and (8.7) [17, 18]:

$$r(t) = m(t)M(t) \tag{8.6}$$

$$M(t) = \sqrt{\frac{1}{2w} \int\limits_{t-w}^{t+w} r^2(\tau)d\tau} \tag{8.7}$$

where $r(t)$ is the received signal envelope, $m(t)$ is the short-term fading envelope, and $M(t)$ is the local root-mean-square (RMS) value of the envelope constituting the long-term fading. The averaging window size $2w$ is a critical parameter and should be within the 5λ to 20λ interval, where λ is the free space wavelength [17, 18].

The measured fast fading and slow fading channel data was tested against some well-known distribution models using KS-test for the goodness of fit [19]. The goodness of fit test reveals that the short-term fading envelope can be modelled best with Rician distribution, as shown in Figure 8.2, with parameters given in Table 8.4 for the two on-body channels. The data with dynamic movements were used only since the static posture data was not fitting any model. The long-term fading envelope best

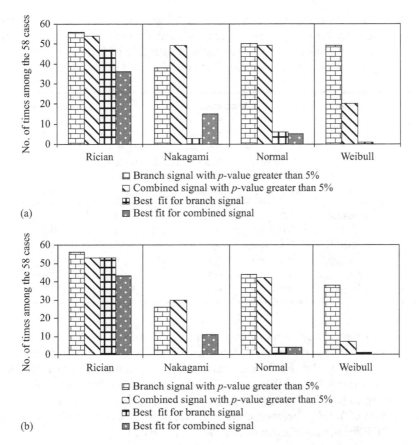

(a)

(b)

Figure 8.2   Graphs showing the no. of times short-term fading data sets of branch and combined signals fitted the four dominant distributions with p values higher than 5% and the best fit (highest p-value) among the 58 cases for (a) belt-head and (b) belt-wrist channel. © 2010 IEEE. Reprinted with permission from Reference 14

fits the log-Normal distribution, as shown in Figure 8.3, with parameters given in Tables 8.5 and 8.6 for various scenarios. It is evident that for almost 85% of the cases for the short-term fading envelope, Rician is the best fit with its probability density function (PDF) given by [20]:

$$p(m, s, \sigma) = I_0 \left(\frac{ms}{\sigma^2} \right) \frac{m}{\sigma^2} \exp \left(-\frac{m^2 + s^2}{2\sigma^2} \right) \tag{8.8}$$

where m is the random variable representing the short-term fading envelope $m(t)$, $I_0(x)$ is the modified Bessel function of the first kind and zeroth order, and s and σ are the two parameters characterizing the strongest ray power, s^2, and the average scattering power, $2\sigma^2$, respectively. Rician distribution can also be expressed in terms of its parameters K and p_o. The Rician K-factor represents the ratio of the strongest

Table 8.4 *Short-term fading parameters of the Rician branch and combined signals*
for the belt-head channel, K_b and K_C are the K-factor values of branch
and combined signals respectively. © 2010 IEEE. Reprinted with
permission from Reference 14

Frequency (GHz)	Antenna	Belt-head channel			Belt-wrist channel		
		Average Rician K-factor (dB)		Difference between branches (dB)	Average Rician K-factor (dB)		Difference between branches (dB)
		K_b	K_C	$\lvert\delta K_b\rvert$	K_b	K_C	$\lvert\delta K_b\rvert$
2.45	Monopole	3.3	6.8	1.9	2.4	7.3	2.3
	PIFA	3.0	6.6	0.5	1.6	5.6	0.2
	Printed-IFA	−0.8	3.6	0.3	−0.2	5.1	1.8
5.8	Monopole	2.4	6.4	0.8	1.9	5.4	0.7
	PIFA	2.4	6.1	0.3	1.8	6.2	0.2
	Printed-IFA	2.7	7.0	0.3	1.7	5.5	0.1

ray power to the average scattered power, i.e. $K = s^2/2\sigma^2$. p_o is the total power, i.e. $p_o = s^2 + 2\sigma^2$. For the model described in Table 8.4, $p_o = 1$ or 0 dB for all the short-term fading envelopes and the Rician distribution is completely characterized by the K-factor. Table 8.4 shows the average K-factor values of the branch signal, K_b, and combined signal (MRC), K_C, for the two channels. The table also shows the maximum difference (in dB), $\lvert\delta K_b\rvert$, between average values of the two diversity branches.

The PDF for the log-Normal distribution is given by [20]:

$$p(x, \mu, \sigma_s) = \frac{1}{\sqrt{2\pi}\sigma_s x} \exp\left(-\frac{(\ln x - \mu)^2}{2\sigma_s^2}\right) \tag{8.9}$$

μ and σ_s are the mean and the standard deviation of the natural logarithm of x, where x is a random variable representing the long-term fading envelope $M(t)$. Tables 8.5 and 8.6 show the average parameters of the fitted model for the two channels along with the difference between the two diversity branches. Parameters μ and σ_s are presented here in dB to make them consistent with the rest of the results.

The results suggest a Rician fit to the short-term fading but with K-factor values not as high as expected in short range LOS communication. The reason for low K-factor values is the attenuation of the direct ray while it propagates along the surface of the body. This results in scattering power comparable to that of direct ray power. The very low K-factor values with printed-IFA support this statement since attenuation of direct ray along the body surface is much more by virtue of its polarization. Furthermore, the mean received power, represented by the μ parameter, is highest for monopole antennas with lower standard deviation, σ_s. Thus, monopole antennas provide better link budget followed by PIFA but are less convenient to use for wearable devices due to its high profile shape.

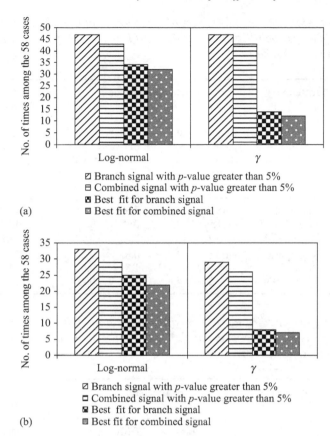

Figure 8.3   *Graphs showing the no. of times the long-term fading data sets of combined signal fitted the two dominant distributions with p-values higher than 5% and the best fit (highest p-value) among the 58 cases for (a) belt-head and (b) belt-wrist channel. © 2010 IEEE. Reprinted with permission from Reference 14*

It can be concluded from the results that the two-branch signals for on-body diversity channels may not be identically distributed. The difference between the two-branch signals, which effectively shows the power imbalance, is noticeable for monopole antennas at 2.45 GHz. The Rician K-factor increases significantly for the combined signals compared to the K-factors of the corresponding branch signals, as shown in Table 8.4. This clearly suggests an improvement of the coherent component over the scattering power with diversity. This improvement can also be seen by comparing the μ and σ_s parameters for branch and combined signals.

Along with the first-order statistics and probability model, the second-order statistics, like average fade duration (AFD) and level-crossing rate (LCR) of the channel also give useful information and are significant in characterizing the channel completely [21]. These statistics give an idea about the severity of fading. AFD is

Table 8.5   *Long-term fading parameters of the log-normal branch and combined*
signals for belt-head channel. © 2010 IEEE. Reprinted with permission
from Reference 14

Frequency (GHz)	Antennas	Branch signals				Combined (MRC) signal	
		Log-normal parameters (dB)		Difference between branches (dB)		Log-normal parameters (dB)	
		μ	σ_s	$|\delta\mu|$	$|\delta\sigma_s|$	μ	σ_s
2.45	Monopole	−45.8	3.0	3.3	0.2	−43.9	2.6
	PIFA	−51.9	2.9	0.2	0.1	−48.3	2.5
	Printed-IFA	−53.1	5.2	0.5	0.2	−49.5	4.9
5.8	Monopole	−60.5	3.2	0.8	0.3	−57.7	3.3
	PIFA	−65.3	3.7	1.8	0.6	−62.7	3.8
	Printed-IFA	−67.2	4.4	1.0	0.4	−64.1	3.8

Table 8.6   *Long-term fading parameters of the log-normal branch and combined*
signals for belt-wrist channel. © 2010 IEEE. Reprinted with permission
from Reference 14

Frequency (GHz)	Antennas	Branch signals				Combined (MRC) signal	
		Log-normal parameters (dB)		Difference between branches (dB)		Log-normal parameters (dB)	
		μ	σ_s	$|\delta\mu|$	$|\delta\sigma_s|$	μ	σ_s
2.45	Monopole	−46.3	3.8	3.2	0.7	−41.1	4.0
	PIFA	−52.1	3.4	1.9	0.0	−49.6	2.8
	Printed-IFA	−61.4	5.2	0.6	0.2	−57.7	5.1
5.8	Monopole	−58.5	4.8	0.7	0.7	−54.7	4.1
	PIFA	−64.4	4.0	1.8	0.4	−62.0	3.7
	Printed-IFA	−68.2	4.8	1.4	0.0	−65.3	4.3

the average amount of time for which the signal remains under a given fade level F
[20]. The LCR, N_R, signifies the rapidity of fading, and is the rate at which the fading
signal envelope, r, crosses the threshold fade level, F, in one direction, either positive
or negative going [20]:

$$N_R \triangleq \int_0^\infty \dot{r} f(F, \dot{r}) \, d\dot{r} \qquad (8.10)$$

where $f(F, \dot{r})$ is the joint PDF of r and $\dot{r}$ at $r = F$, and $\dot{r}$ is the time derivative of $r(t)$. AFD, T_D, can be estimated from P_r and N_R as [20]:

$$T_D \triangleq \frac{P_r}{N_r} \tag{8.11}$$

where P_r is the probability of fade (PF) and is the probability that the fading signal remains under the threshold fade level F. The PF can be determined for a given threshold F as [20]:

$$P_r(r \leq F) = \frac{N_{r \leq F}}{N_T} \tag{8.12}$$

where $N_{r < F}$ is the number of times the sample magnitude of the fading envelope, r, is below F. N_T is the total number of samples in the data set.

These statistics were calculated for the measured data of the belt-head channel for the two diversity branch signals and combined MRC signal. In addition to the 2.45 GHz and 5.8 GHz, an additional measurement and analysis is presented at 10 GHz to see the trend at higher frequencies. Figure 8.4 shows a comparison of LCR and AFD values of the branch and diversity combined signal and compares these parameters at the two frequencies as well. The figure shows comparison of channels using monopole antennas but the same trend was observed for the printed-IFA and PIFA as well. The higher values of LCR for the higher frequency suggest that the fading is more severe at 5.8 GHz. In addition, the signals at low frequency remain under a given fade level for more time compared to high-frequency signals, as shown in the AFD curves of Figure 8.4. In other words, the fading is more rapid at higher frequency. The figure also clearly suggests the improvement offered by the diversity. At lower fade levels, both LCR and AFD for the diversity combined signal are lower compared to the values at the branch signal before combining. For fair comparison, the branch and combined signals were both normalized to the median of the branch signals. An improvement in the mean power level can also be observed for the diversity combined signal since its LCR profile is slimmer than that of the branch signal. This signifies that the signal values remain close to the median after combining. Figure 8.5 presents a comparison of the LCR and AFD for the three antennas used at 5.8 GHz. Similar conclusions regarding the link budget can be drawn as given above in the first-order statistics. Printed-IFA is more prone to fading due to its weak link budget compared to the monopole.

8.2.2.2 Spectral analysis

The motion of transmitter and receiver in on-body channels is restricted and hence their small relative movement may result in low Doppler shifts. However, to complete the narrowband characterization of the channel, the Doppler spectrum is required. The Doppler frequency shift, referred to here as the body Doppler shift, f_d, can be calculated as [20]:

$$f_d = \frac{v}{\lambda} \cos \alpha \tag{8.13}$$

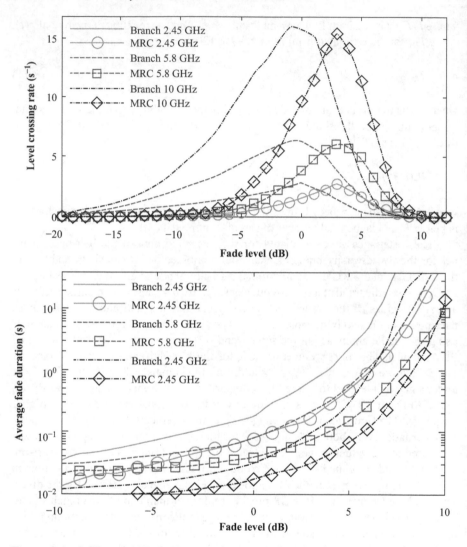

*Figure 8.4   LCR and AFD for Monopole antennas at the three frequencies. © 2009
John/Wiley & Sons, Inc. Reprinted with permission from Reference 16*

where v is the velocity of motion, α is the angle between the direction of arrival of the
signal and the direction of receiver movement, and λ is the wavelength. The channel
coherence time, t_c, can be estimated from the maximum Doppler shift, $f_m = v/\lambda$,
as [20]:

$$t_c = \sqrt{\frac{9}{16\pi f_m^2}}$$ (8.14)

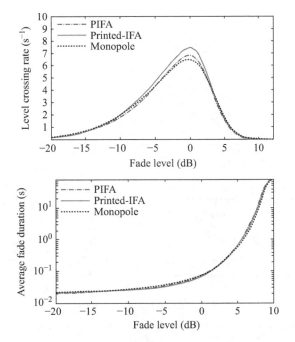

Figure 8.5 *Comparison of LCR and AFD for the three antennas at 5.8 GHz.*
© 2009 John/Wiley & Sons, Inc. Reprinted with permission from
Reference 16

Doppler spectra for the received branch signal envelope and diversity combined signal were calculated by taking the Fourier transform of the autocorrelation function of the signal envelope. The spectra of branch and the corresponding diversity combined signals were identical. Figure 8.6 shows the Doppler spectra of the two channels for walking posture using Hamming window with side lobe level of −40 dB. Additional measurements at 10 GHz were carried out just to observe the trend with higher frequencies. Any frequency point with an ordinate value below −40 dB may represent a side lobe and hence should be ignored. The maximum Doppler shift, f_m, can be approximated from these plots since the frequency where the slope of the curve changes and drops down rapidly is $2f_m$ [22]. It can be identified for the 2.45 GHz case as approximately 4 Hz for belt-head and 10 Hz for belt-wrist channel but cannot be identified for the other frequency due to the limitations of the measurement system on the sampling time and the noise level. These values of maximum Doppler shift can also be verified using (8.13) with an approximate speed of motion of 1.3 m/s, which was the approximate average speed of movement of Rx antenna with respect to the Tx antenna. Figure 8.7 shows a portion of the average Doppler spectra of the complex signal which are asymmetric about the centre frequency. The asymmetry may be due to the variable speed of motion and random movement of the scatterers around the antennas due to the random movements. Besides this, most scatterers (equipment

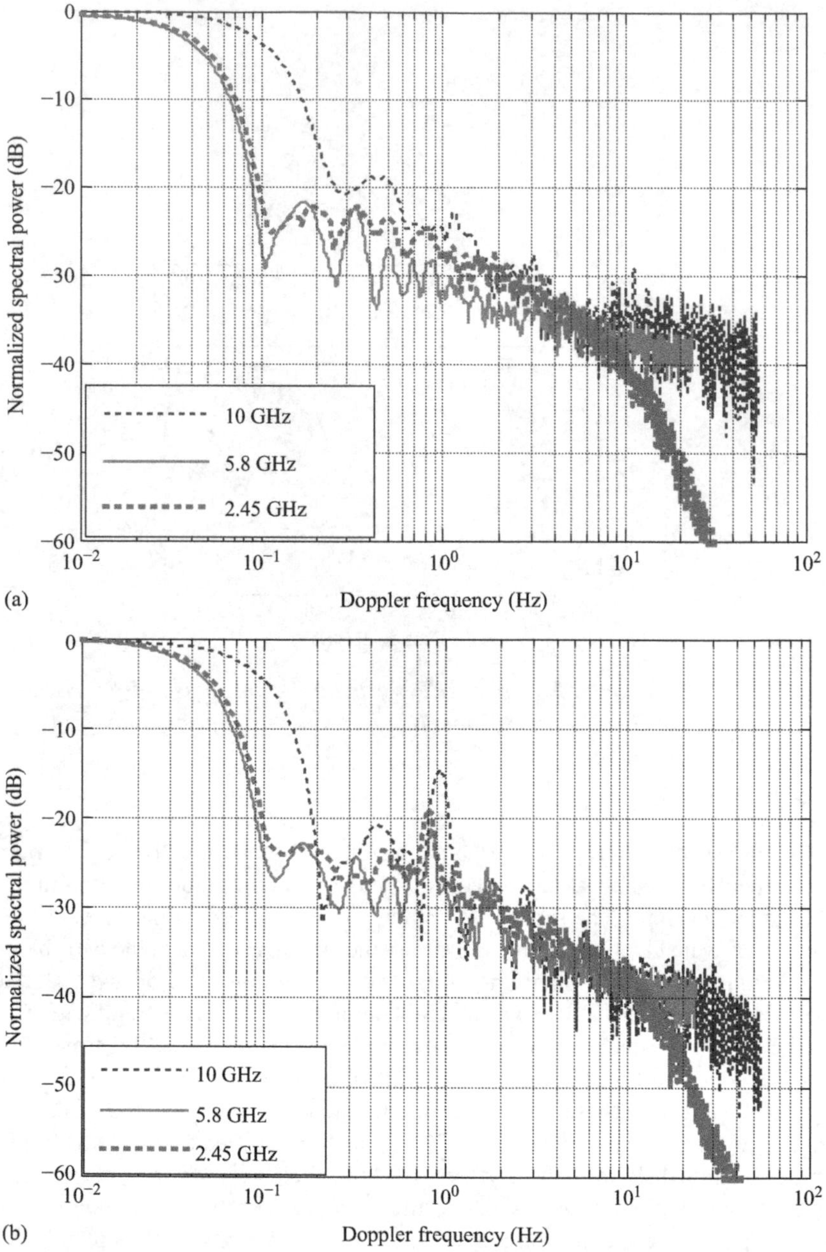

Figure 8.6 Average Doppler spectrum. (a) Belt-head channel and (b) belt-wrist channel. © 2010 IEEE. Reprinted with permission from Reference 14

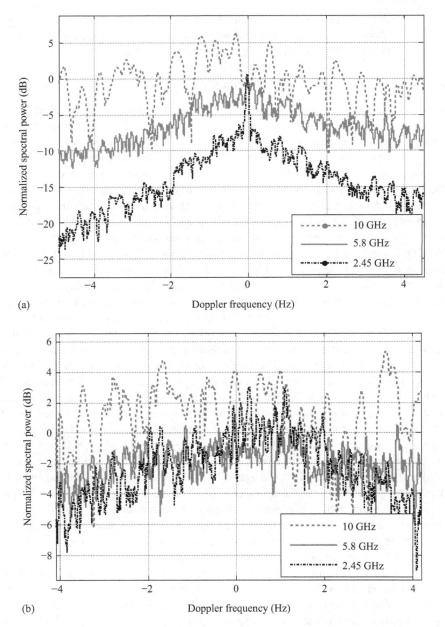

Figure 8.7 Average Doppler spectrum with walking posture using complex signal (a) Belt-head channel and (b) belt-wrist channel. © 2010 IEEE. Reprinted with permission from Reference 14

and furniture) were positioned in the horizontal plane. So a large proportion of the directions of arrival was in the horizontal plane and are shielded by the body on which the antennas were mounted, making the scattering non-isotropic.

8.2.3 Diversity for interference cancellation

Co-channel interference is a serious issue in BAN, where the nodes, operating in the same frequency band, operate and move close to each other. The interference and the desired signal power in this case can be the same. Interference cancellation hence becomes unavoidable. The challenge is that signal to interference plus noise ratio (SINR) becomes very low and moreover, more than one interferer move randomly in the close vicinity. This makes the angle of arrival of interference signal spread over the entire space domain and makes interference cancellation more difficult and challenging. Another challenge is the complexity of the receiver while using the conventional interference rejection techniques like optimum combining (OC) [23] and Weiner–Hopf (WH) method [24, 25] with smart antenna arrays. This complexity adds up to the cost and size of the receiver, which is undesirable for wearable devices. Receive diversity can play a significant role in interference cancellation by employing interference rejection combining (IRC) algorithms [23]. A new low complexity interference rejection technique with receive diversity is proposed in Reference 26, which is suitable for BAN applications. The technique works on the interference signal estimates at the two receive diversity branches. The interference signals are combined in such a way that they cancel each other. This can be accomplished by scaling and phase shifting one branch interference signal to make it equal and out of phase with the second branch interference signal. To make it possible, the interference signal needs to be separated from the desired signal. It can be done by interrupting the desired signal transmitter to stop signal transmission for a short interval. In this interval, the received signal is thus interference plus noise only. The signal received in this short interval can be used as an estimate of interference plus noise for the forthcoming time interval until the transmission is interrupted again. The performance of this scheme is mainly dependent upon the duration and frequency of occurrence of the interval. A long gap in the interruption interval outdates the estimate of the interference signal, whereas, a short gap affects the data rate of the desired signal. The performance of the proposed algorithm in rejecting interference is compared to the performance of the conventional OC and WH algorithms applied to the same channel data. Real-time measurements were carried out by mounting the diversity antennas and the desired signal transmitter on the same human body and the interfering signal transmitter on another human body. The two subjects were moving around in close vicinity doing random movements. Further details of the measurement setup can be found in Reference 26. System model and results are discussed below.

8.2.3.1 Systems model

Consider a narrowband m-branch diversity combiner, as shown in Figure 8.8. The received signal, $x_i(t)$ at the ith antenna of the diversity combiner is thus:

$$x_i(t) = d_i(t) + i_i(t) + n_i(t) \qquad (8.15)$$

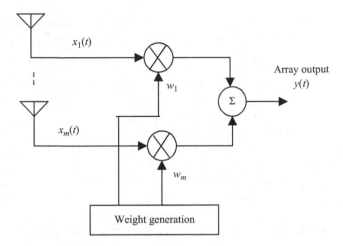

Figure 8.8 Simplified diversity combiner. © 2010 IEEE. Reprinted with permission from Reference 26

where $d_i(t)$ is the received desired signal, $i_i(t)$ is the received interference signal, and $n_i(t)$ is the additive white Gaussian noise (AWGN), at the *ith* branch antenna.

IRC combines the two-branch signals with an aim to suppress the interference signal and improve the output SINR. This objective is achieved by calculating an optimum weight, w_i, for each branch [23]. For a two-branch diversity receiver, the received signal vector, X, and the weight vector, W, can be defined as:

$$X = \begin{bmatrix} x_1(t) & x_2(t) \end{bmatrix}^T \tag{8.16}$$

$$W = \begin{bmatrix} w_1 & w_2 \end{bmatrix}^T \tag{8.17}$$

where $[\ldots]^T$ represents the transpose of the vector. The array output, $y(t)$, is then:

$$y(t) = W^T X \tag{8.18}$$

The desired signal vector, D, interference signal vector, I, and RMS noise vector, N_{rms}, are given as:

$$D = \begin{bmatrix} d_1(t) & d_2(t) \end{bmatrix}^T, \quad I = \begin{bmatrix} i_1(t) & i_2(t) \end{bmatrix}^T, \tag{8.19}$$

$$N_{rms} = \begin{bmatrix} \sqrt{<|n_1(t)|^2>} & \sqrt{<|n_2(t)|^2>} \end{bmatrix}^T \tag{8.20}$$

where $<\ldots>$ represents the mean value.

In case of QC, the weight vector, W, is generated as [23]:

$$W = R^{-1}H \tag{8.21}$$

H is the desired channel transfer gain vector and R is the covariance matrix of interference plus noise ($U = I + N$), called the error covariance matrix [23, 24], i.e.

$$R = E(UU') \tag{8.22}$$

U can be estimated by using a known transmitted training sequence and the channel response knowledge. In the WH solution, W can be generated as [24, 25]:

$$W = \Phi^{-1}Z \tag{8.23}$$

where Φ is the covariance matrix of the received signals, $x_1(t)$ and $x_2(t)$, Z is the correlation matrix of the received signal, $x_i(t)$, and a reference signal $z(t)$, that must be correlated to the desired signal and uncorrelated to the interference signal.

For the proposed Interference Cancellation with Interrupted Transmission (ICIT) scheme described above and given in detail in Reference 26, the estimated value of the interference signal is used to calculate the weight vector, W, for the forthcoming time interval. Such a signal, however, is subject to fast fading and thus should be averaged over time to achieve an accurate estimate. The larger the averaging window, the more is the time for desired signal blockage and more is the throughput degradation. Thus at any time instant, k, the average estimated interference signal at the ith branch is:

$$\bar{i}_i(k) = |\bar{i}_i(k)| \angle (\bar{\psi}_i(k)) \tag{8.24}$$

and W at that instant, k, is calculated as:

$$W = \left[1 \quad \frac{|\bar{i}_1(k)|}{|\bar{i}_2(k)|} \angle (\Delta\psi) \right] \tag{8.25}$$

where $\Delta\psi = \bar{\psi}_1(k) - \bar{\psi}_2(k) + 180° \tag{8.26}$

The input and the output SINR can be calculated as [27]:

$$\text{SINR}_{in(i)}(t) = \frac{|d_i(t)|^2}{|i_i(t)|^2 + \left\langle |n_i|^2 \right\rangle} \tag{8.27}$$

$$\text{SINR}_{out}(t) = \frac{|W^T D|^2}{|W^T I|^2 + |W^T N_{rms}|^2} \tag{8.28}$$

Interference rejection gain (IRG) is the improvement offered by the IRC in output SINR over the input SINR of the branch having highest SINR value among the branches. The IRG can be calculated from the CDFs of the input and output SINR by taking the difference at a certain level of outage probability, usually 1%.

8.2.3.2   Interference cancellation performance analysis

In order to quantify the interference rejection by all the above mentioned algorithms, the CDFs of the input and output SINR are plotted, as shown in Figure 8.9, and IRG is calculated in each case at 1% outage probability. Table 8.7 shows the calculated values of the IRG and the mean desired signal and interference signal powers. To evaluate the IRG offered by each scheme, various averaging window sizes (for an average estimate of the covariance matrices and the interference signal) were tried and the window size giving the best IRG was used. For the proposed ICIT scheme, the SINR improvement in the form of IRG was evaluated as a function of the interruption period (time interval between two shut downs of the desired signal transmitter) and is plotted for the three on-body channels in Figure 8.10. The figure reveals that in

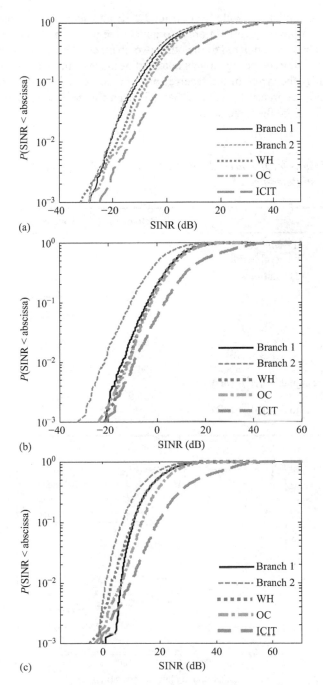

Figure 8.9   SINR CDFs. (a) Belt-head channel, (b) belt-wrist channel, and
(c) belt-chest channel. © 2010 IEEE. Reprinted with permission from
Reference 26

order to obtain some reasonable IRG (around 2 dB and above), the transmitter must be interrupted more frequently in order to update the interference signal in the case of belt-chest channel compared to the other two channels. This may probably be due to the fact that interference signal gets outdated because of more body shadowing of the interferer by the virtue of the receiver position compared to the other channels. For consistency, an interval period of 60 ms is used for the results of Table 8.7 and Figure 8.10 for all of the three channels.

Table 8.7   Results for the three channels. © 2010 IEEE. Reprinted with permission from Reference 26

		Belt-head	**Belt-chest**	**Belt-wrist**
IRG (dB)	ICIT with 60 ms interval period	6.4	1.2	4.7
	WH solution	2.2	−3	2.2
	OC	4.5	0.2	3.0
Mean power, desired signal, $<|s_1|^2>$ (dB)		−53.7	−37.4	−48.1
Mean power, desired signal, $<|s_2|^2>$ (dB)		−56.6	−41.3	−51.4
Mean power, interference signal, $<|i_1|^2>$ (dB)		−54.3	−52.8	−51.9
Mean power, interference signal, $<|i_2|^2>$ (dB)		−55.9	−53.4	−48.5
SIR_{avg} (dB)		8.13	21.8	10.9

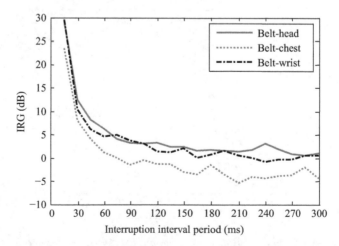

Figure 8.10   IRG vs. interval period for ICIT with interference estimated at a single instant. © 2010 IEEE. Reprinted with permission from Reference 26

Table 8.7 also gives the IRG values calculated using OC and WH algorithms. ICIT shows better rejection capability. OC and WH use covariance and correlation matrices to evaluate the weight vector. Relatively high correlation and covariance between the branch signals for on-body channels result in degraded performance of these schemes. This fact is pronounced more in the case of belt-chest channel having high correlation values. The comparison shows that ICIT can perform better if a proper interruption interval is selected. Moreover, it provides a simple and robust solution for interference rejection compared to the complex and expensive smart antenna techniques. It also provides a handle to control the IRG. For high interference prone scenarios where high IRG is desired for better rejection, interruption interval can be reduced to increase IRG. This comes at the expense of throughput degradation. Thus there is a trade-off between the IRG and throughput of the system. In packet radio systems, the gap between two transmitted packets can be used to estimate the interference and the throughput degradation can be avoided. However, the disadvantage will be an increased power consumption at the receiver end, as the sleep time for the receiver will be decreased and utilized in estimating the interference signal.

The average desired signal to interference ratio (SIR_{avg}), can be calculated as:

$$SIR_{avg} = 10 \times \log_{10} \left(\left\langle \frac{|s_1(t)|^2 + |s_2(t)|^2}{|i_1(t)|^2 + |i_2(t)|^2} \right\rangle \right) \tag{8.29}$$

By plotting IRG against the average SIR value (SIR_{avg}), as shown in Figure 8.11 for belt-head case as an example, the performance of the IRC schemes with increasing level of interference can be observed. The figure shows that the performance of IRC algorithms degrade in suppressing the interference when the interference signal is

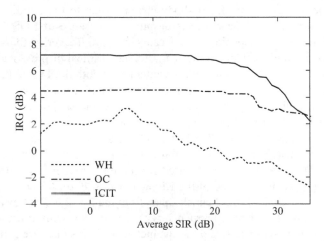

Figure 8.11   IRG vs. average SIR for belt-head channel with interval length for ICIT = 60 ms. © 2010 IEEE. Reprinted with permission from Reference 26

quite weak compared to the desired signal level. The figure also gives a comparison of the three schemes for various values of average SIR.

8.3   MIMO channels and capacity of on-body channels

The emergence of 4G and 5G wireless communications technology has extensively increased the demand of high data rate and reliability. MIMO systems are well known to increase the channel capacity by providing significant amount of multiplexing gain. These systems are also effective to increase the reliability by providing diversity gain as well. Being an integral part of the Personal Communications System (PCS), BAN also demands the use of multiple antennas to increase the channel capacity and reliability. MIMO exploits the multipath richness in providing multiplexing gain. The performance of MIMO is limited by the presence of LOS link, and correlation among the spatial subchannels as well as by the mutual coupling between the spatially separated antennas [27–30]. Section 8.2 shows that the short-term fading envelope for the on-body channels is Rician distributed with varying K-factor values for various channels and antenna types. The capacity increase in Rician fading channels depends on the K-factor which is a measure of the multipath richness and receiver SNR. High K-factor values indicate the presence of a strong direct ray, and hence high receive SNR values are expected. The high SNR values may increase the channel capacity compared to a NLOS link with the same configuration. On the other hand, this results in high correlation among the spatial subchannels thus decreasing the multiplexing gain. Thus, a trade-off exists between the effect of increased SNR and increased correlation on the channel capacity. In the case of an on-body MIMO link, body movements may cause changes to both these parameters, thus leading to a temporal variation of the channel capacity. The diversity results presented above encourage the use of MIMO for on-body channels due to the low channel correlation values for some of the channels despite the presence of LOS. This can be exploited to achieve reasonable channel capacity gain. To verify the significance of MIMO for on-body channels, the measurement campaign presented above for receive diversity was extended to multiple antennas at both the Tx and Rx sides. The performance of a 2×2 MIMO system was investigated through real measurement data which was collected by the same measurement setup as described above at the 2.45 GHz frequency. Instead of a single Tx antenna, a 2-element MIMO PIFA was used as Tx to make a 2×2 MIMO channel, as shown in Figure 8.12. The two transmitting antennas were connected through an RF switch to a signal generator operating at 2.45 GHz. The antennas used for the measurements are also shown in Figure 8.12. The switching time of the RF switch was 40 μs which was much less than the coherence time of the channel. Apart from the 2×2 MIMO measurements, 2×1, 1×2, and 1×1 measurements, with $n \times m$ representing n Rx and m Tx antennas, were done separately, and the channel capacity results were compared. The MIMO subchannel matrix, as described below in the channel model, was obtained from the measured data. Further details of the measurement procedure and antenna characteristics are given in Reference 31. The following sections describe the MIMO

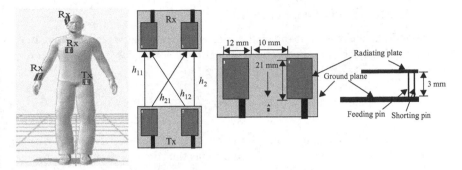

*Figure 8.12   Placement of the antennas on the body and the MIMO channel and
the top view of the PIFA array. The Rx antenna array was placed at the
three positions separately for the three on-body channels. Tx antenna
remained at the waist position. © 2010 IEEE. Reprinted with
permission from Reference 31*

channel capacity, analysis of spatial channel correlation matrices, and the transmit–
receive diversity gains with the 2 × 2 MIMO setup. Furthermore, the stochastic
MIMO channel model is presented and validated with the measured results. The
channel model is presented below first, and then the channel capacity analysis is
done by showing the CDFs of the capacity and its relationship with the SNR. The
transmit–receive diversity and stochastic model are then presented in the subsequent
subsections.

8.3.1   MIMO channel model

For a narrowband, single-user MIMO channel with m transmit and n receive antennas,
if X is the [$m \times 1$] transmitted vector, Y is the [$n \times 1$] received vector, the input–output
relationship between the Tx and Rx is given as:

$$Y = HX + N \qquad (8.30)$$

where N is receive AWGN vector, and H is the $n \times m$ channel matrix. For a 2 × 2
MIMO channel, H can be written as:

$$H = \begin{bmatrix} h_{11} & h_{12} \\ h_{21} & h_{22} \end{bmatrix} \qquad (8.31)$$

where h_{ij} is the complex random variable representing the channel-fading coefficients
or the complex subchannel gains from transmitting antenna j to receiving antenna i, as
shown in Figure 8.12. The H matrix, representing the measured channel, includes the
effect of mutual coupling between the antenna elements and the correlation among
the subchannels since the subchannel gains were measured at the actual antenna ports
with antenna elements placed next to each other.

8.3.2   *MIMO for channel capacity*

For a MIMO channel with uniform distribution of the transmitted power among the m transmitting antennas, if channel state information is not available at the transmitter, the channel capacity can be expressed by [3, 28–30]:

$$C = \log_2 \left(\det \left[I_n + \frac{\xi}{m} HH* \right] \right) \ bps/Hz \tag{8.32}$$

where I_n is $n \times n$ identity matrix, and ξ is the average SNR per receive antenna. Here H is the normalized channel matrix, and * represents the complex conjugate transpose. The H matrix is normalized such that its average Frobenius Norm (averaged over all instances of H matrix) is equal to nm [3, 32, 33], i.e.

$$\langle \|H\|_F^2 \rangle = nm \tag{8.33}$$

The normalization assumes a fixed transmitted power, and hence the average SNR at the receiver for each realization of the channel changes with variation in pathloss [33]. A high correlation among the subchannels, h_{ij}, may result in reducing the rank of the H matrix which reduces the channel capacity. Hence, a knowledge of the subchannel correlation is required. The complex signal correlation coefficients (ρ_s) among the subchannels can be calculated as [34]:

$$\rho_s = \frac{\sum\limits_{i=1}^{N} V_1(i)V_2^*(i)}{\sqrt{\sum\limits_{i=1}^{N} V_1(i)V_1^*(i)}\sqrt{\sum\limits_{i=1}^{N} V_2(i)V_2^*(i)}} \tag{8.34}$$

where N is the total no. of samples of received envelope. V_1 and V_2 represent the zero-meaned complex voltage signals. The MIMO spatial correlation matrix, given by (8.35) for the 2×2 MIMO channel, gives a comprehensive view of the degree of correlation among the subchannels [35]:

$$\rho = \begin{bmatrix} 1 & \rho_{11}^{12} & \rho_{11}^{21} & \rho_{11}^{22} \\ \rho_{12}^{11} & 1 & \rho_{12}^{21} & \rho_{12}^{22} \\ \rho_{21}^{11} & \rho_{21}^{12} & 1 & \rho_{21}^{22} \\ \rho_{22}^{11} & \rho_{22}^{12} & \rho_{22}^{21} & 1 \end{bmatrix} \tag{8.35}$$

where ρ_{ab}^{cd} is the correlation coefficient between subchannel h_{ab} and h_{cd}. The symmetry $\rho_{ab}^{cd} = \rho_{cd}^{ab*}$ reduces the total number of significant coefficients to 6. The calculated spatial correlation matrices for the measured data of the three on-body channels are given in Table 8.8.

Figure 8.13 presents the CDF plots of the channel capacity calculated using (8.32) at various values of average receive SNR, ξ, for the three on-body channels. The figure also shows the CDFs of channel capacity for other configurations, i.e. MISO, SIMO,

Table 8.8   *Spatial correlation matrices for the three on-body channels. © 2010*
             IEEE. Reprinted with permission from Reference 31

Channel	Spatial correlation matrix (ρ_s)
Belt-head	$\begin{bmatrix} 1 & 0.67 + 0.39i & 0.31 - 0.27i & 0.27 - 0.02i \\ 0.67 - 0.39i & 1 & 0.13 - 0.36i & 0.17 - 0.24i \\ 0.31 + 0.27i & 0.13 + 0.36i & 1 & 0.56 + 0.33i \\ 0.27 + 0.02i & 0.17 + 0.24i & 0.56 - 0.33i & 1 \end{bmatrix}$
Belt-chest	$\begin{bmatrix} 1 & 0.92 + 0.22i & 0.14 - 0.63i & 0.23 - 0.54i \\ 0.92 - 0.22i & 1 & 0.00 - 0.60i & 0.12 - 0.61i \\ 0.14 + 0.63i & 0.00 + 0.60i & 1 & 0.88 + 0.19i \\ 0.23 + 0.54i & 0.12 + 0.61i & 0.88 - 0.19i & 1 \end{bmatrix}$
Belt-wrist	$\begin{bmatrix} 1 & 0.78 - 0.01i & -0.26 + 0.08i & -0.24 + 0.10i \\ 0.78 + 0.01i & 1 & -0.13 + 0.01i & -0.23 + 0.11i \\ -0.26 - 0.08i & -0.13 - 0.01i & 1 & 0.81 + 0.22i \\ -0.24 - 0.10i & -0.23 - 0.11i & 0.81 - 0.22i & 1 \end{bmatrix}$

and SISO at $\xi = 25$ dB, which are represented by thick lines. By observing the plots closely, it can be noticed that in the low SNR regime, the capacities, at a fixed outage probability, are almost the same for the three channels despite the difference in the subchannel correlation values. It shows that the direct ray at the low SNR values, in the case of belt-chest channel, is not strong enough to introduce significant correlation among the subchannels. At higher SNR values, the direct ray is strong, and the high correlation can influence the channel capacity. This is reflected in the belt-chest channel case where the channel capacity decreases at high SNR compared to the other two channels. Moreover, the curves for the belt-chest channel have steeper slope signifying a less spread in the channel capacity. This may be by the virtue of less pathloss variation for belt-chest channel compared to the dynamic belt-wrist and belt-head channels. Nevertheless, it is noticeable that despite the correlation among the subchannels and the presence of direct link due to LOS for the belt-chest channel, the capacity improvement offered by MIMO over the same channel with SISO, MISO, and SIMO links is encouraging. The other two channels offer significant amount of channel capacity improvement with MIMO.

The channel capacity improvement can be measured in terms of the spatial multiplexing gain which is typically measured as $\min(n, m)$ bps/Hz for independent $n \times m$ MIMO system in ideal case [36]. It can be calculated at a certain outage level (usually 1%) and fixed SNR as the improvement offered by MIMO over SISO at the same fixed SNR. For the 2×2 measured channels presented, the multiplexing gain is close to 2. The amount of decrease in gain from ideal value depends upon the correlation among the subchannels. Furthermore, Figure 8.14 shows the ergodic capacity as a

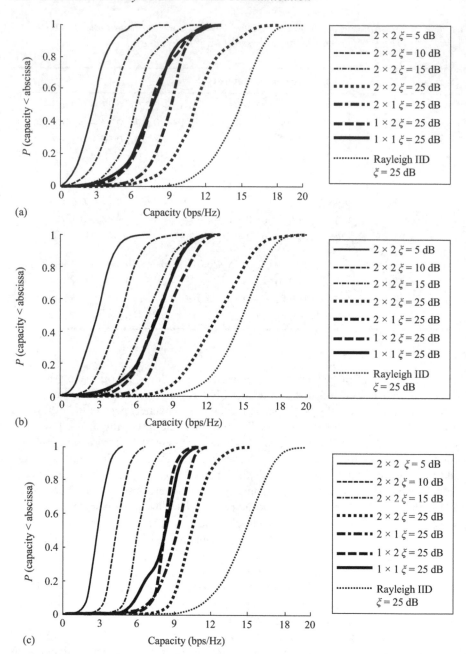

Figure 8.13 *Capacity CDFs. © 2010 IEEE. Reprinted with permission from Reference 31. (a) Belt-wrist channel, (b) belt-head channel, and (c) belt-chest channel*

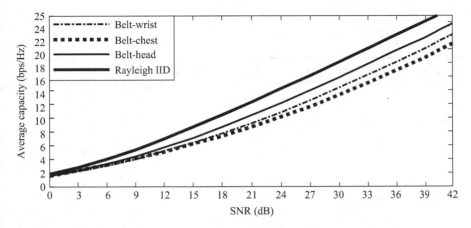

Figure 8.14 Average capacity vs. SNR. © 2010 IEEE. Reprinted with permission from Reference 31

function of SNR for the three channels, and the average capacity increase with each 3 dB increase in SNR can be observed to be less than but close to 2 bps/Hz, in the high SNR regime. Ideally, the increase in capacity with every 3 dB increase in SNR for independent $n \times m$ MIMO system is $\min(n, m)$ bps/Hz [36].

8.3.3 Transmit–receive diversity with MIMO

For any communication system, a fundamental trade-off exists between the two key performance metrics, i.e. the frame-error rate (FER) and the transmission rate offered by the system. Intuitively, an increase in SNR results in reduced FER for a fixed transmission rate. Similarly, an increase in SNR may result in increased transmission rate at a fixed target FER. This trade-off is often referred to as the diversity-multiplexing trade-off in the context of MIMO systems with diversity signifying the FER reduction and multiplexing signifying an increase in transmission rate.

Transmit diversity system, with number of transmitting antenna, $M_T = 2$, and number of receive antennas M_R, can be analysed for diversity gains using the well-known Alamouti scheme [37]. It realizes a diversity gain of the order of $2M_R$ at a fixed transmission rate equivalent to fourth-order diversity in a 2×2 MIMO channel. It has been shown by Alamouti that 2×2 MIMO configurations are equivalent to traditional four-branch MRC after decomposing the MIMO channel into parallel subchannels. For an ideal case with independent diversity branches having no power imbalance, a diversity gain of around 20 dB can be achieved with four-branch MRC [36]. Transmit–receive combined diversity performance analysis is presented in Reference 38 for the 2×2 MIMO measure on-body channels. The four subchannels of the measured MIMO on-body channel were taken as the four diversity branches in a fourth-order receive diversity system. The diversity combined signals were achieved using extensions of (8.3)–(8.5). Table 8.9 presents the DG values with

Table 8.9   Results for 2 × 2 MIMO. © 2011 IET. Reprinted with permission from Reference 38

Channel	DG MRC (dB)	Max power diff (dB)	ρ_e matrix
Belt-head	11.5	8.3	$\begin{bmatrix} 1.0 & 0.5 & 0.4 & 0.2 \\ 0.5 & 1.0 & 0.3 & 0.2 \\ 0.4 & 0.3 & 1.0 & 0.4 \\ 0.2 & 0.2 & 0.4 & 1.0 \end{bmatrix}$
Belt-wrist	10.6	6.4	$\begin{bmatrix} 1.0 & 0.6 & 0.2 & 0.1 \\ 0.6 & 1.0 & 0.1 & 0.2 \\ 0.2 & 0.1 & 1.0 & 0.5 \\ 0.1 & 0.2 & 0.5 & 1.0 \end{bmatrix}$
Belt-chest	4.2	6.8	$\begin{bmatrix} 1.0 & 0.6 & 0.6 & 0.3 \\ 0.6 & 1.0 & 0.3 & 0.5 \\ 0.6 & 0.3 & 1.0 & 0.6 \\ 0.3 & 0.5 & 0.6 & 1.0 \end{bmatrix}$

MRC, the maximum power difference among the four subchannels, and the envelope correlation values. The results reveal that belt-head and belt-wrist channels give reasonable diversity gains while belt-chest channel again is the worst performer by virtue of the high correlation and high power imbalance. The diversity gains are not as high as the ideal case but still significant in the context of on-body propagation environment.

8.3.4   MIMO stochastic channel modelling

The characterization of on-body channels reveals that the on-body channel behaviour is dependent upon not only the environment but also the antenna type, type of movements, body shape and postures, location of antenna on the body, and polarization of the antenna [14]. The channel may become highly non-stationary as a consequence of random movements of the body where antennas are mounted. This makes the channel very difficult to understand and requires a complete model which can describe most of the mentioned complex effects. In the frequency band of 2.45 GHz, a realistic pathloss model was presented in Reference 12 where it was shown that the slow fading can be modelled by the lognormal distribution. The results presented above in Section 8.2 also validate this fact along with the findings that the fast fading envelope's distribution is Rician. However, these studies concern with either SISO or SIMO systems. A MIMO channel model is presented in Reference 39 based on the measurement campaign described above. Furthermore, for the channel capacity evaluation, equal power distribution was solely considered. In the subsequent section, the achievable capacity gain is evaluated using water filling as well and is

Table 8.10 *Average pathloss, shadowing and K values used in Model 1. © 2012 IEEE. Reprinted with permission from Reference 39*

On-body channel	K value	Mean pathloss (dB)	Standard deviation (dB)
Belt-head	3.90	58.501	1.004
Belt-chest	11.42	39.38	1.217

compared to the uniform power distribution. Herein, the Rician K-factor, received power, and shadowing deviation are used as channel parameters. Proposed Model 1 uses the average values of these parameters for the different spatial subchannels of a given on-body channel. A second model, Model 2, is presented to incorporate the various on-body propagation mechanisms perceived at the receive antenna elements.

The evaluated statistical parameters for belt-chest and belt-head on-body channels are given in Table 8.10 using Model 1. As depicted in the diversity channel model above, the high K-factor values (K) and low channel attenuation for belt-chest channel suggest that the LOS component is significantly higher than the scattered component. Belt-head channel is also Ricain but with low K-factor values. The parameters presented in Table 8.10 are averages recorded for the different types of body postures over the different MIMO subchannels, thus the same values are assumed for each receive antenna.

The received signal thus is a combination of a constant LOS signal and a Rayleigh distributed time-varying component. So for a MIMO system, the spatial subchannel linking the ith receive element and the jth transmit element can be given as:

$$h_{ij}(t) = \sqrt{\frac{P_r}{K+1}}\left[\sqrt{K}e^{j\varphi_{ij}} + z_{ij}(t)\right] \tag{8.36}$$

where φ_{ij} is the phase of the j–ith subchannel in the constant component and is uniformly distributed over $[0, 2\pi]$ due to time variant orientation of transmit and receive antennas, $z_{ij}(t)$ is the correlated NLOS component. The received power or the path-gain term P_r– can be modelled for a given transmitter-receiver separation as:

$$P_r(d) = P_r(d_0) - 10n\log_{10}\left(\frac{d}{d_0}\right) + X_{shad}(d) \tag{8.37}$$

where X_{shad} is the log-normally distributed shadowing term. Following these definitions, the $(N \times M)$ MIMO channel matrix H in the Model 1 can be formulated as:

$$\mathbf{H} = \sqrt{\frac{P_r}{K+1}}\left[\sqrt{K}\begin{bmatrix} e^{j\varphi_{11}} & \cdots & e^{j\varphi_{1M}} \\ & \vdots & \\ e^{j\varphi_{N1}} & & e^{j\varphi_{NM}} \end{bmatrix} + \begin{bmatrix} z_{11}(t) & \cdots & z_{1M}(t) \\ & \vdots & \\ z_{N1}(t) & \cdots & z_{NM}(t) \end{bmatrix}\right] \tag{8.38}$$

Table 8.11   *K-factor, mean pathloss and standard deviation of the four MIMO subchannels. © 2012 IEEE. Reprinted with permission from Reference 39*

Channel	Subchannel	K-factor	Mean pathloss (dB)	Standard deviation (dB)
Belt-head	h_{11}	6.38	52.65	0.61
	h_{12}	7.33	55.81	0.919
	h_{21}	1.052	62.10	1.467
	h_{22}	0.82	63.43	1.019
Belt-chest	h_{11}	10.68	36.67	0.75
	h_{12}	15.02	40.48	0.93
	h_{21}	7.688	38.14	1.417
	h_{22}	12.30	43.25	1.767

where the first $N \times M$ term of the right hand part of (8.38) corresponds to the LOS component and the second $N \times M$ term corresponds to the NLOS component.

The MIMO subchannels may not have identical distribution due to random postures, orientation and different shadowing effects on the various elements of the MIMO receive antenna. Table 8.11 summarizes the statistical parameters for the four MIMO subchannels while performing averaging over different body postures. The channels parameters in this case clearly suggest the disparity in the propagation experienced in each subchannel specifically in the belt-head channel where the h_{11} and h_{12} exhibit lower mean pathloss and higher K values compared to h_{21} and h_{22} subchannels. An index i is attached to the receive power $(p_r)_i$ and to the Rician K-factor in Model 2 to indicate their dependence on the receive antenna. It follows that the MIMO channel matrix in Model 2 can be formulated as:

$$
\mathbf{H} = \left[\begin{bmatrix} \sqrt{\dfrac{K_1\,(p_r)_1}{K_1+1}}e^{j\varphi_{11}} \cdots \sqrt{\dfrac{K_1\,(p_r)_1}{K_1+1}}e^{j\varphi_{1M}} \\ \vdots \\ \sqrt{\dfrac{K_N\,(p_r)_N}{K_N+1}}e^{j\varphi_{N1}} \quad \sqrt{\dfrac{K_N\,(p_r)_N}{K_N+1}}e^{j\varphi_{NM}} \end{bmatrix} \right.
$$

$$
\left. + \begin{bmatrix} \sqrt{\dfrac{(p_r)_1}{K_1+1}}z_{11}(t) \cdots \sqrt{\dfrac{(p_r)_1}{K_1+1}}z_{1M}(t) \\ \vdots \\ \sqrt{\dfrac{(p_r)_N}{K_N+1}}z_{N1}(t) \cdots \sqrt{\dfrac{(p_r)_N}{K_N+1}}z_{NM}(t) \end{bmatrix} \right]
$$

$$(8.39)$$

The correlated channel coefficients z_{ij} are generated from the spatial correlation matrix ρ, given above in (8.35), and the zero-mean complex *i.i.d.* random variables a_{nm} as follows:

$$\mathbf{A} = \mathbf{Ca} \tag{8.40}$$

where $\mathbf{A}_{NM\times 1} = [z_{11} \dots z_{N1}, \cdots, z_{21} \dots z_{NM}]^T$, $\mathbf{a}_{NM\times 1} = [a_1, a_2, \cdots, a_{NM}]^T$, and $\mathbf{C}$ is a symmetric matrix extracted from the standard Cholesky factorization of the matrix ρ. It is worth mentioning that the Kronecker product method to evaluate z_{ij} does not hold for the on-body channels since its assumptions of independent spatial correlation coefficients at the receiving and transmitting side is not valid. This is clear from the spatial correlation matrices presented in the above section.

Assuming uniform power allocation among transmitting antennas, as explained in the above section, the channel capacity can be obtained using (8.32). Now considering water filling instead of equal power allocation, and assuming the existence of K eigenchannels, the capacity with water filling is given by:

$$C = \sum_{k=1}^{K} \log_2 \left(1 + \lambda_k \frac{P_k}{\sigma^2} \right) \tag{8.41}$$

where P_k is the power allocated to the kth MIMO eigenchannel and λ_k is the corresponding eigenvalue. To reach this capacity, each eigenchannel is iteratively filled up to a common level.

8.3.4.1 Simulation results

Figure 8.15 shows the CDFs of the maximum and minimum eigenvalues for the measured, simulated Model 1 and simulated Model 2 cases for the two on-body 2×2 MIMO channels using 2×2 Rayleigh channel as a reference. It is clear from the figure that due to the presence of correlation and LOS one eigenmode is activated more compared to the other as opposed to the full rank 2×2 Rayleigh reference where both the eigenmodes are active. The even higher values of spatial correlation and stronger LOS in the measured belt-chest channel moves the minimum eigenvalue away from the reference and even lower than the belt-head channel values. The maximum eigenvalues of the measured, and both simulated models of the belt-head channel are comparable to the reference case. Model 2 shows good agreement with the measured results in case of minimum eigenvalues. It also suggests that only one eignemode is almost solely active in belt-chest case.

Channel capacity CDFs with equal power allocation are shown for various SNR values in Figure 8.16 for belt-head and belt-chest channels, respectively. The plots show the capacities for both the models and measured channels. Model 2 is in close agreement with the measured channel while Model 1 is close to the measured values at low SNR values but deviates and is no longer valid at high SNR values for the belt-head channel.

Figure 8.17 shows the achievable ergodic capacities at 10% outage for the measured and simulated (Model 1) channels for belt-head and belt-chest cases along

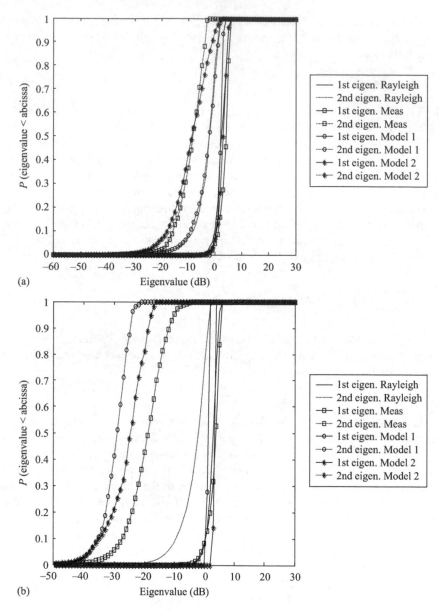

Figure 8.15   CDFs of the eigenvalues using the measurements, Model 1 and Model 2. © 2012 IEEE. Reprinted with permission from Reference 39. (a) Belt-head and (b) belt-chest

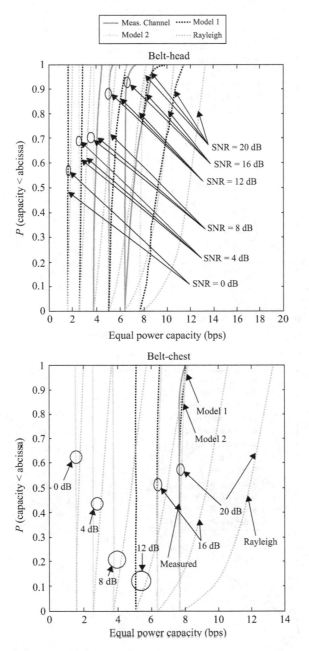

Figure 8.16　*CDF of the equal power capacity of the belt-head and belt-chest channel using the measured channel, and the simulated ones using Model 1 and Model 2. © 2012 IEEE. Reprinted with permission from Reference 39*

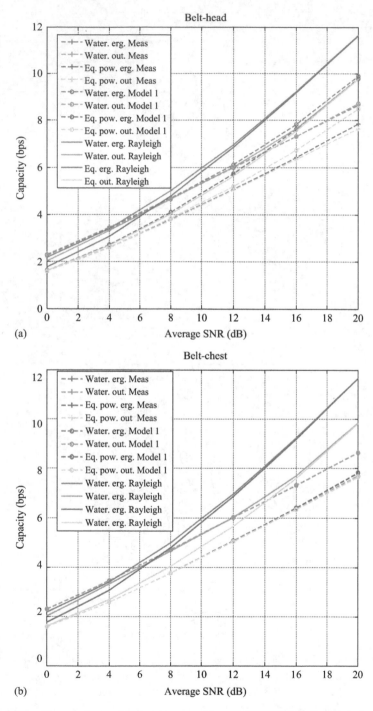

Figure 8.17   Comparison of different capacities. © 2012 IEEE. Reprinted with permission from Reference 39

with the Rayleigh reference case. The capacities are calculated with equal power and water filling and compared. Up to 4 dB in SNR values, the on-body and Rayleigh channels behave similarly when using water filling scheme. Around the cut-off value of 4 dB in SNR and above, the Rayleigh channel outperforms the on-body channel. Water filling improves the capacity at low SNRs due to the fact that it optimally distributes the power among eigenchannels according to their state and balances the impairment of low SNR.

8.4 Summary

The contents of the chapter focused upon the use of multiple antennas for a more efficient and reliable front-end design for the body-worn receivers. It is obvious from the results and discussions that both MIMO and diversity can play a significant role in making the BAN receivers more robust and suitable for high-speed communications. Significant improvement in SNR may result in belt-wrist and belt-head on-body channels by using two-branch receive diversity system. This may also help in reducing the transmit power to the on-body devices thus reducing the risk of exposure of human body to high power electromagnetic waves. Antenna diversity in all the three forms, i.e. space, pattern, and polarization diversity, is effective for belt-head and belt-wrist channels. Belt-chest channel by virtue of strong LOS may not be benefited much by diversity and MIMO. Receive diversity was also found to be an effective tool in BAN–BAN interference cancellation. A new interference cancellation scheme was proposed and was found to be more suitable and efficient for BAN applications. The use of MIMO for on-body channels was also investigated, and the results were very encouraging in terms of channel capacity improvement offered for various on-body scenarios. Stochastic channel models were proposed for on-body diversity and MIMO channels. The fast fading envelope of both diversity and MIMO channels was found to be Rician distributed, and the long-term fading envelope was found to be log-normal distributed. The parameters of the two models were presented for various scenarios, and channel correlation based MIMO channel model was presented and compared to the measured data. These stochastic models can be used by the front-end designers to design efficient, reliable, and robust body-worn devices.

References

[1] R. G. Vaughan and J. B. Andersen, 'Antenna diversity in mobile communications', *IEEE Transactions on Vehicular Technology*, vol. VT-36, no. 4, November 1987, pp. 149–172.

[2] D. Bladsjo, A. Furuskar, S. Javerbring, and E. Larsson, 'Interference cancellation using antenna diversity for EDGE-enhanced data rates in GSM and TDMA/136', *Proceedings of the 50th IEEE Vehicular Technology Conference*, Fall 1999.

[3] E. Biglier, R. Calderbank, A. Constantnides, A. Goldsmith, A. Paulraj, and H. V. Poor, *MIMO Wireless Communications*. New York: Cambridge University Press, 2007.

[4]   W. C. Jakes, *Microwave Mobile Communications*, New York: Wiley, 1974.

[5]   A. M. D. Turkmani, A. A. Arowojolu, P. A. Jefford, and C. J. Kellett, 'An experimental evaluation of the performance of two-branch space and polarization diversity schemes at 1800 MHz', *IEEE Transactions on Vehicular Technology*, vol. 44, no. 2, May 1995, pp. 318–326.

[6]   I. Khan, M. R. Kamarudin, L. Yu, Y. I. Nechayev, and P. S. Hall, 'Comparison of space and pattern diversity for on-body channels', *Proceeding of 5th European Workshop on Conformal Antennas*, Bristol, UK, 10–11 September 2007, pp. 47–50.

[7]   I. Khan, P. S. Hall, A. A Serra, A. R. Guraliuc, and P. Nepa, 'Diversity performance analysis for on-body communication channels at 2.45 GHz', *IEEE Transactions on Antennas and Propagation*, vol. 57, no. 4, April 2009, pp. 956–963.

[8]   I. Khan and P. S. Hall, 'Multiple antenna reception at 5.8 and 10 GHz for body-centric wireless communication channels', *IEEE Transactions on Antennas and Propagation*, vol. 57, no. 1, January 2009, pp. 248–255.

[9]   L. Akhoondzadeh-Asl, I. Khan, and P. S. Hall, 'Polarization diversity performance for on-body communication applications', *IET Microwaves, Antennas and Propagation Journal*, vol. 5, no. 2, January 2011, pp. 232–236.

[10] A. A. Serra, A. R. Guraliuc, P. Nepa, G. Manara, I. Khan, and P. S. Hall, 'Dual-polarization and dual-pattern planar antenna for diversity in body-centric communications', *IET Microwaves, Antennas and Propagation Journal*, vol. 4, no. 1, January 2010, pp. 106–112.

[11] P. S. Hall and Y. Hao, Eds., *Antennas and Propagation for Body-Centric Wireless Communications*, London: Artech House, 2006.

[12] Y. I. Nechayev and P. S. Hall, 'Multipath fading of on-body propagation channels', *Proceedings of IEEE Int. AP-S Symp.—USNC/URSI National Radio Science Meeting* San Diego, CA, 2008.

[13] S. L. Cotton and W. G. Scanlon, 'Characterization and modeling of the indoor radio channel at 868 MHz for a mobile body-worn wireless personal area network', *IEEE Antennas and Wireless Propagation Letters*, vol. 6, 2007, pp. 51–55.

[14] I. Khan, Y. I. Nechayev, and P. S. Hall, 'On-body diversity channel characterization', *IEEE Transactions on Antennas and Propagation*, vol. 58, no. 2, February 2010, pp. 573–580.

[15] S. L. Cotton and W. G. Scanlon, 'Channel characterization for single- and multiple-antenna wearable systems used for indoor body-to-body communications', *IEEE Transactions on Antennas and Propagation*, vol. 57, no. 4, April 2009, pp. 980–990.

[16] I. Khan, Y. I. Nechayev, and P. S. Hall, 'Second-order statistics of measured on-body diversity channels', *Microwave and Optical Technology Letters*, vol. 51, no. 10, October 2009, pp. 2335–2337.

[17] R. Steele, Ed., *Mobile Radio Communications*, London, New York: Pentech Press, 1994.

[18]   N. L. Scott and R. G. Vaughn, 'The effect of demeaning on signal envelope cor-relation analysis', *Proceedings of 4th International Symposium on Personal, Indoor and Mobile Radio Communications*, Yokohama, Japan, September 1993.

[19]   R. B. D'Agostino and M. A. Stephens, *Goodness-of-Fit Techniques*, New York: Marcel Dekker, Inc., 1986.

[20]   R. Prasad, *Universal Wireless Personal Communications*, Boston, MA: Artech House, 1998, pp. 1464–1479.

[21]   M. D. Yacoub, C. R. C. M. da Silva, and J. E. V. Bautista, 'Second-order statistics for diversity-combining techniques in Nakagami-fading channels', *IEEE Transactions on Vehicular Technology*, vol. 50, 2001.

[22]   W. C. Y. Lee, *Mobile Communications Engineering*, New York: McGraw-Hill, 1982.

[23]   J. H. Winters, 'Optimum combining in digital mobile radio with co-channel interference', *IEEE Journal on Selected Areas in Communications*, vol. SAC-2, July 1984, pp. 528–539.

[24]   R. T. Compton, *Adaptive Antennas, Concepts and Performance*, Upper Saddle River, NJ: Prentice-Hall, Inc., 1988.

[25]   X. N. Tran, T. Taniguchi, and Y. Karasawa, 'Subband Adaptive Array for Multirate Multicode DS-CDMA', *Proceedings of IEEE Tropical Conference on Wireless Communication Technology*, Honolulu, HI, October 15–17, 2003.

[26]   I. Khan, Y. I. Nechayev, K. Ghanem, and P. S. Hall, 'BAN-BAN interference rejection with multiple antennas at the receiver', *IEEE Transactions on Antennas and Propagation*, vol. 58, no. 3, March 2010, pp. 927–934.

[27]   G. J. Foschini and M. J. Gans, 'On limits of wireless communications in fading environment when using multiple antennas', *Wireless Personal Communications*, vol. 6, March 1998, pp. 311–335.

[28]   D. Neirynck, C. Williams, A. Nix, and M. Beach, 'Exploiting multiple-input multiple-output in the personal sphere', *IET Microwaves, Antennas and Propagation*, vol. 1, no. 6, December 2007, pp. 1170–1176.

[29]   H. Ozcelik, M. Herdin, R. Prestros, and E. Bonek, 'How MIMO capacity is linked with single element fading statistics', *Proceedings of International Conference on Electromagnetics in Advanced Applications*, Torino, Italy, September 8–12, 2003, pp. 775–778.

[30]   L. Garcia, N. Jalden, B. Lindmark, P. Zetterberg, and L. Haro, 'Measurements of MIMO indoor channels at 1800 MHz with multiple indoor and outdoor base stations', *EURASIP Journal on Wireless Communications and Networking*, vol. 2007, Article ID 28073.

[31]   I. Khan and P. S. Hall, 'Experimental evaluation of MIMO capacity and correlation for body-centric wireless channels', *IEEE Transactions on Antennas and Propagation*, vol. 58, no. 1, January 2010, pp. 195–202.

[32]   T. Svantesson and J. Wallace, 'On signal strength and multipath richness in multi-input multi-output systems', *Proc. IEEE International Conference on Communications*, May 2003, vol. 4, pp. 2683–2687.

[33]   H. Carrasco, R. Feick, and H. Hristov, 'Experimental evaluation of indoor MIMO channel capacity for compact arrays of planar inverted-F antennas', *Microwave and Optical Technology Letters*, vol. 49, no. 7, July 2007, pp. 1754–1756.

[34]   J. S. Colburn, Y. Rahmat-Samii, M. A. Jensen, and G. J. Pottie, 'Evaluation of personal communications dual-antenna handset diversity performance', *IEEE Transactions on Vehicular Technology*, vol. 47, August 1998, pp. 737–746.

[35]   R. E. Jaramillo, O. Fernandez, and R. P. Torres, 'Empirical analysis of 2 × 2 MIMO channel in outdoor-indoor scenarios for BFWA applications', *IEEE Antennas and Propagation Magazine*, vol. 48, no. 6, December 2006, pp. 57–69.

[36]   A. Goldsmith, *Wireless Communications*, New York: Cambridge University Press, 2005.

[37]   S. M. Alamouti, 'A simple transmit diversity technique for wireless communications', *IEEE Journal on Selected Areas in Communications*, vol. 16, no. 8, 1998, pp. 1451–1458.

[38]   I. Khan, Irfanullah, P. S. Hall, 'Transmit–receive diversity for 2 × 2 MIMO channel in body area networks', *IET Microwaves, Antennas and Propagation Journal*, vol. 5, no. 13, October 2011, pp. 1589–1593.

[39]   K. Ghanem, I. Khan, and P. Hall, 'MIMO stochastic model and capacity evaluation of on-body channels', *IEEE Transactions on Antennas and Propagation*, vol. 60, no. 6, June 2012, pp. 2980–2986.

Chapter 9

On-body antennas and radio channels for GPS applications

Masood Ur-Rehman*, Xiaodong Chen† and Zhinong Ying‡

The Global Navigation Satellite System (GNSS) is providing navigation and positioning services worldwide. The European GNSS Agency (GSA) has reported usage of an estimated 4.5 billion GNSS devices worldwide in 2015. It is forecasted to increase to over 7 billion by 2019 having a market size of approximately 300 billion euros [1]. Being the only fully functional system till date, these navigation devices rely primarily on the Global Positioning System (GPS).

Ever-growing demand of navigation and positioning facilities to be available in portable devices has made the GPS antennas an essential part of the modern Wireless Personal Area Networks (WPAN) and Wireless Body Area Networks (WBAN) applications. A Federal Communications Commission (FCC) adoption to enhance the provision of emergency services by tracking the user's location through his mobile also necessitates the integration of the GPS to the cellular phones [2]. The WPAN/WBAN works in collaboration with the GPS to provide these services. GPS enabled wearable devices range from Personal Navigation Devices (PNDs), wrist watches, wristband activity trackers, headband fitness trackers, smart shoe soles and most importantly the smart phones. These devices typically operate either in on-body positions (held by the user) or in near-body positions (working in the proximity of the human body). In either case, the presence of the human body in the vicinity, degrades the performance of the embedded GPS antennas of such devices.

It is now a well-established phenomenon that the human body is a very lossy medium that affects the performance of the antenna in three ways: reduction in efficiency due to electromagnetic absorption in the tissues, degradation of the radiation pattern and variation in the feed point impedance [3–9]. Characterisation of these effects is a challenging but necessary task to provide guidelines for the design of an optimal performance antenna resilient to these degrading factors [10–13].

*Centre for Wireless Research, University of Bedfordshire, Luton (UK)
†School of Electronic Engineering and Computer Science, Queen Mary University of London
‡Network Technology Lab., Research and Technology, Sony Mobile Communications Mobilvägen, Lund, Sweden

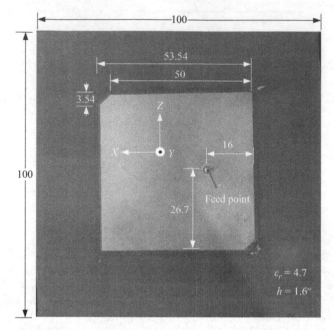

Figure 9.1 *Schematic layout of the truncated corner microstrip patch antenna for GPS operation at 1575.42 MHz fed by a coaxial port ©2012 IEEE. Reprinted with permission from Reference 16*

9.1 GPS antennas in the presence of human body

A simple truncated corner microstrip patch (TC patch) antenna has been used to study the fundamental phenomena inflicting performance changes due to the human body presence [14, 15]. The antenna design has a ground plane of 100 mm × 100 mm with a printed square radiating patch of 53.4 mm × 53.4 mm and fed by a coaxial port. A FR4 substrate of 1.6-mm thickness and $\varepsilon_r = 4.7$ is used. Figure 9.1 shows the geometry of the antenna.

Antenna performance is analysed via simulations and validated through measurements. The comparison of the simulated and measured S_{11} curves of the antenna are shown in Figure 9.2. A good agreement between the two has been observed. The highlighted area shows ±5 MHz impedance bandwidth region typically desired for a good performing GPS antenna [17, 18]. The antenna performs well in L1 band with centre frequency at 1578 MHz. The antenna has a −10 dB bandwidth of 25 MHz covering frequencies in the range of 1566–1591 MHz.

Since, the microstrip patch antenna radiates normal to its patch surface, the gain patterns for both *XY* plane and *YZ* plane are of importance. The simulated and measured 2-D gain patterns in *XY* plane and *YZ* plane are shown in Figure 9.3. A good agreement can be found between the simulated and measured patterns. The small differences are due to the fabrication imperfections. Figure 9.4 illustrates the

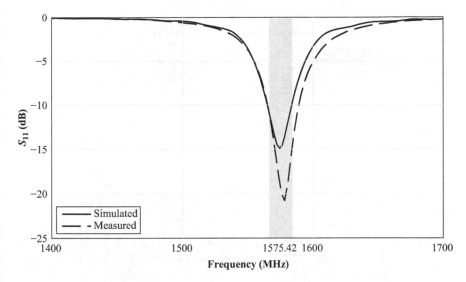

Figure 9.2   Simulated and measured S_{11} curves for the truncated corner microstrip patch GPS antenna in free space

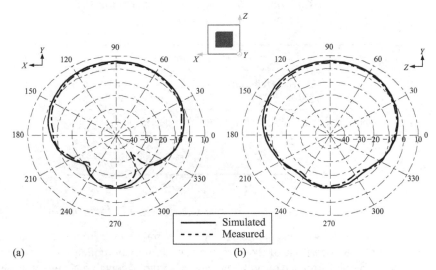

(a)   (b)

Figure 9.3   Simulated and measured 2-D gain patterns in XY and YZ planes of the truncated corner microstrip patch GPS antenna in free space at 1575.42 MHz. (a) XY plane and (b) YZ plane

simulated 3-D RHCP and LHCP gain patterns of the antenna. These results confirm that the antenna performance is excellent for the GPS operation with good RHCP and a small back side radiation of order −20 dB, in accordance with the specified requirements in classic literature.

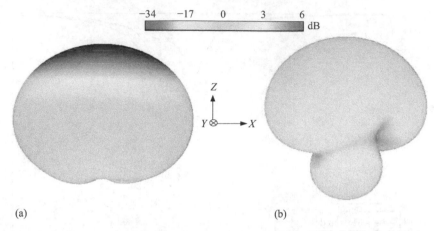

Figure 9.4 *Simulated 3-D RHCP and LHCP gain patterns of the truncated corner microstrip patch GPS antenna in free space at 1575.42 MHz. (a) RHCP and (b) LHCP*

Table 9.1 *Electric properties of specific human body tissues used within the constructed homogeneous body model at 1575.42 MHz*

Tissue	Electric properties	
	Dielectric constant (ε_r)	Tissue conductivity (σ) (S/m)
Bone	19.65	0.52
Fat	5.37	0.07
Muscle	53.83	1.22
Skin	39.28	1.09

9.1.1 *Experimental set-up for on-body antenna performance*

The internal organs play a very insignificant role at microwave frequencies due to small skin depth. It allows use of the homogeneous numerical models of the human body to study the performance of on-body antennas with acceptable accuracy levels [7, 9, 19–23]. Therefore, electromagnetic interaction between the human body and the GPS antenna is studied using a single-layer numerical model of the human body. Simulations were carried out in CST Microwave Studio®. The weighted averaged tissue properties have been adopted to develop the model. The homogeneous human body model is therefore, considered as a compound with 15% skin, 20% fat, 45% muscle and 20% bone, which resulted in an averaged relative permittivity of 35.12 and conductivity of 0.83 S/m at 1575.42 MHz. The dielectric properties of the human body tissue are taken as described in References 24–26. The values for the four types of tissues at 1575.42 MHz used in this study are given in Table 9.1.

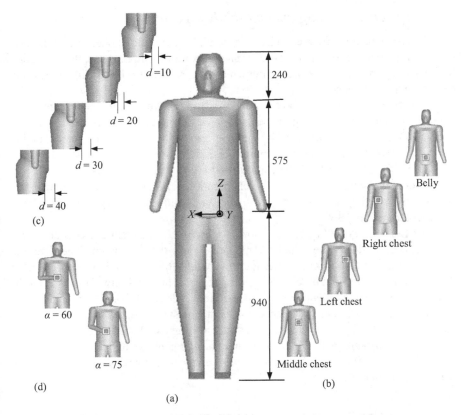

Figure 9.5   Human body model and different on-body configurations of the truncated corner microstrip patch antenna used to study the effects of the human body presence on the GPS antenna (all lengths are in mm). (a) Human body model, (b) antenna with varying on-body positions, (c) antenna with varying distances along the chest and (d) antenna held-in-hand

A realistic complete body model with a height of 1755 mm representing an average physique human user is used in this study. This model offers not only faster computations as a result of reduced complexity but also gives flexibility in terms of re-positioning the body parts [27, 28]. Effects of the presence of the human body on the GPS receiver antenna are studied for different configurations considering varying separations of the body and the antenna, different positions of the antenna on the body and different body postures shown in Figure 9.5.

An adaptive mesh is employed with different cell sizes that has reduced the number of volume cells (voxels) in the computational domain significantly. The Perfectly Matched Layer (PML) absorbing boundary conditions [29] are used with a maximum cell size of 10 mm near the boundaries and a minimum size of 0.08 mm at the edges of the solids in the computational region.

9.1.2   *Effects of varying antenna-body separation*

The GPS antennas in portable devices, especially the mobile phones, are often integrated with other radio communication systems. For the navigation use, the separation between the portable device and the human body can change which results in varying the antenna performance.

The effects of the human body on the performance of the microstrip patch GPS antenna are studied by varying the separation (d) along the chest, illustrated in Figure 9.5. Different gaps between the antenna and the body are considered ranging from $d = 10$ mm to $d = 120$ mm. Figures 9.6 and 9.7 provide a comparison of the antenna performance in free space and in the presence of the human body at these separations. A summary of various antenna parameters for different simulated configurations including f_c, bandwidth, total efficiency and maximum gain (in XY plane) is presented in Table 9.2.

The S_{11} curves in Figure 9.6 show a shift in the resonance frequency from 1575.42 MHz to lower frequencies depending upon the separation. The detuning of the antenna is caused due to the fact that while the antenna is placed on-body, the electromagnetic field produced in the space near human body contains both the fields induced by the antenna itself and the fields reflected from the body surface. These reflected fields induce currents on the antenna surface disturbing the free space distribution. It changes the antenna impedance and, hence, detunes the resonance frequency. A maximum drop of 49% in the antenna efficiency (simulated), compared to that in free space (93%), has been observed when the antenna is placed at $d = 10$ mm from the human body.

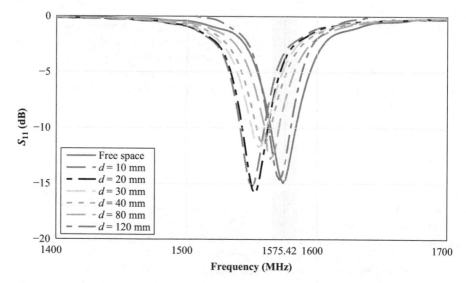

Figure 9.6   *Comparison of the simulated S_{11} responses of the truncated corner microstrip patch GPS antenna for various antenna-body separations (d) along the chest*

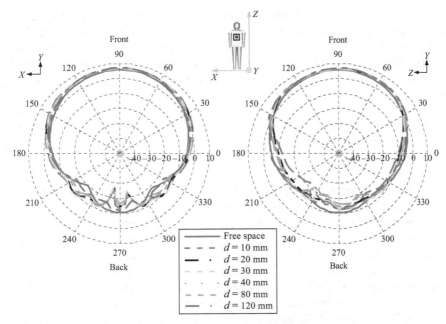

Figure 9.7   Comparison of simulated 2-D gain patterns in XY and YZ planes of the truncated corner microstrip patch GPS antenna as a function of the antenna distance (d) from the body along the chest at 1575.42 MHz

Table 9.2   Comparison of different simulated parameters of CP patch antenna analysed for various body-worn configurations working at the GPS frequency of 1575.42 MHz

Antenna body-worn configuration		f_c (MHz)	BW (MHz)	η_t (%)	Gain in XY plane dBi
Antenna with no body presence (free space)		1575	19	93	5.8
Varying separation of antenna and body	$d = 10$ mm	1550	19	44	6.3
	$d = 20$ mm	1552	20	48	6.6
	$d = 30$ mm	1555	12	53	6.5
	$d = 40$ mm	1560	11	55	6.7
	$d = 80$ mm	1564	13	68	7.1
	$d = 120$ mm	1571	17	79	6.2
Varying antenna position on-body	Middle chest	1550	19	44	6.3
	Right chest	1549	17	45	6.9
	Left chest	1549	17	46	7.1
	Near belly	1537	13	16	3.7
Varying hand-held antenna position	$\alpha = 60°$	1554	8	51	7.0
	$\alpha = 75°$	1557	9	53	6.4

It appears that greater separations between the antenna and the human body tend to improve the antenna performance. At $d = 40$ mm, the antenna resonates at 1560 MHz as compared to 1550 MHz when $d = 10$ mm. It could be observed from the presented results that further the antenna from the body, the closer is its resonance to that in free space. The antenna efficiency also shows improvement. With $d = 120$ mm, the antenna exhibits very close performance to the free space operation with resonating at 1571 MHz having an efficiency of 79%. This enhanced performance is caused by change in the amount of the reflected fields by changing the antenna-body gap. At larger separations, the effective permittivity of the medium becomes closer to the value for the free space that reduces the extent of frequency detuning and the radiation deformation.

Figure 9.7 demonstrates the antenna performance from radiation perspective in both XY and YZ planes. The antenna radiation is compared for the absence of the human body to that in its presence, as a function of the antenna and the human body separation, along the chest. These results show that the human body presence deforms the antenna radiation patterns substantially in both planes due to power absorption in the lossy tissues. It also causes increased gain levels and reduced antenna efficiency. The antenna pattern shape keeps constant with increasing separations from the body. However, radiated power increases in front direction, away from the body because of greater reflected waves. An increase of 0.5–1.3 dB is noted in the antenna gain in front direction while it reduces by up to 5 dB in backward direction. It is also observed that the antenna radiation patterns have a tendency of improvement by getting closer to free space performance as value of d increases from 10 mm to 120 mm due to reduced losses in the human body tissues. Also, large ground plane used for the matching and optimised performance of the antenna has played a part to reduce the pattern deformations by shielding the antenna from some of the reacting field reflected by the body at closer gaps. These effects would have a greater impact on antennas with smaller or no ground plane.

It could also be noted from Figure 9.7 that antenna radiation characteristics are affected more in YZ plane as compared to XY plane. It is a result of the presence of greater body mass with increased losses in the tissues since, the height of the torso is larger than the width as shown in Figure 9.5.

9.1.3 Dependency on on-body GPS antenna position

The portable devices are commonly placed at different positions near the human body, for example in the pocket of a shirt (near chest) or in the pocket of a jacket (near belly). Difference in the shape of the body parts at different locations could influence the GPS antenna to perform with varying radiation characteristics. The effects of varying on-body placements of the antenna are also analysed to study the change in the antenna performance.

The microstrip patch GPS antenna is placed at different positions on-body while maintaining a gap of 10 mm between the antenna and the body to allow the covering assembly clearance. The considered on-body positions include the antenna placed at

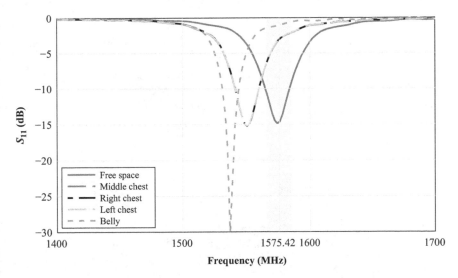

*Figure 9.8   Comparison of simulated S_{11} of the truncated corner microstrip patch
            GPS antenna for different antenna positions on-body*

the middle chest, the right chest, the left chest and right waist near the belly. The
results are compared to the antenna performance in the absence of the human body.

The S_{11} curves in Figure 9.8 depict that resonance shifts from 1575.42 MHz to
lower frequencies depending upon the placement of the antenna on the body. The
effect on the antenna impedance is nearly equal for the three positions along the
chest. An average drop of 48% in the antenna total efficiency, compared to the free
space efficiency, is observed. However, the antenna is detuned to a larger extent and
centre frequency comes down to 1537 MHz when placed near the belly. The antenna
efficiency also experiences a huge drop reaching to 16%. It is because the presence
of discontinuities towards legs and arms near the belly changes size and shape of
the lossy tissue more as compared to the three chest positions. Hence, the resulting
modification of the effective medium causes larger variation in the antenna input
impedance.

The gain patterns shown in Figure 9.9 confirm that the radiation characteristics
of the antenna are also affected but the shielding provided through the ground plane
minimises this effect. The pattern shapes are again similar for the three chest positions
with reduced back lobes. The discontinuities towards legs and arms near the belly have
also caused scattered fields with greater pattern deformations.

9.1.4   Effects of body posture

The portable devices, especially the GPS navigators, are usually used in held-in-hand
scenarios with the user watching the screen. The direct contact of the human hand with
the GPS antenna affects the radiation properties of the antenna resulting in a reduced
performance. The presented microstrip patch GPS antenna is tested to demonstrate

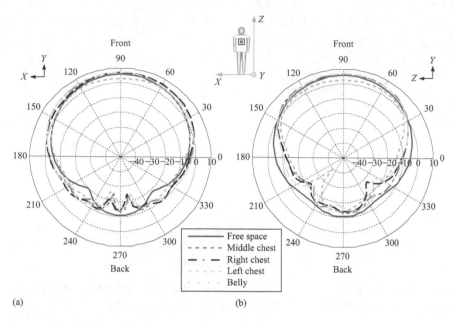

Figure 9.9   Simulated 2-D gain patterns in XY and YZ planes for the truncated corner microstrip patch GPS antenna with effects of variation in on-body antenna position at 1575.42 MHz

the human body effects on its performance for held-in-hand scenarios. Although a realistic hand model (constructed in later studies) could not be added to the designed structure due to the model limitations, the human body model is modified to represent the user's watching position with the antenna held in the left hand. The bending angle of the arm is represented by α. Two held-in-hand configurations with the arm bent at an angle of $\alpha = 60°$ and $\alpha = 75°$ are considered, and results are compared with the antenna in the absence of the human body and antenna placed on-body at the middle chest ($\alpha = 0°$). The three configurations have the separation between the body and the antenna equal to 10 mm, 190 mm and 350 mm, respectively, as illustrated in Figure 9.5.

The antenna impedance performance is compared in Figure 9.10. The presence of the human body has detuned the antenna causing the resonance frequency to shift from 1575 MHz to lower values. The detuning of the antenna again depends upon the separation of the antenna and the human body. The resonance gets closer to the one observed for the antenna in free space scenario with wider bending angles as antenna becomes less affected by the torso. It also improves the antenna efficiency to 51% in comparison to the free space value with $\alpha = 75°$ while it is 53% when $\alpha = 60°$ as drop is minimised. However, the antenna impedance undergoes detuning in the two cases due to the currents induced in hand tissues causing the drop in the efficiency.

The antenna radiation performance is compared in Figure 9.11. The reflected fields again have caused changes in the gain and introduced deformations in the

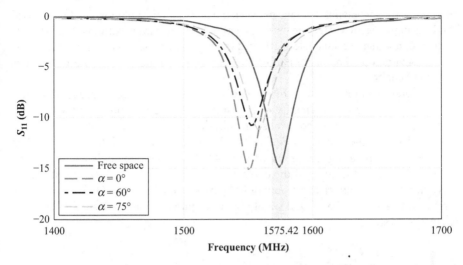

Figure 9.10   Simulated S_{11} responses of the truncated corner microstrip patch GPS antenna for hand-held configurations with arm bent at three different angles (α)

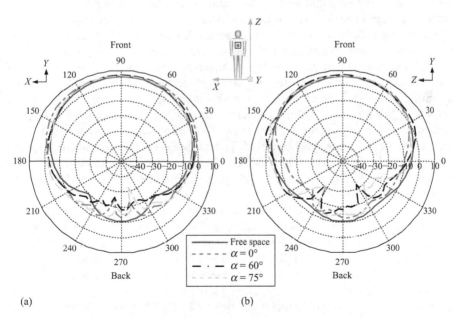

(a)   (b)

Figure 9.11   Comparison of 2-D gain patterns in XY and YZ planes of the truncated corner microstrip patch GPS antenna for hand-held configurations with arm bent at three different angles (α) at 1575.42 MHz. (a) XY plane and (b) YZ plane

polar patterns. Increasing distance between the body and the antenna due to increased bending angle has modified the patterns with high backward radiations. It is because of the fact that the antenna is less affected by the greater body mass of torso and legs at these separations. The major part of the degradation in these configurations comes from the bent arm.

The presented results confirm that the GPS antenna undergoes frequency detuning while operating in the vicinity of the human body depending upon the on-body antenna placement, present body mass and physiological parameters of the body. This tends to decrease the resonant frequency of the antenna, causing it to be mismatched at its intended operating frequency of 1575.42 MHz. The antenna loses the desired ±5 MHz bandwidth in most of the cases whereas its radiation pattern also deforms. It is, therefore, evident that the human body presence affects the GPS antenna performance to a visible extent and should be taken into account to design an efficient navigation system.

9.2   On-body GPS antennas in real working environment

The discussion in the above sections has established that the performance of the GPS antennas tends to deteriorate while placed on/near the human body. This study, however, does not take into account the nature of the surrounding environment. In practical scenarios, these antennas have to operate in cluttered urban/sub-urban areas. The incoming radio wave undergoes reflections, diffractions and scattering from the surrounding objects including buildings, vehicles, vegetation and ground in these difficult environments resulting in the multipath phenomenon, as illustrated in Figure 9.12 [20, 21]. It increases the magnitude of degradations in the GPS antennas operating in the vicinity of the human body significantly resulting in attenuation, delay and distortion of the communication link. To guarantee a reliable navigation system that can also meet the FCC's mandate E911 of provision of precise location of the mobile user to ensure public safety [2] necessitates that the GPS antennas should cope with both of these degrading factors. Therefore, these antennas must be tested not only for the effects of the human body presence but also for their performance in the multipath environment.

In ideal conditions, the GPS antennas are required to exhibit good Right Hand Circular Polarisation (RHCP) with a uniform radiation pattern over entire upper hemisphere in order to receive the incoming GPS signal efficiently. Theoretically, multipath is needed to be avoided through good rejection of Left Hand Circular Polarisation (LHCP) [30–32]. Fulfilment of these requirements in portable devices is, however, a very challenging task which have to allow maximum mobility of the user and flexibility of use offering multiple functions such as Wi-Fi, Bluetooth, FM radio, digital camera, mobile TV and GPS [18, 33].

Portable GPS antennas are commonly used in cluttered environments such as city streets, vehicles and indoors. In these working scenarios, the incoming Line-of-Sight (LOS) GPS signals are weak while the polarisation of the reflected signals is arbitrary and unknown. Furthermore, these hand-held or on-body devices are rarely used in a fixed position varying the "up" direction of the embedded GPS antenna

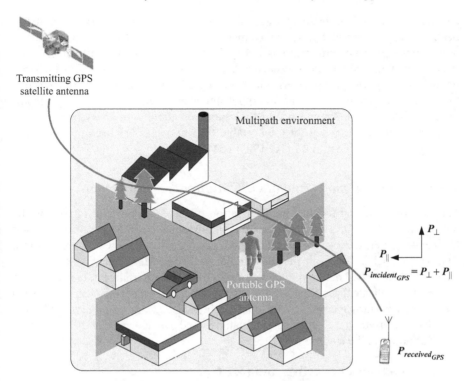

*Figure 9.12   GPS environment and reception of multipath signal by GPS mobile
terminal antenna operating near human user*

with the orientation of the device. Presence of the human body also increases the
losses due to electromagnetic absorptions and blocking the clear sky view of the
antenna. Paramount task of establishing the GPS link for these on-body antennas is
therefore, a very challenging task. Despite the potential losses due to polarisation
mismatch, linear polarised GPS antennas with wide-beam coverage have found to
be more efficient in overcoming these problems as compared to the conventional
RHCP antennas [34–36]. The multipath signal can be used constructively to establish
a quick GPS link. The resulting positioning errors can be removed using widely
available software approaches [37–40]. It requires detailed characterisation of the
GPS antenna performance in the portable devices.

The performance assessment of a portable GPS antenna through field tests suffers
from longer test durations that have uncontrollable factors of weather, temperature
and location. It causes lack of accuracy due to poor repeatability and efficiency.
The statistical modelling of the multipath environment provides a good alternative
to the field tests. It can estimate the antenna performance efficiently with realistic
replication of the actual working conditions while avoiding all these drawbacks.

The impact of the antenna on the wireless link budget in response to the nature
of the surrounding environment can be described statistically using Mean Effective
Gain (MEG). It estimates the antenna performance considering Angle of Arrival

and polarisation of the incident waves and antenna gain patterns [41–44]. It has been used as an important performance metric for mobile handsets in multipath land mobile propagation environments [45, 46]. This method can efficiently incorporate the effects of the human body presence as the calculations of antenna MEG are based on the antenna power gain patterns, where the degradations in gain and efficiency due to radiation pattern deformations and input impedance variations are easily accommodated [45, 47–49]. An enhanced MEG approach along with a new parameter of antenna coverage efficiency for the portable GPS antennas has been proposed and developed by Ur-Rehman *et al.* [16, 19, 50, 51].

9.2.1   Statistical modelling of GPS multipath environment

The statistical method for the characterisation of the environmental factors on the performance of the GPS antennas uses the parameters of GPS MEG (MEG_{GPS}) and GPS coverage efficiency (η_c) to evaluate the performance of the GPS antennas [16, 51]. The model is based on the re-formulation of the MEG equation derived by Taga [41]. The RHCP nature of the incoming waves and environmental reflections is included to replicate the multipath GPS environment. The model efficiently estimates the expected performance of the GPS antennas in realistic working conditions through calculations based on 3-D antenna gain patterns and average angular distribution of incident power in the environment.

9.2.1.1   GPS mean effective gain (MEG_{GPS})

The average gain of the antenna performance in multipath radio environment is termed as MEG. It is ratio of total mean received power of the antenna and total mean incident power in the same environment [41]:

$$MEG = \frac{\text{Mean received power }(P_{received})}{\text{Total mean incident power }(P_{incident})} \tag{9.1}$$

For the GPS operation, the incident radio wave is circularly polarised. Therefore, it could be split into perpendicular and parallel polarised components. Using the spherical coordinates (Figure 9.13), $P_{received}$ can be expressed as follows:

$$P_{receivedGPS} = \int_0^{2\pi} \int_0^{\pi} \left[P_\perp G_\perp(\theta,\phi) p_\perp(\theta,\phi) + P_{||} G_{||}(\theta,\phi) p_{||}(\theta,\phi) \right] \sin\theta \, d\theta \, d\phi \tag{9.2}$$

where $P_\perp$ and $P_{||}$ are the mean received powers in the perpendicular and parallel polarisations with respect to the ground plane while the angular density functions of the incident waves in the two polarisations (defined by AoA_{GPS}) are denoted by $p_\perp(\theta,\phi)$ and $p_{||}(\theta,\phi)$ (as shown in Figure 9.12), respectively. $G_\perp$ and $G_{||}$ represent the antenna power gains for the perpendicular and the parallel polarisations, respectively. The total mean incident power ($P_{incident}$) arriving at the antenna would be the summation of the mean powers in the two polarisations.

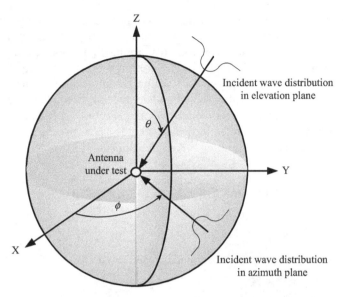

Figure 9.13   Spherical coordinates system and representation of a hypothetical incident wave distribution model for the GPS multipath environment ©2012 IEEE. Reprinted with permission from Reference 16

The ratio between the powers received in the two polarisations is referred as cross polarisation ratio (*XPR*) and expressed as:

$$XPR = \frac{P_\perp}{P_\parallel} \tag{9.3}$$

Based on this discussion, the MEG expression for the GPS multipath environment has been formulated as [16, 51]:

$$MEG_{GPS} = \int_0^{2\pi} \int_0^\pi \left[\frac{XPR}{1+XPR} G_\perp(\theta,\phi) p_\perp(\theta,\phi) \right.$$
$$\left. + \frac{1}{1+XPR} G_\parallel(\theta,\phi) p_\parallel(\theta,\phi) \right] \sin\theta d\theta d\phi \tag{9.4}$$

In this model, *XPR* governs the polarisation of the incoming wave. Since the transmission of two perpendicular and parallel polarised waves simultaneously, out of phase by $\pi/2$ (radian), generates a circularly polarised wave, *XPR* = 0 *dB* has been used to accommodate the circular polarisation of the incident GPS wave [16, 19].

9.2.1.2   GPS angle of arrival (*AoA*_{GPS})

Formulation of the GPS multipath environment requires statistical definitions of the direction (angle) of arrival (*AoA*_{GPS}) of the incoming waves in azimuth and elevation planes.

The statistical model assumes a uniform angular density function in azimuth plane due to the random occurrence of the incident waves caused by reflections, diffractions and scattering from geometrically varying objects located in the vicinity of the receiving antenna [16, 42, 43]. The elevation plane has been divided into incident and reflection regions. Incident region assumes uniform angular density function while the reflection region accommodates the ground reflections by reducing the angular density function in accordance with the reflection coefficients. Sum of the received powers in the incident and reflected regions has been considered to use the multipath signal favourably for quick link establishment [16]:

$$p_\perp(\theta, \phi) = \begin{cases} 1 & 0 \le \theta \le \pi/2 \\ A(\theta)_\perp & \pi/2 \le \theta \le \pi \end{cases} \tag{9.5}$$

$$p_\parallel(\theta, \phi) = \begin{cases} 1 & 0 \le \theta \le \pi/2 \\ A(\theta)_\parallel & \pi/2 \le \theta \le \pi \end{cases} \tag{9.6}$$

Here, $A(\theta)$ depends upon the reflection coefficients for the perpendicular and parallel components as [52]:

$$A(\theta_i)_\perp = \frac{\cos\theta_i - \sqrt{(\varepsilon_2/\varepsilon_1) - \sin^2\theta_i}}{\cos\theta_i + \sqrt{(\varepsilon_2/\varepsilon_1) - \sin^2\theta_i}} \tag{9.7}$$

$$A(\theta_i)_\parallel = \frac{(\varepsilon_2/\varepsilon_1)\cos\theta_i - \sqrt{(\varepsilon_2/\varepsilon_1) - \sin^2\theta_i}}{(\varepsilon_2/\varepsilon_1)\cos\theta_i + \sqrt{(\varepsilon_2/\varepsilon_1) - \sin^2\theta_i}} \tag{9.8}$$

The model calculations are based on the consideration of an open field ground to be of a semi-grassy semi-concrete type with a relative permittivity of 4.5 [53, 54].

9.2.1.3   GPS coverage efficiency (η_c)

The GPS antennas receive signals from all directions that lie within its coverage of the open sky. Conventional techniques, however, consider the performance of a GPS antenna by its capability to receive the signals for elevation angles higher and lower than 10° (from the horizon) [31]. To overcome this problem, the statistical model also evaluates the performance of the GPS antenna in terms of its coverage efficiency that defines the ability of the antenna to receive the direct satellite signals in whole upper hemisphere. It is calculated as [16, 51]:

$$\eta_c = \frac{\text{Coverage area}}{\text{Total area}} \tag{9.9}$$

Here, the coverage area represents the solid angle subtended by the directions in which the GPS signal is above −13 dBi threshold (for a GPS receiver sensitivity of 145 dBi) in the incident region ($0 \le \theta \le \pi/2$). Signals below this level are considered too weak to make an impact and hence, wasted. The maximum coverage that can be obtained by a reference GPS antenna has been considered as the total area. It represents the half hemispherical solid angle of 2π for an isotropic antenna. Figure 9.14 illustrates η_c calculations. The box encloses the incident region (upper hemisphere)

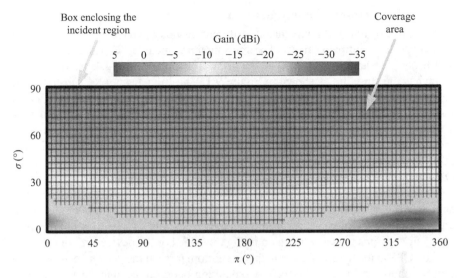

Figure 9.14 *Illustration of η_c calculations based on RHCP radiation pattern of a GPS antenna with cross-hatched regions indicating coverage area (where signal is above $-13\,dBi$) ©2012 IEEE. Reprinted with permission from Reference 16*

with the horizon at $0°$ and the zenith at $90°$. Cross-hatched part indicates the coverage area. These calculations are based on RHCP gain pattern of the antenna-under-test to suit the RHCP incoming radio waves.

9.2.2 On-body GPS antennas in multipath environment

The statistical model discussed in the preceding sections is employed to investigate the performance of the on-body GPS antennas in multipath environment. Different common possibilities of the on-body positioning of the GPS antennas are considered including antenna placed in the user's pocket, antenna held-in-hand used in watching position and antenna held-in-hand placed beside the ear in talking-on-phone position.

The performance of the on-body GPS antennas is characterised in terms of MEG_{GPS} and η_c. The 3-D power gain patterns required to perform the calculations using the statistical model only considers the power absorptions in the human body tissues [29]. To include the mismatch losses caused by the antenna detuning in the on-body configurations, the "Realised Gain" is considered calculated using the following expression: using the following equation [55]:

$$Gain_{realised} = Gain \times \eta_m \tag{9.10}$$

where *Gain* is the simulated gain taking the material losses into account, and η_m represents the antenna mismatch efficiency accounting for the antenna detuning losses. η_m is calculated by:

$$\eta_m = 1 - |S_{11}|^2 \tag{9.11}$$

9.2.2.1   Experimental configuration

Performance of four GPS antennas in different on-body placements is evaluated to investigate the multipath effects. The four antennas include TC patch, PIFA, DRA and helix. The antennas operate at the GPS frequency of 1575.42 MHz. The TC patch has the structure as shown in Figure 9.1. The PIFA has a FR4 substrate with 1.6 mm of thickness and PCB size of 100 mm × 40 mm. The DRA antenna is loaded with a dielectric of $\varepsilon_r = 21$ which is covered with lossy silver. The ground plane is 100 mm × 40 mm of lossy copper type. The helix is mounted on the left side of the metallic ground plane of 100 mm × 40 mm × 0.45 mm. The antennas are fed using 50 Ω coaxial port feed. Figures 9.15–9.17 show the schematic layout of the PIFA, DRA and helix antennas.

The realistic homogeneous human body model discussed in Section 9.1.1 with similar mesh scheme is used in this study. The flexibility of the model allowed to add a realistic hand model to realise the antenna held-in-hand scenarios and represent different body postures as illustrated in Figure 9.18. For the device kept in the user's pocket, GPS antennas are placed on the left bottom torso at the pocket position on the body as shown in Figure 9.18(b). The separation between the antenna and the body is kept at 8 mm to allow the clearance for the mobile casing.

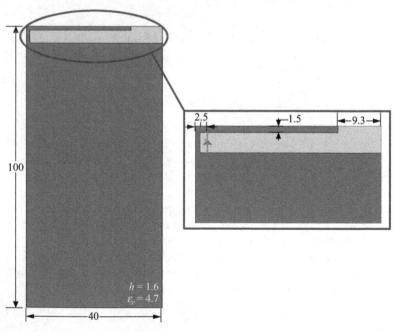

Figure 9.15   Geometrical structure of GPS mobile terminal PIFA antenna used to study the effects of human body presence on the antenna performance in multipath environment (all lengths are in mm) ©2012 Wiley. Reprinted with permission from Reference 51

The talking-on-phone scenario is configured by re-positioning the arm in such a way that the body model now has the antenna held in the user's hand beside the head, as illustrated in Figure 9.18(c). The separation between the antenna and the head is maintained at 8 mm to allow the covering assembly, while the antennas are inclined at 45° with respect to the z-axis. For the configuration where the navigation device is being watched by the user, the GPS antennas are modelled to be held in the user's hand in front of the body, at a distance of 175 mm from the body surface, depicted in Figure 9.18(d). Figure 9.19 shows the details of position and orientation of the radiating elements of the three antennas in the considered on-body placements.

9.2.2.2 Evaluation of GPS antenna performance

The performance of the on-body GPS antennas operating in the multipath environment is analysed by observing the simulated S_{11} responses and statistical model's calculations of MEG_{GPS} and η_c values.

The S_{11} response of the four antennas working without the human body presence and in different on-body placements is plotted and compared in Figures 9.20–9.23. The frequency detuning caused by the human body presence in the vicinity of the

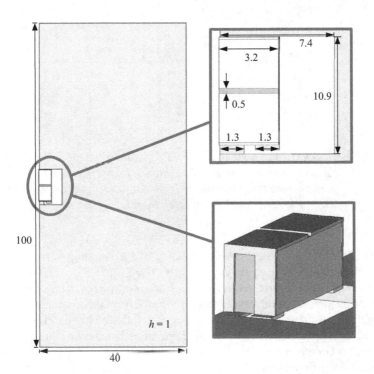

Figure 9.16 *Schematic layout of GPS mobile terminal DRA antenna to study the effects of human body presence on the antenna performance in multipath environment (all lengths are in mm) ©2012 Wiley. Reprinted with permission from Reference 51*

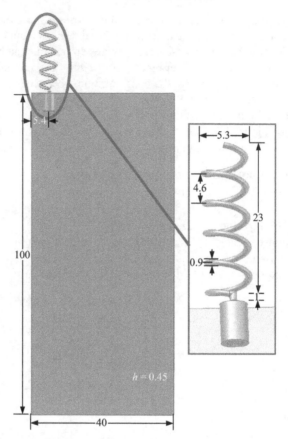

Figure 9.17   *Geometry and dimensions of GPS mobile terminal helix antenna for*
the study of the presence of the human body effects on the antenna
performance in multipath environment (all lengths are in mm) ©2012
Wiley. Reprinted with permission from Reference 51

antennas is evident from these results. Amount of the detuning varies for the four
antennas depending on the on-body placement. The TC patch is worst-hit due to the
human body presences in all three on-body positions considered. It has lost the −10 dB
impedance bandwidth of ±5 MHz (desired for efficient GPS operation) completely
with highest shift in the resonance frequency. It is followed by the PIFA, which also
has failed to exhibit the desired −10 dB impedance bandwidth, though the extent
of detuning is comparatively less than that observed for the TC patch. Although the
DRA has also suffered from completely missing the required −10 dB impedance
bandwidth in the watching and talking positions, it managed to achieve a −10 dB
impedance bandwidth of 5 MHz in the frequency range of 1570–1575 MHz in the
placed at pocket position. The helix antenna has appeared to be the best performer
having least detuning for the on-body operation. It has not only achieved the ±5 MHz

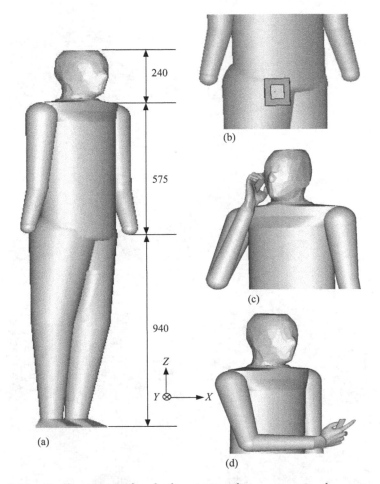

Figure 9.18   *Configuration of on-body antenna placements using homogeneous realistic human body model (all lengths are in mm). (a) Human body model, (b) at pocket, (c) talking and (d) watching*

impedance bandwidth for the GPS operation for placed at pocket and held in watching positions but also managed to attain a comparatively reasonable −8 dB impedance bandwidth of ±5 MHz in the talking position.

The greater detuning of the four antennas in watching and talking positions is primarily due to electromagnetic absorptions in the human head, palm of the holding hand and the gripping fingers. Hence, the way a user holds the antenna plays a vital role to define the extent of antenna detuning. Therefore, these GPS antennas could perform better if re-designed in such a way that the radiating element is placed on the PCB in a way that it is cleared from the palm of the hand and the gripping fingers, for example, placed at the top edge of the PCB.

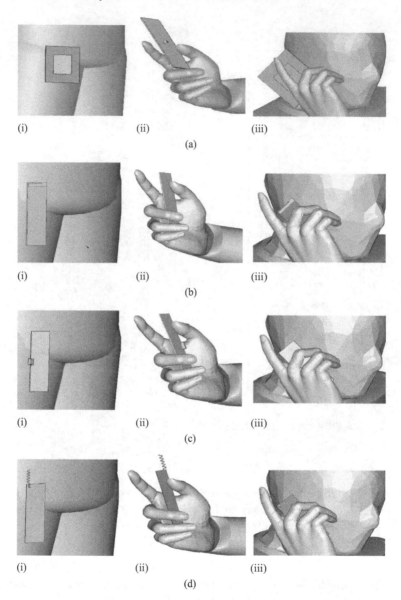

Figure 9.19 *Different on-body placements of TC patch, PIFA, DRA and helix antennas for GPS operation. (a) TC patch: (i) pocket position, (ii) watching position and (iii) talking position; (b) PIFA: (i) pocket position, (ii) watching position and (iii) talking position; (c) DRA: (i) pocket position, (ii) watching position and (iii) talking position and (d) helix: (i) pocket position, (ii) watching position and (iii) talking position*

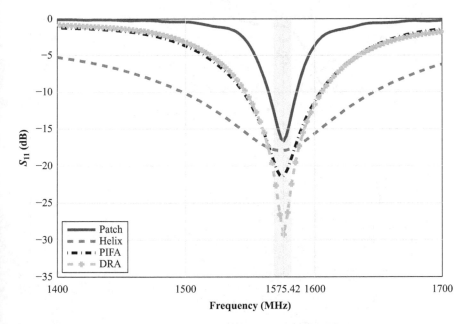

Figure 9.20 S_{11} *response of TC patch, PIFA, helix and DRA GPS antennas operating in the absence of the human body*

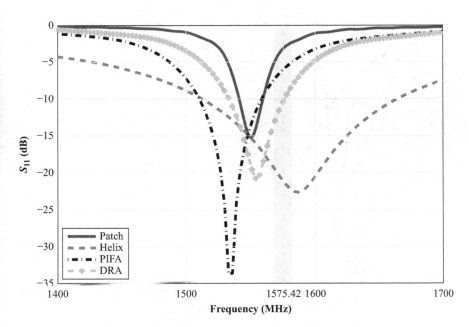

Figure 9.21 S_{11} *response of TC patch, PIFA, helix and DRA GPS antennas operating on-body placed at the pocket position*

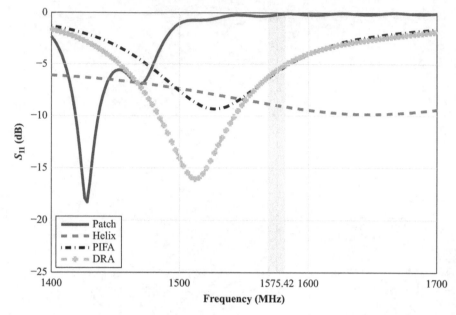

*Figure 9.22 S_{11} response of TC patch, PIFA, helix and DRA GPS antennas
operating on-body held in hand in talking-on-phone position*

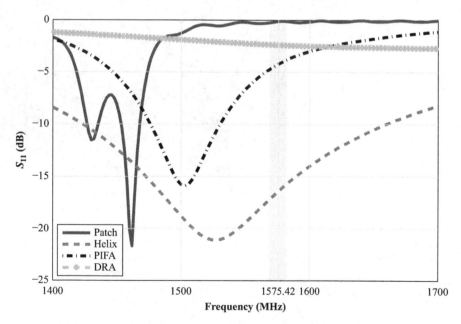

*Figure 9.23 S_{11} response of TC patch, PIFA, helix and DRA GPS antennas
operating on-body held in hand in watching position*

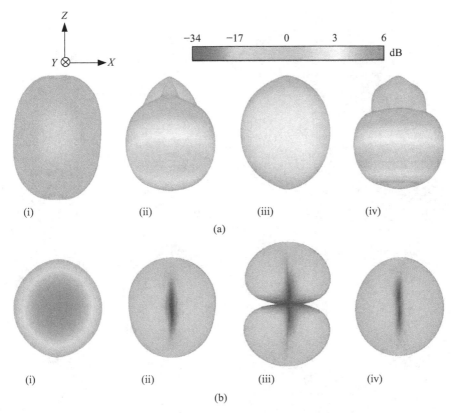

Figure 9.24 *Simulated 3-D gain patterns for perpendicular and parallel polarisations of TC patch, PIFA, DRA and helix GPS antennas operating in the absence of the human body. (a) $G_\perp$: (i) TC patch, (ii) PIFA, (iii) DRA and (iv) helix and (b) $G_\parallel$: (i) TC patch, (ii) PIFA, (iii) DRA and (iv) helix*

The 3-D simulated gain patterns of the four GPS antennas operating in the absence of the human body and in different on-body positions are also observed (Figures 9.24–9.27). These results show that the presence of the lossy human body tissues causes significant deformation of the antenna gain patterns for both perpendicular and parallel polarisations due to electromagnetic absorptions in the tissues and field reflections from the surface of the human body. When the antennas are held in hand in talking and watching positions, the radiating element of the PIFA is in the negative x-direction while that of TC patch and DRA is in the positive x-direction (Figure 9.19). It causes the PIFA to be affected more by the gripping fingers while the TC patch and DRA suffers more by the absorptions in the palm. For the helix in free space, the ground plane is a major contributor in the radiation. Therefore, it is also affected by the gripping fingers. In the talking position, the antennas are placed beside the head, and

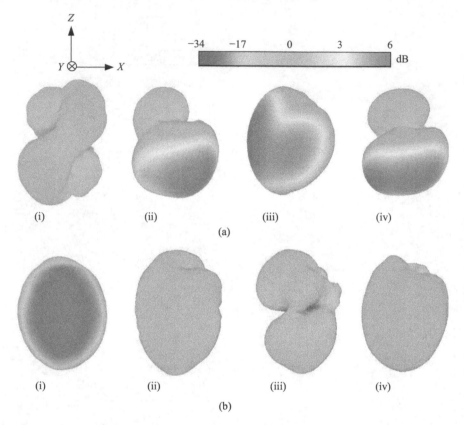

Figure 9.25   *Simulated 3-D gain patterns for perpendicular and parallel*
polarisations of TC patch, PIFA, DRA and helix GPS antennas placed
at user's pocket position. (a) $G_\perp$: (i) TC patch, (ii) PIFA, (iii) DRA and
(iv) helix and (b) $G_\parallel$: (i) TC patch, (ii) PIFA, (iii) DRA and (iv) helix

a greater body mass reduces the radiation of the four antennas in positive x-direction. It results more directive gain patterns in negative x-direction as shown in Figure 9.27. Also, presence of both the hand and the head on the two sides of the antenna in this on-body position increases electromagnetic absorptions while reflected fields are very low. It degrades the antenna radiation greatly and causes poor gain levels in all the directions. Moreover, in the three on-body positions, the gain levels are less in the upper hemisphere as compared to the lower hemisphere for the PIFA, DRA and helix antennas. TC patch has highest level of gain in parallel polarisation.

Table 9.3 presents η_c and MEG_{GPS} values (calculated based on realised gain to incorporate both the power absorption and the detuning effects caused by the human body presence). Figure 9.28 compares η_c and MEG_{GPS} for the four antennas.

An optimally performing antenna needs to exhibit a balanced performance in terms of its η_c and MEG_{GPS} [16, 19]. In the absence of the human body, TC patch

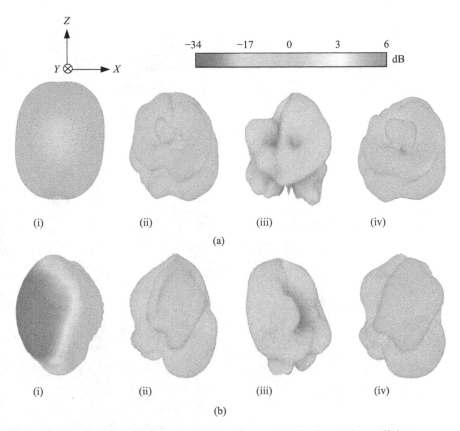

Figure 9.26 *Simulated 3-D gain patterns for perpendicular and parallel polarisations of TC patch, PIFA, DRA and helix GPS antennas held in user's hand at watching position. (a) $G_\perp$: (i) TC patch, (ii) PIFA, (iii) DRA and (iv) helix and (b) $G_\parallel$: (i) TC patch, (ii) PIFA, (iii) DRA and (iv) helix*

gives an overall best performance of the four tested antennas with η_c of 81% and of −4.9 dB. It is followed by the DRA with η_c and MEG_{GPS} values of 95% and −5.7 dB, and helix with values of 88% and −5.8 dB. PIFA has shown the weakest GPS operation with 92% of η_c and −8.2 dB of MEG_{GPS}.

The results show that frequency detuning and deterioration of the gain patterns reduce the coverage and MEG_{GPS} of the on-body GPS antennas operating in actual working environment affecting their ability to pick up the GPS satellite signal drastically. The reduction in η_c depends mainly on the extent of available clear sky view to the radiating element of the antenna. Higher shielding of the radiating element by the user gives lesser η_c. The amount of present body mass in the vicinity of the GPS antenna also plays a key role. The GPS antennas have a minimum drop of 3% in η_c and 1.5 dB in MEG_{GPS} due to the presence of the human body.

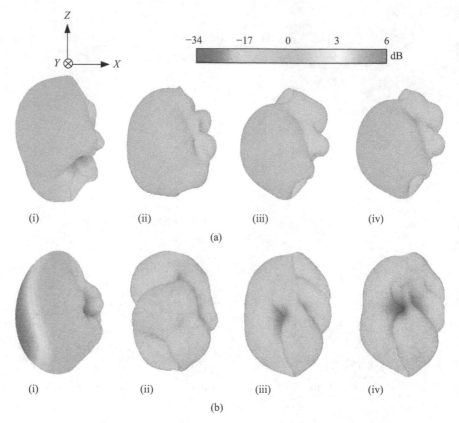

Figure 9.27   Simulated 3-D gain patterns for perpendicular and parallel polarisations of TC patch, PIFA, DRA and helix GPS antennas held in user's hand beside head in talking position. (a) $G_\perp$: (i) TC patch, (ii) PIFA, (iii) DRA and (iv) helix and (b) $G_\parallel$: (i) TC patch, (ii) PIFA, (iii) DRA and (iv) helix

The four antennas show their best on-body performance when they are placed at pocket position. A greater availability of clear view of the sky alone with enhanced gain levels in directions away from the body due to reflections from the surface of the body helps to minimise the degradations caused by the presence of the human body in this position for the PIFA, DRA and helix. On contrary, despite exhibiting higher gain levels, the TC patch suffers from the greatest level of detuning that makes it to give lowest realised gain values in this experiment. It results in a very poor performance in the three on-body placements with a maximum η_c of 15% and MEG_{GPS} of -18 dB when it is placed at the pocket position. The DRA antenna has shown the best performance in this on-body placement having a stronger radiation in upper hemisphere with η_c of 58% and MEG_{GPS} of -7.2 dB.

Table 9.3   Calculated GPS coverage efficiency and GPS MEG of four
GPS antennas working on-body in multipath environment

Antenna working scenario	Antenna type	Antenna performance in multipath environment	
		$\eta_c(\%)$	MEG_{GPS}(dB)
Antenna without human body presence	TC patch	81	−4.9
	PIFA	92	−8.2
	DRA	95	−5.7
	Helix	88	−5.8
Antenna placed on-body at pocket position	TC patch	15	−18
	PIFA	52	−10.7
	DRA	58	−7.2
	Helix	53	−7.3
Antenna held in watching position	TC patch	8	−20
	PIFA	69	−13.2
	DRA	3	−24.8
	Helix	85	−8.0
Antenna held in talking-on-phone position	TC patch	3	−22
	PIFA	34	−17.5
	DRA	17	−18.5
	Helix	20	−13.8

In the watching position, larger gap between the torso and the antenna has lessened the tissue losses but the presence of the holding hand has proved the major source of degradation in this case. The PIFA and helix have exhibited an improved η_c of 69% and 85%, respectively, as a combined result of less shielded radiating element and increased gain in the upper hemisphere because of the reflections from the palm and arm. However, this increased coverage is made less effective by a reduced MEG_{GPS} of −13.2 dB and −8.0 dB, respectively, due to pattern deformations. On the other hand, the performance of the TC patch and DRA has suffered greatly since, the radiating element is blocked to a larger extent by the palm and gripping fingers of the holding hand. The two antennas managed to provide a η_c of just 8% and 3%, respectively, with MEG_{GPS} values of −20 dB and −24.8 dB, respectively.

The four antennas have again shown a poor performance in the talking position. The electromagnetic shielding by the lossy head and hand tissues on both sides limits the GPS signal reception from all directions. The η_c values are observed to be 3%, 34%, 17% and 20% while the MEG_{GPS} is noted to be −22 dB, −17.5 dB, −18.5 dB and −13.8 dD for the TC patch, PIFA, DRA and helix antennas, respectively.

Overall, the helix has shown better capability to operate in the vicinity of the human body in the three tested scenarios, exhibiting reasonable levels of MEG_{GPS} and η_c with lesser detuning in comparison to the PIFA and DRA. The operation of the GPS on-body antennas depends not only upon the clear view of the sky, shielding body mass and antenna placement but also on the size and position of the radiating

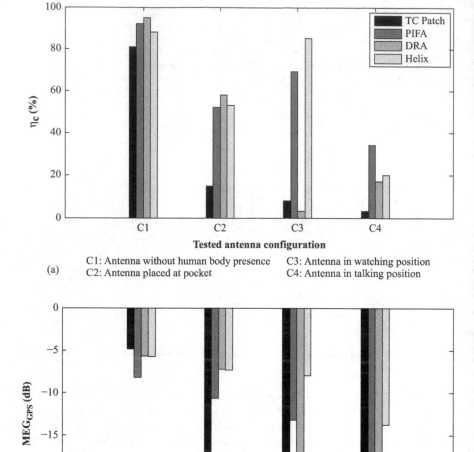

(a)

C1: Antenna without human body presence   C3: Antenna in watching position
C2: Antenna placed at pocket   C4: Antenna in talking position

(b)

C1: Antenna without human body presence   C3: Antenna in watching position
C2: Antenna placed at pocket   C4: Antenna in talking position

*Figure 9.28   Multipath environment performance of GPS antennas in the absence
of the human body and in different on-body scenarios. (a) Coverage
efficiency of GPS antennas and (b) MEG of GPS antennas*

element within the device. Moreover, CP antennas are not necessarily be the best
choice for portable navigation devices and linearly polarised antennas can outperform
them in cluttered environments. Therefore, efficient antenna designs for these devices
need to be characterised for not only free space parameters of impedance matching,

resonance frequency and radiation pattern but also for their η_c and MEG_{GPS} to predict performance in real working scenarios.

9.3 Summary

This chapter has presented the characterisation of the GPS antennas operating in different on-body scenarios. Performance of these antennas is severely affected by the multipath arrival of the GPS signal in realistic operational conditions. These effects have been investigated using a statistical model replicating the multipath GPS environment. It has been shown that the linear polarised antennas can work better for on-body GPS links in multipath environment as compared to the conventional choice of CP antennas. It has also been shown that the antenna performance depends hugely on the body posture and on-body antenna position. The model also reduces the antenna design and testing durations by removing the need of detailed, uncontrolled and lengthy open field test procedures and providing efficient and accurate predictions of GPS antenna performance in cluttered environment.

References

[1]   Z. Fu, A. Hornbostel, J. Hammesfahr, and A. Konovaltsev, "GNSS market report," *The European GNSS Agency (GSA)*, vol. 4, 2015.

[2]   "FCC's wireless 911 rules," *Federal Communication Commission (FCC)*, URL: *http://www.fcc.gov/guides/wireless-911-services*.

[3]   M. A. Jensen and Y. Rahmat-Samii, "EM interaction of handset antennas and human in personal communications," *IEEE Transactions on Antennas and Propagation*, vol. 83, no. 1, pp. 7–17, January 1995.

[4]   J. Toftgard, S. Hornsleth, and J. B. Anderson, "Effects on portable antennas of the presence of a person," *IEEE Transactions on Antennas and Propagation*, vol. 41, no. 6, pp. 739–746, June 1993.

[5]   M. Okoniewski and M. A. Stuchly, "A study of the handset antenna and human body interaction," *IEEE Transactions on Microwave Theory and Techniques*, vol. 44, no. 10, pp. 1855–1864, October 1996.

[6]   P. A. Mason, W. D. Hurt, T. J. Walters, *et al.*, "Effects of frequency, permittivity and voxel size on predicted specific absorption rate values in biological tissue during electromagnetic field exposure," *IEEE Transactions on Microwave Theory and Techniques*, vol. 48, no. 11, pp. 2050–2058, November 2000.

[7]   P. S. Hall and Y. Hao, "Antennas and propagation for body-centric wireless networks," *Artech House Publishers, London, UK*, 2012.

[8]   J. Wang and O. Fujiwara, "EM interaction between a 5 GHz band antenna mounted PC and a realistic human body model," *IEICE Transactions on Communications*, vol. E88-B, no. 6, pp. 2604–260, 2005.

[9]   G. A. Conway and W. G. Scanlon, "Antennas for over-body-surface communication at 2.45 GHz," *IEEE Transactions on Antennas and Propagation*, vol. 57, no. 4, pp. 844–855, April 2009.

[10]   M. Sanad, "Effect of the human body on microstrip antennas," *IEEE Antennas and Propagation Society International Symposium (AP-S)*, vol. 1, pp. 298–301, June 1994.

[11]   H. R. Chuang, "Human operator coupling effects on radiation characteristics of a portable communication dipole antenna," *IEEE Transactions on Antennas and Propagation*, vol. 42, no. 4, pp. 556–560, April 1994.

[12]   J. S. Colburn and Y. Rahmat-Samii, "Human proximity effects on circular polarized handset antennas in personal satellite communications," *IEEE Transactions on Antennas and Propagation*, vol. 46, no. 6, pp. 813–820, January 1998.

[13]   "Considerations for the evaluation of human exposure to electromagnetic fields (EMFs) from mobile telecommunication equipment (MTE) in the frequency range from 30MHz-6GHz," *CENELEC, European Specification, Ref. No. ES-59005:1998 E*, 1998.

[14]   L. Boccia, G. Amendola, and G. Di Massa, "Design a high-precision antenna for GPS," *Microw. RF*, vol. 42, no. 1, pp. 91–93, 2003.

[15]   M. Ur Rehman, Y. Gao, X. Chen, and C. G. Parini, "Effects of human body interference on the performance of a GPS antenna," *European Conference on Antennas and Propagation (EuCap)*, November 2007.

[16]   M. Ur Rehman, X. Chen, C. Parini, and Z. Ying, "Evaluation of a statistical model for the characterization of multipath affecting mobile terminal GPS antennas in sub-urban areas," *IEEE Transactions on Antennas and Propagation*, vol. 60, no. 2, pp. 1084–1094, February 2012.

[17]   N. Padros, J. I. Ortigosa, J. Baker, M. F. Iskander, and B. Thornberg, "Comparative study of high-performance GPS receiving antenna designs," *IEEE Transactions on Antennas and Propagation*, vol. 45, no. 4, pp. 698–706, April 1997.

[18]   R. B. Langley, "A primer on GPS antennas," *GPS World*, pp. 50–55, July 1998.

[19]   M. Ur Rehman, Y. Gao, X. Chen, C. Parini, and Z. Ying, "Environment effects and system performance characterisation of GPS antennas for mobile terminals," *IET Electronics Letters*, vol. 45, no. 5, pp. 243–245, February 2009.

[20]   M. Ur Rehman, Y. Gao, X. Chen, C. G. Parini, and Z. Ying, "Impacts of human body on built-in GPS antennas for mobile terminal in multipath environment," *European Conference on Antennas and Propagation (EuCap)*, April 2010.

[21]   M. Ur Rehman, Y. Gao, X. Chen, C. G. Parini, and Z. Ying, "Mobile terminal GPS antennas in multipath environment and effects of human body presence," *Loughborough Antennas and Propagation Conference (LAPC)*, November 2009.

[22]   M. Ur Rehman, Q. Abbasi, X. Chen, and Z. Ying, "Numerical modelling of human body for bluetooth body-worn applications," *Progress in Electromagnetics Research*, vol. 143, pp. 623–639, 2013.

[23] W. Whittow, C. Panagamuwa, R. Edwards, and J. Vardaxoglou, "On the effects of straight metallic jewellery on the specific absorption rates resulting from face illuminating radio communications devices at popular cellular frequencies," *Physics in Medicine and Biology*, vol. 53, pp. 1167–1182, 2008.

[24] C. Gabriel, "Compilation of the dielectric properties of body tissues at RF and microwave frequencies," *Brooks Air Force Technical Report, AL/OE-TR-1996-0037*, 1996.

[25] "Body tissue dielectric properties," *Federal Communication Commission (FCC)*, URL: http://www.fcc.gov/oet/rfsafety/dielectric.html.

[26] "Calculation of the dielectric properties of body tissues," *Institute of Applied Physics, Italian National Research Council*, URL: http://niremf.ifac .cnr.it/tissprop.

[27] M. Ur Rehman, Y. Gao, Y. Alfadhl, *et al.*, "Study of human body exposure to RF signal at UHF frequencies," *Joint Meeting of Bioelectromagnetics Society and the European Bioelectromagnetics Association (BioEM)*, June 2009.

[28] M. Ur Rehman, Y. Gao, Z. Wang, *et al.*, "Investigation of on-body bluetooth transmission," *IET Microwaves, Antennas and Propagation*, vol. 4, no. 7, pp. 871–880, July 2010.

[29] *CST Microwave Studio®*, 2015 *User Manual, Computer System Technology*, Germany.

[30] R. Bancroft, "Microstrip and printed antenna design (2nd edition)," *SciTech Publishing, Inc., Raleigh, NC, USA*, 2009.

[31] G. Moernaut and D. Orban, "GNSS antennas," *GPS World*, vol. 20, no. 2, pp. 42–48, February 2009.

[32] L. Boccia, G. Amendola, and G. Di Massa, "A shorted elliptical patch antenna for GPS applications," *IEEE Antennas and Wireless Propagation Letters*, vol. 2, pp. 6–8, 2003.

[33] G. Miller, "Adding GPS applications to an existing design," *RF Design*, pp. 50–57, March 1998.

[34] V. Pathak, S. Thornwall, M. Krier, S. Rowson, G. Poilasne, and L. Desclos, "Mobile handset system performance comparison of a linearly polarized GPS internal antenna with a circularly polarized antenna," *IEEE Antennas and Propagation Society International Symposium (AP-S)*, vol. 3, June 2003.

[35] S. Kingsley, "GPS antenna design for mobile phones," *Electronics Weekly*, 11 April 2007.

[36] T. Haddrell, N. Ricquier, and M. Phocas, "Mobile-phone GPS antennas: can they be better?" *GPS World*, February 2010.

[37] D. Betaille, P. Cross, and H.-J. Euler, "Assessment and improvement of the capabilities of a window correlator to model GPS multipath phase errors," *IEEE Transactions on Aerospace and Electronic Systems*, vol. 42, no. 2, pp. 705–717, April 2006.

[38] J. Juan, M. Hernandez-Pajares, J. Sanz, *et al.*, "Enhanced precise point positioning for GNSS users," *IEEE Transactions on Geoscience and Remote Sensing*, vol. 50, no. 10, pp. 4213–4222, October 2012.

[39] Q.-H. Phan, S.-L. Tan, and I. McLoughlin, "GPS multipath mitigation: a nonlinear regression approach," *GPS Solutions*, vol. 17, no. 3, pp. 371–380, 2013.

[40] M. R. Azarbad and M. R. Mosavi, "A new method to mitigate multipath error in single-frequency gps receiver with wavelet transform," *GPS Solutions*, vol. 18, no. 2, pp. 189–198, April 2014.

[41] T. Taga, "Analysis for mean effective gain of mobile antennas in land mobile radio environments," *IEEE Transactions on Vehicular Technology*, vol. 39, no. 2, pp. 117–131, May 1990.

[42] K. Kalliola, K. Sulonen, H. Laitinen, O. Kivekas, J. Krogerus, and P. Vainikainen, "Angular power distribution and mean effective gain of mobile antenna in different propagation environments," *IEEE Transactions on Vehicular Technology*, vol. 51, no. 5, pp. 823–838, September 2002.

[43] P. Carro and J. de Mingo, "Mean effective gain of compact WLAN genetic printed dipole antennas in indoor-outdoor scenarios," *International Conference on Personal Wireless Communications (PWC)*, pp. 275–283, September 2006.

[44] A. Ando, T. Taga, A. Kondo, K. Kagoshima, and S. Kubota, "Mean effective gain of mobile antennas in line-of-sight street microcells with low base station antennas," *IEEE Transactions on Antennas and Propagation*, vol. 56, no. 11, pp. 3552–3565, November 2008.

[45] K. Fujimoto and J. R. James, "Mobile antenna systems handbook (2nd edition)," *Artech House, Inc., Norwood, MA, USA*, 2001.

[46] Z. N. Chen, "Antennas for portable devices," *John Wiley and Sons, Inc., Chichester, UK*, 2007.

[47] G. F. Pedersen and J. B. Andersen, "Handset antennas for mobile communications: integration, diversity, and performance," *Radio Science Review 1996 1999, Oxford University Press, Oxford, UK*, 1999.

[48] K. Ogawa and T. Uwanao, "Mean effective gain analysis of a diversity antenna for portable telephones in mobile communication environments," *Electronics and Communication in Japan (Part I: Communications)*, vol. 83, no. 3, pp. 88–96, December 2000.

[49] J. Krogerus, C. Ichelun, and P. Vainikainen, "Dependence of mean effective gain of mobile terminal antennas on side of head," *European Conference on Wireless Technology (ECWT)*, pp. 467–470, October 2005.

[50] M. Ur Rehman, Y. Gao, X. Chen, C. Parini, and Z. Ying, "Characterisation of system performance of GPS antennas in mobile terminals including environmental effects," *European Conference on Antennas and Propagation (EuCap) Proceedings*, March 2009.

[51] X. Chen, C. G. Parini, B. Collins, Y. Yao, and M. U. Rehman, "Antennas for global navigation satellite systems," *John Wiley and Sons, Chichester, UK*, 2012.

[52] D. Cheng, "Field and wave electromagnetics (2nd edition)," *Addison Wesley, Inc., Boston, MA, USA*, 1989.

[53] J. Jemai, T. Kurner, A. Varone, and J. F. Wagen, "Determination of the permittivity of building materials through WLAN measurements at 2.4GHz," *IEEE International Symposium on Personal, Indoor and Mobile Radio Communications*, September 2005.

[54] G. Klysza, J. P. Balayssaca, and X. Ferrièresb, "Evaluation of dielectric properties of concrete by a numerical FDTD model of a GPR coupled anten-naparametric study," *NDT & E International*, vol. 41, no. 8, pp. 621–631, December 2008.

[55] C. Balanis, "Antenna theory analysis and design (2nd edition)," *John Wiley and Sons, Inc., Hoboken, NJ, USA*, 1997.

Chapter 10

Textile substrate integrated waveguide technology for the next-generation wearable microwave systems

Sam Agneessens, Sam Lemey*, Riccardo Moro[†],
Maurizio Bozzi[†], and Hendrik Rogier**

10.1 Introduction: the contribution of wearable technology to ubiquitous computing

Ever since the transistor replaced the vacuum tube by the end of the 1950s and, thereafter, the invention of the integrated circuits (ICs), the influence of computers on society has increased hand over fist. Whereas, in the 1990s, we used to rely on keyboard and mouse to browse our first web pages and send our first emails, most people nowadays carry around, in their pocket, a device that lets them contact anyone, anywhere in the world, or access any piece of information from a data source more vast than all of the world's libraries combined. And all of this by simply swiping a screen or by using voice commands. But what does the future have in store for us? Where will this evolution bring us in the next 5–10 years?

In the vision of the Internet-of-things (IoT) [1–3] and "ubiquitous computing" [4, 5] it is expected that, with the continued miniaturization of electronic sizes and their ever increasing capabilities, technology will retract further and further into the background, becoming more and more interwoven with our everyday lives.

In this evolution toward total symbiosis between man and machine, an important role is reserved for wearable devices. These body-centric systems, dealing with a plethora of tasks, offer a lot of potential owing to the combination of inconspicuous integration of functionality, extending the user's capabilities while still being comfortable enough to carry around all day long.

The possibilities offered by wearables have not gone unnoticed, and the subject attracts a lot of attention from researchers in academia, industry, and designer groups. Numerous application domains can be conceived that may benefit from wearable technology, such as medical systems for patient monitoring [6–11], epidemiologic studies [12–14], rehabilitation and geriatric care [15–17], in-home healthcare [18, 19].

*Department of information technology, Ghent University-Minds
[†]Department of Electrical, Computer and Biomedical Engineering, University of Pavia

Rescue workers, first responders, and security personnel [20, 21] will certainly benefit from low-weight, high-performance, and smart systems. In addition, wearables may enhance the experiences of leisure activities, such as sports & fitness monitoring [22, 23] and augmented reality [24]. Another interesting topic concerns body-centric energy harvesting [25–31] and some researchers have even focused on the use of wearable electronics to improve human–animal interaction [32, 33].

Clearly, the future looks bright for wearable devices, given the abundance of application areas and their user-friendly potential. But what is a wearable system, from a technical point of view, really? To engineers, a wearable system is a wireless body-centric system, consisting of a number of functional blocks that should be integrated without diminishing the user's comfort. There is a hardware part, with sensors (gathering the data), actuators (providing interaction with the user), a computing core (such as microcontroller that takes care of data processing), an energy source (batteries or power-harvesting unit), and communication capabilities (RF front-end and antenna). The software part, running on the microcontroller, handles the available data streams from/toward the hardware components and makes the system "smart." And finally, what makes this combination of separate building blocks into a full blown wearable is the fact that it is integrated, in some way, such that it can easily be worn by a user, inconspicuously, without hindering him or her during the execution of his/her tasks or day to day activities.

An interesting type of wearable with strong emphasis on integration is the smart textiles [34, 35]. These are fabrics in which electronics are embedded, exploiting as much of the space provided by the textile to achieve optimal integration, to extend its functionality. The electronics could be off-the-shelf components, such as microcontrollers, sensors, actuators, or could be realized by using the fabrics of the smart textile itself, such as RFID tags [36–38], stretchable interconnects [39–41], or antennas.

10.2   Conventional wearable antennas

10.2.1   On-body considerations

In any wireless system, the antenna is a key component. Its function is to facilitate the wireless transfer of data by converting an electric signal, carrying information, into an electromagnetic wave and launching it into space. At the receiving end, its task is comparable, but reversed: the electromagnetic wave carrying the information is converted into an electric signal that can be handled by the system's "data processing core." It stands to reason that the quality of the antenna will severely influence the overall performance of the wireless system. A decent antenna converts the signal into EM waves in a power-efficient way, giving the system a long communication range, high bit rate, and long-lasting battery life. A bad one, on the other hand, results in short range, bad signal quality, low bit rates, and high power consumption, inevitably leading to quickly drained batteries.

Sometimes the design of an efficient antenna is straightforward: all you need is a bulky geometry, and a large keep out area such that there are no objects in the direct vicinity. Overall best performance is achieved when the size of the design is about half the wavelength. For example, this would mean that the optimal antenna size for Wi-Fi is about 6 cm, and for 900-MHz GSM (mobile telephone) ± 17 cm, and with no nearby objects (i.e. no laptop screen or human body parts). Obviously, this is not always possible, especially not in scenarios where integration is important (as for wearable systems), and, therefore, this approach needs to be abandoned in such cases. Dedicated design and implementation methods are needed for wearable antennas to be able to cope with a large number of – often conflicting – objectives.

The design should be low profile, compact, and easy to integrate inconspicuously. The person using the wearable system should be able to take advantage of the additional functionality offered, without being hindered, and having to pay the price of reduced comfort. There are also specific challenges related to the deployment near the human body. Mechanical stress occurs, such as bending, crumpling, and stretching. This can lead to failure, such as broken connectors, cracked interconnections, or delamination of the different layers. Electromagnetic issues are associated to on-body deployment as well. In today's technologically advanced societies, exposure to all types of radio-frequency (RF) sources [42–44] is inescapable and additional absorption caused by a radiating antenna, placed near the human body, should be avoided as this could lead to health issues. The presence of the body also has the tendency to affect the propagation characteristics [45–50], and careless design could result in poor link quality, depolarization, and augmented power consumption. The body might also detune the antenna's operating frequency and deteriorate its matching characteristics and/or radiation efficiency, rendering it useless for wearable applications. On top of this, the characteristics should be robust and stay reliable under changing environmental parameters, such as air humidity and heat.

A good strategy to achieve stable and efficient on-body performance is to make high body-antenna isolation a design priority. In other words, the antenna has to operate in such a way that the amount of radiation toward, and near field coupling with, the body is minimized (and ideally equal to zero). Placing the body in a "blind spot" of the antenna will permit stable performance, comparable to free-space results, and will make the design independent of the position on the human body, the spacing between the antenna and the user, and the person's morphology (child or adult, female or male, skinny or fat, etc.).

10.2.2 Topologies and fabrication methods

In the previous section, it became clear that a wearable antenna needs to meet two different design aspects. First of all, good free-space and on-body performances (i.e. high body-antenna isolation) are essential. Second, it should be possible to integrate the design inconspicuously, without causing hinder to the user, and with stable performance in various operating conditions. This is where textile antennas become interesting. Many of these designs are microstrip line or coplanar waveguide (CPW) topologies, and they can be fabricated in the same materials as the garments into which they are integrated by combining them with conductive textiles. As an antenna substrate, a thicker textile, with a low permittivity, is ideally used. Wearable antenna conductors can be realized by using conductive textiles [51], such as copper-plated nylon taffeta and silver-plated stretchable fabric, by using laminations of metal film onto a polymer carrier [52, 53], such as copper-on-polyimide, or by using conductive patterns defined by screen printing conductive ink [54–56] or embroidering conductive yarns [57, 58].

To determine the suitability of (conductive) textile materials for antenna fabrication, the electromagnetic parameters have to be known. Several methods have been developed to extract the permittivity, losses, and conductivity of these materials, such as microstrip line methods with connector de-embedding [59], surrogate-based methods [60], or in-situ methods [61, 62]. The use of textile materials, in combination with a dedicated design strategy, results in flexible, low weight, high performance on body components. This benefits the system and facilitates longer communication ranges, lower power consumption, and higher bitrates.

10.3 Substrate integrated waveguide technology for a new class of wearable microwave components

10.3.1 Substrate integrated waveguides: fundamentals

This section briefly introduces the concepts behind substrate integrated waveguides (SIWs) and their relation to rectangular waveguides. For more in-depth information, the reader is directed to literature such as References 63 and 64.

SIW technology yields a planar implementation of rectangular waveguides and all microwave components that can be derived from it, such as filters, couplers, and antennas. It is realizable with standard printed circuit board (PCB) manufacturing techniques by using two rows of conductive cylinders, vias, or slots to connect two conductive plates on opposing sides of a dielectric substrate. This structure yields a performance comparable to that of a rectangular waveguide in terms of dispersion characteristics and field pattern. The modes that exist in the SIW coincide with the $TE_{n0}, n = 1, 2, \ldots$ modes of the rectangular waveguide, TM modes are not supported due to the gaps between the metal posts.

To facilitate easy design, empirical relations between the geometrical dimensions of an SIW waveguide and the effective width, w_{eff}, of the equivalent rectangular waveguide with the same propagation characteristics have been developed:

$$w_{\text{eff}} = w - 1.08\frac{d^2}{s} + 0.1\frac{d^2}{w} \tag{10.1}$$

where w is the width of the waveguide, d the diameter of the vias, and s the spacing between them. This relation can be used to quickly obtain initial dimensions of a design by substituting the w_{eff} in well-known formulas regarding rectangular waveguide structures.

The losses in SIW structures can be subdivided into three groups: dielectric losses, conductor losses, and radiation losses. Dielectric losses are caused by the losses in the dielectric substrate ($\tan \delta$). The conductor losses are proportional to the height of the used substrate and the surface roughness of the conductors. The radiation loss, caused by leakage through the apertures between the posts, can be controlled by choosing the right dimensions for the posts and the corresponding spacing between them. A rule of thumb which should be used is:

$$d < \frac{\lambda_g}{5}, \quad p \le 2d \tag{10.2}$$

10.3.2 SIW techniques and textile materials

The SIW implementation of microwave structures has some very interesting advantages over microstrip or CPW, such as higher isolation owing to the vertical metal posts, low radiation losses for thicker substrates (such as textile materials), no need for a large ground plane, and the miniaturization options by exploiting the symmetry planes of the modes.

10.3.2.1 Fabrication technique

Fabrication of SIW technology on textile materials has the same basis as the production steps for conventional textile antennas, with some additional steps and considerations. As for conventional textile antennas, metalized layers are realized by conductive textiles, which are often copper or silver based. They can be patterned manually, e.g. with a scalpel, or by laser cutters, when higher precision and better repeatability are required. As for the dielectric substrates, serving as non-conductive textiles, a stable height, low moisture retention, and resilience against permanent deformation are required, in addition to low losses and limited inhomogeneity in electromagnetic parameters (mainly permittivity and loss tangent). The stable height is important for reliable and repeatable performance. Thicker substrates are preferred, as they yield lower losses and larger bandwidth. A suitable height may be obtained by applying one thick substrate or a stack of thinner textile layers. Lamination is typically used to assemble the different layers of the textile stack. A popular choice of adhesive is thermally activated glue, which enables fast realization of prototypes.

The main difference in fabrication procedure and material selection between conventional textile antennas and SIW microwave components relates to the realization of the interconnection between the two metallization layers. Currently, two techniques are popular to create the vias: stitching and eyelets.

The stitching methods generate very closely stitched, "nearly-sold walls" of conductive thread that connect the parallel metal layers. This method realizes very good electrical contact between top and bottom planes. There is a risk of fabric tearing due to the very close stitch of the thread. Furthermore, compression of the substrate due to the tension on the stitching wire can result into local substrate compression, which cannot be ignored in the design process.

Another implementation method for the SIW technology on textiles is by relying on eyelets. They can be found in many different shapes, sizes, and materials and are very commonly used in all types of textile-related products. For SIW implementation, cylindrical brass eyelets are often used, which are spaced according to the common SIW rules of thumb. Due to the resemblance between cylindrical eyelets in textile materials and metalized vias in PCB technology, empirical rules that describe the relationship between rectangular waveguides and SIW waveguides can be used without modification, simplifying the design process. Care needs to be taken in selecting the substrate material, as the quality of the electrical connection between eyelet and e-textile depends on mechanical pressure. Good contact can be obtained if the eyelets are fixed very tightly, leading to substrate compression, which needs to be taken into account in the design, or by selecting a substrate material that is very resilient to compression. Another strategy that can be used is applying conductive glue or paste between the eyelet and the e-textile.

10.3.2.2 State of the art

The first implementation of SIW technology on textile materials was published in August 2012 [65]. The proposed design is a cavity-backed dog-bone slot antenna, completely covering the 2.45-GHz ISM band and boasting excellent radiation characteristics. This study demonstrated that SIW techniques can be applied in textile by means of simple, low-cost materials, and manufacturing techniques.

The SIW technique is very versatile and can be used to meet very distinctly different design goals. As such, topologies aiming at wideband operation or covering multiple frequency bands [27, 66] have been reported. The miniaturization of antennas by relying on the symmetry of the mode distributions has also been exploited. Half-mode SIW [66, 67] and quarter-mode SIW (QMSIW) [68] topologies were proposed that combine very high body-antenna isolation with excellent performance and compact dimensions. Another option to combine good shielding and compact dimensions is the multilayered cavity-backed stacked patch design presented in References 62 and 69.

When eyelets are used for the implementation of SIWs on textile, wires (e.g. power active electronics) can be routed through the eyelets from one side of the substrate to the other without affecting the performance of the microwave component, which is not feasible with microstrip line or CPW designs. Based on this principle, SIW antennas have been presented that act as an energy-harvesting platform for solar energy and thermal energy [27, 70].

SIW techniques can also be adopted to realize other microwave components, and studies have been presented that focus on the realization of SIW waveguides and filters implemented on textile materials [62, 71].

10.4 Textile substrate integrated waveguide designs

10.4.1 Textile SIW cavity-backed slot antennas

The cavity-backed antenna topology was selected for the implementation of SIW antennas on textile. Cavity-backed antennas offer several advantages, including the suppression of undesired surface waves and a high front-to-back ratio, resulting in a low sensitivity for on-body operation. In addition, the SIW technology allows for a simple and cost-effective fabrication process [63], well developed in the case of PCBs and suitable to apply to fabrics. Furthermore, the SIW structure guarantees the easy integration of active components on the antenna [72, 73], thus permitting the realization of active antennas or complete transceivers on a textile substrate.

10.4.1.1 Cavity-backed SIW antenna

The first implementation of an SIW antenna on textile consisted in a cavity-backed slot antenna [65]. The antenna was designed for operation in the 2.45-GHz industrial, scientific and medical (ISM) band, for short-range communication between rescue workers.

This antenna comprises a square SIW cavity, with a radiating slot etched in the top metal layer and a feed line consisting of a 50-Ω grounded CPW in the bottom metal layer (Figure 10.1). The SIW cavity was designed to operate with the TM_{120} cavity mode at 2.45 GHz. The slot has a dog-bone shape, to reduce its length and fit it inside the cavity. The length of the CPW was determined to optimize the input matching. Metal eyelets were adopted to implement the metal vias, to guarantee the flexibility of the structure, required for wearable applications.

The substrate adopted for the fabrication of the antenna has a thickness of 3.94 mm, dielectric permittivity $\varepsilon_r = 1.575$ and loss tangent $\tan \delta = 0.0238$ at 2.45 GHz. The top and bottom conductive layers are realized by using Flectron®, a conductive fabric with surface resistivity $R_s = 0.18\ \Omega$/sq at 2.45 GHz.

The prototype of the antenna was experimentally characterized under different operation conditions. Preliminarily, the antenna was measured in nominal conditions inside an anechoic chamber: the comparison between simulated and measured reflection coefficients is shown in Figure 10.2 (solid lines). The antenna exhibits a bandwidth of 165 MHz around the central frequency of 2.45 GHz, meeting the design specification of 10-dB input matching over the entire ISM band. The radiation pattern was measured in the E-plane and in the H-plane, and the comparison between simulation and experimental data is reported in Figure 10.3: the measured antenna gain resulted 3.21 dBi in the simulation and 3.9 dBi in the measurement, and the measured front-to-back ratio was 19.7 dB. The antenna efficiency resulted 52% in the simulation and 68% in the measurement.

Subsequently, to verify the proper operation of the antenna in wearable conditions, on-body measurements were performed, with the antenna integrated on the backside of a fire-fighter jacket worn by a person. To avoid the direct contact with the skin,

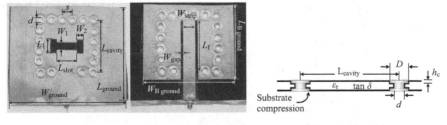

Figure 10.1   *Geometry and dimensions of the cavity-backed SIW textile antenna (dimensions in mm: $L_{cavity} = 71.5$, $L_{slot} = 26.2$, $L_1 = 19.1$, $W_1 = 10.5$, $W_2 = 8.7$, $L_f = 64.6$, $W_{strip} = 15$, $W_{gap} = 3$, $d = 7$, $D = 13$, $s = 14$, $h_c = 1$, $L_{ground} = 126.5$, $W_{ground} = 131.5$, $L_{B\ ground} = 91.5$, $W_{B\ ground} = 111.5$) (© 2012 IET from Reference 65)*

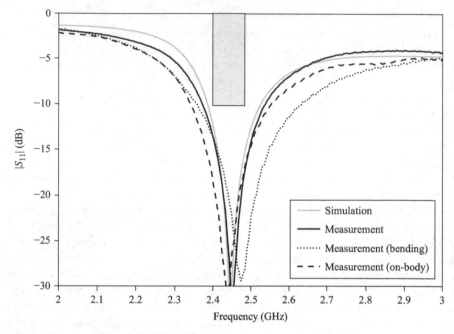

Figure 10.2   *Amplitude of the input reflection coefficient* $|S_{11}|$ *of the cavity-backed SIW textile antenna vs. frequency, under different operation conditions (© 2012 IET from Reference 65)*

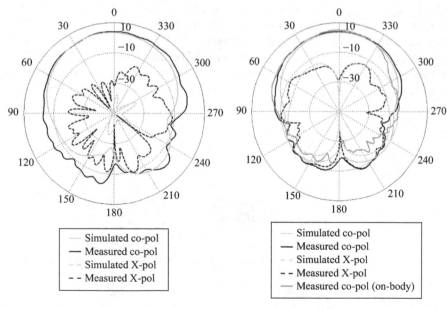

Figure 10.3   *Simulated and measured radiation patterns of the cavity-backed SIW textile antenna at the frequency of 2.45 GHz, under different operation conditions (© 2012 IET from Reference 65)*

the antenna was mounted between a waterproof aramid inner liner of the jacket and a moisture/thermal barrier. Measurement results of the reflection coefficient and of the radiation pattern are shown in Figures 10.2 (dashed line) and 10.3, respectively. The presence of human body does not affect significantly the input matching and the radiation characteristics. Moreover, an increase of the measured gain at broadside was observed (4.9 dBi), attributed to the larger ground plane due to the presence of the body. The measured efficiency of the antenna mounted on-body was 60%.

The final experimental verification aimed to investigate the influence of bending on the antenna performance, in the hypothesis to locate the antenna on the arm of a fire-fighter jacket. The antenna was bent along the *H*-plane around a cylinder with a diameter of 10 cm. The only effect was a small shift of the resonance frequency (Figure 10.2, dotted line). Under this operation condition, however, the antenna still met the specifications of 10-dB input matching in the ISM band.

10.4.1.2   Folded cavity-backed SIW antenna

A folded configuration was adopted to reduce the size of the cavity-backed SIW antenna and improve its integration in the garments [62]. In this antenna topology, the SIW cavity is folded around a metal patch, which is connected to the bottom ground plane by a metal via (Figure 10.4a). A square ring slot, cut out in the top metal layer, is responsible for the radiation. The dimensions of the radiating slot were selected to maximize the radiation efficiency at the frequency of 2.45 GHz and to broaden the operating bandwidth of the antenna. The antenna is fed by a coaxial probe in the back side, whose position was selected for optimal input matching. Due to the position of the coaxial probe, the antenna radiates an electric field linearly polarized along the diagonal of the cavity. This folded topology allows reducing the area of the antenna of approximately 50%, compared to the classical cavity-backed antenna described above, at the cost of the additional complexity in the fabrication of a double-layer structure.

A prototype of the folded cavity-backed SIW antenna was manufactured, and photographs of the back and front sides are shown in Figure 10.4b and c, respectively.

The antenna was subsequently characterized to determine its electromagnetic performance. The simulated and measured values of the input reflection coefficient of the antenna are shown in Figure 10.5. The measurements confirm that the antenna meets the design specifications of 10 dB matching in the ISM band and exhibits a 10 dB bandwidth of 130 MHz around the frequency of 2.45 GHz. Bending and compression tests were also performed, and in all cases the antenna meets the design specifications [62].

The radiation pattern of the antenna was subsequently measured at the frequency of 2.45 GHz. Due to the polarization properties of the antenna, the radiation pattern was measured along the planes $\Phi = 45°$ (co-polarization) and $\Phi = 135°$ (cross-polarization). Figure 10.6 shows the comparison between simulation and measurement of the radiation pattern, which exhibit a good agreement. The measured antenna efficiency was 74%, with a measured gain of 5.93 dBi, close to the simulated value of 5.9 dBi. The front-to-back ratio of the antenna was approximately 18 dB. On-body measurements are also reported in Figure 10.6, confirming the suitability of SIW components for on-body applications.

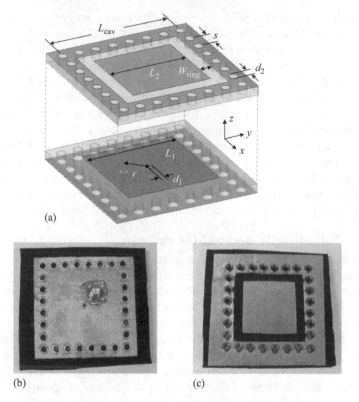

*Figure 10.4   Textile antenna based on a folded SIW cavity-backed antenna
configuration: (a) geometry of the antenna (dimensions in mm:
$L_{cav} = 54.1$, $L_2 = 35$, $W_{ring} = 5.5$, $L_1 = 41.2$, $d_1 = 2$, $d_2 = 4$, $s = 8$,
$r = 11.3$); (b) photograph of the back side (feed input); and
(c) photograph of the front (radiating) side (©2015 IEEE. Reprinted
with permission from Reference 62)*

10.4.2   Half-mode SIW textile antenna

10.4.2.1   Introduction

Previously presented antenna designs have shown that it is possible to implement SIW antennas relying solely on off-the-shelf textile (related) materials. The techniques have as additional benefit that they augment body-antenna isolation, resulting in robust and safe on-body deployment. The main advantage is the achievement of high body-antenna isolation without the need for an extended ground plane. This, however, has as a downside that the actual antenna occupies a slightly larger area. Solutions to further reduce size can be found by applying miniaturization techniques. Another improvement for this antenna topology consists of implementing multi-band operation, making it a more powerful candidate if an on-body system requires flexible operation in multiple frequency bands.

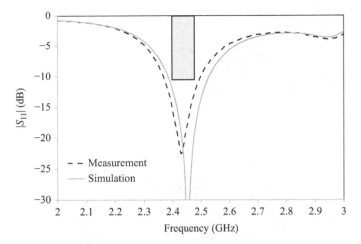

Figure 10.5   Amplitude of the input reflection coefficient $|S_{11}|$ of the folded SIW cavity-backed antenna: simulation and measurement (©2015 IEEE. Reprinted with permission from Reference 62)

In this section, a dedicated SIW design is presented that is developed to exhibit good body-antenna isolation, reduced dimensions, and, optionally, providing dual-band operation. The novel design, discussed in detail in Reference 66, yields a signification size reduction with respect to conventional textile patch antennas owing to the lack of large ground plane.

10.4.2.2   Design and material selection

SIWs provide an interesting option to reduce dimensions by exploiting the symmetry in the magnetic field distribution. If a magnetic field has no tangential component w.r.t. a certain symmetry plane, then there is no current flow (on the top and bottom metal walls of the waveguide) across this plane. Since there is no current flow, it is possible to replace the plane by a virtual magnetic wall, which is implemented by removing half of the structure. This is analogous to the virtual electric wall, where a null in the electric field can be replaced by a shorting wall without altering the operation of the structure.

The design evolution of this antenna is depicted in Figure 10.7. The strategy uses two different miniaturization techniques that rely on the symmetrical distribution of the electromagnetic fields inside a resonant cavity: a row of shorting vias, acting as a virtual electric wall, and a virtual magnetic wall, which both help to reduce the dimensions of the design significantly and an additional slot is used to facilitate impedance matching to and tune a higher order mode.

In the first step, the magnetic wall is implemented to reduce the dimensions a full-size cavity back slot antenna, which has a symmetrical magnetic field distribution w.r.t. the y-axis, almost by half. This results in half-mode SIW operation. The

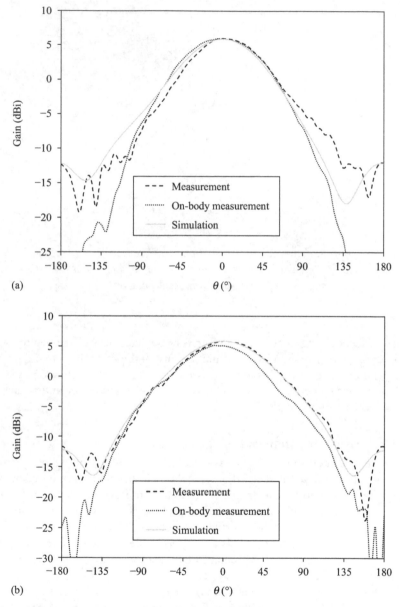

Figure 10.6 *Simulated and measured radiation pattern of the folded SIW cavity-backed antenna: (a) in the plane* $\Phi = 45°$ *and (b) in the plane* $\Phi = 135°$ *(©2015 IEEE. Reprinted with permission from Reference 62)*

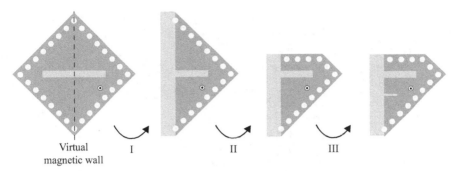

Virtual
magnetic wall   I   II   III

Figure 10.7 Design evolution of the wearable half-mode dual band SIW antenna

following step introduces the electric wall above the slot. The magnitude of the electric field above the slot varies only slightly around zero, so a shorting wall could be introduced, resulting in smaller dimensions and very limited performance loss. In the final step, III in the Figure 10.7, an additional slot is added to the design to allow for good impedance matching to 50 Ω in a high- and low-frequency band.

This structure is fed by a coaxial probe. The probe, the top conductor, the shorting vias, and the bottom conductor form a current loop, which couples to the magnetic field of the resonant cavity, thereby exciting the resonant modes. The choice of probe feeding has an additional advantage for on-body antennas, since it does not introduce any extra back radiation, which is not the case for microstrip or CPW feeding techniques. In addition, it allows for easy integration of the antenna below behind the ground plane. This makes it possible to design very compact systems, where the active electronics and antenna share the same area.

The design realizes antenna operation in the 2.4- and 5.8-GHz ISM bands, making the device suitable for multiple applications. The substrate material of the antenna is a closed-cell rubber fabric with permittivity = 1.495, loss tangent = 0.016, and height 3.94 mm. The antenna is manufactured as described in Section 10.3.2. Computer-aided performance optimization is done by relying on CST microwave studio. The final layout and dimensions of the optimized design is shown in Figure 10.8.

10.4.2.3 Validation

To validate the design, simulation results are compared to the outcomes of a measurement campaign. Different measurement scenarios were investigated, as to guarantee optimal performance in a variety of operating conditions: free-space (when the antenna is not near the human body, this could be due to a deliberate spacers or to movements of the user, yielding a significant spacing between antenna and body) and on-body scenarios (when the antenna is very close to the wearer, only separated by a thin shirt that avoids direct contact between antenna and the user's skin).

Two criteria, vital for decent antenna operation, are verified in all measurement scenarios: the impedance matching (measurement of the *S*-parameters) and the radiation performance. For on-body deployment, an additional performance characteristic

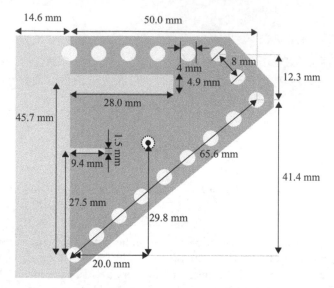

Figure 10.8   Geometry and dimensions of the wearable half-mode, dual-band antenna

is checked: the specific absorption rate (SAR). This will provide information about the amount of radiation that is absorbed by the human body and is a good indication whether the antenna is safe to use in close proximity of the wearer. A network analyzer (Agilent N5242 PNA-X) is used to determine the reflection coefficient in an accurate way, and 3D radiation pattern measurements in an anechoic chamber are conducted to assess the radiation performance of the half-mode SIW design.

Free-space measurements
Figure 10.9 shows the reflection coefficient at the input port of the antenna w.r.t. 50 Ω for the simulated case and the measurement in free space. Very good agreement between both scenarios is observed. The antenna has a fractional bandwidth of 4.9% in the lower band and 5.1% in the higher band.

By comparing the simulated and measured gain patterns in the 2.4- and 5.8-GHz ISM bands, as depicted in Figure 10.10, it is seen that the antenna also behaves as expected in terms of radiation performance. The measured maximal gain and efficiency in both bands are: 4.1 dBi and 72.8% (@2.4 GHz) and 5.8 dBi and 85.6% (@5.8 GHz). These results validate the antenna as a good design in terms of free-space performance.

On-body performance
After the validation in free space, it is necessary to look at on-body performance. Robust operation is required in the on-body scenario if reliable communication at all times is to be guaranteed. The same measurement procedure is conducted as for the free-space scenarios, but the setup now includes the presence of the human body: the antenna is positioned directly on the human torso, separated by a t-shirt.

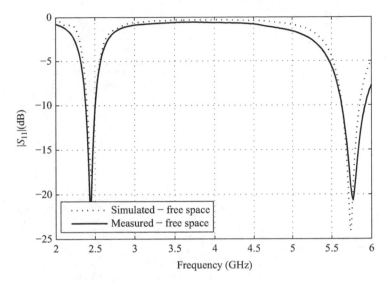

Figure 10.9   Free-space reflection coefficient of the half-mode SIW antenna

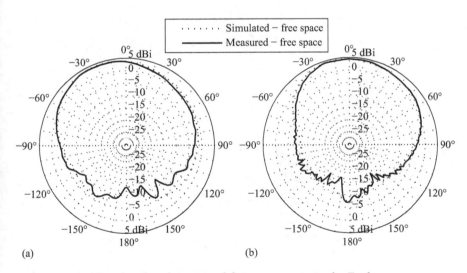

*Figure 10.10   Simulated and measured free-space gain in the E-plane at
(a) 2.45 GHz and (b) 5.8 GHz*

The reflection coefficient for the free-space scenario and the on-body scenario are depicted in Figure 10.11. Good agreement is observed. The bandwidth in the on-body scenario remains unchanged, and no frequency shifts occur.

Comparing the free-space radiation performance with the on-body case, shown in Figure 10.12, indicates that the gain in boresight remains largely unaffected, despite

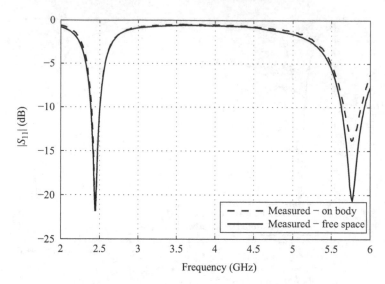

Figure 10.11   Stable, measured, on-body, impedance matching for the half-mode SIW antenna

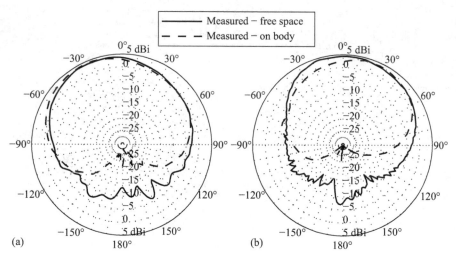

Figure 10.12   Measured free-space gain and on-body gain in the E-plane at (a) 2.45 GHz and (b) 5.8 GHz

the presence of the human body. Observe that the radiation that is directed toward the human body is largely absorbed.

To assess the influence of this absorbed radiation on the user, the SAR values are calculated to get a clear view on how much radiation the human body will be exposed to. CST microwave studio is used to simulate the SAR values when 0.5 W of input power is applied to the antenna, which is placed at 2 mm distance from the human body (emulated by the three-layer human body model). This results in values of 0.55 mW/g @2.45 GHz and 0.9 mW/g @5.8 GHz, both well below the allowed level of 1.6 mW/g.

Performance under deformation

Due to the dynamic nature of the body environment, deformation of the antenna of the antenna is likely to occur. Such an alternation of the antenna's shape will result in a perturbation of the EM-field in the resonant cavity of the antenna and the associated current distributions on its surface, leading to different radiation characteristics and impedance matching. A well-designed antenna is able to cope with deformation, exhibiting robust performance even under significant levels of bending.

The presented design is therefore subjected to bending scenarios, for which the reflection coefficient of the design is measured. The antenna is bent around two cylinders with different radii (75 and 40 mm), along different axes: $\vartheta = 0°$, $\vartheta = 90°$. The outcome of these measurements is shown in Figure 10.13. Although the −10 dB matching criteria are slightly violated in the higher frequency band for certain combinations of bending angles and radii, the antenna performance remains remarkably stable, with limited detuning and conservation of bandwidth. The combinations of bending angles and radii for which the $|S_{11}| < -10$ dB matching criterion are not obtained over the entire frequency band, only generate a slight violation of the impedance matching, which is unlikely to pose problems when connected to a signal source.

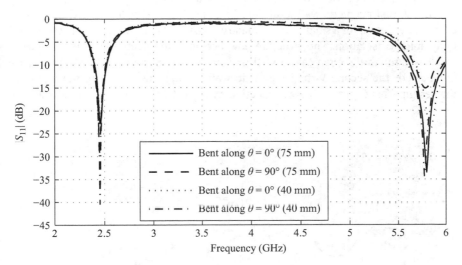

Figure 10.13   *Reflection coefficient of the half-mode SIW antenna under different bending situations*

10.4.3   Quarter-mode SIW antenna

10.4.3.1   Introduction

As seen in Section 10.4.2, SIW technology provides some interesting options to miniaturize designs, while maintaining good body-antenna isolation and robust performance. The exploitation of mode symmetry in a resonant cavity does not need to be limited to a single symmetry plane. If other symmetry planes exist, this technique can be reapplied to further reduce dimensions. This section shows how this can be done, by implementing a quarter-mode SIW (QMSIW) antenna. An in-depth discussion of the design can be found in Reference 68.

The goal is a body-worn antenna, for off-body communication, which is an excellent candidate for integration into a garment. In terms of specifications, this requires maximal performance, high body-antenna isolation, and minimal size. Size reduction is achieved by combining different techniques. As mentioned, the main size-reduction strategy focuses on exploiting the potential of virtual magnetic walls. Additional slots are added afterward in order to lengthen the current paths and resulting in a more compact design.

10.4.3.2   Design evolution

To allow the antenna to be used for body-worn applications, the 2.4-GHz ISM band is selected as operating band. This is a good candidate because, at this frequency, compact antennas can be fabricated, which are easy to integrate into garments, and many popular radio protocols operate in this unlicensed band (Wi-Fi, Bluetooth, ZigBee, etc.)

The design evolution is presented in Figure 10.14 and consists of three steps. An initial full-size resonant cavity has a symmetrical H-field w.r.t. the y-axis (or in fact, any plane that divides the structure into two equal parts). Step one implements the first virtual magnetic wall along this axis: the left size of the cavity is removed completely, except for a small ground plane extension, which is kept to maintain good body-antenna isolation. The structure now becomes an effective radiator, as power can leak through the open cavity wall into free space. Subsequently, the procedure is repeated: the second virtual magnetic wall is implemented along the x-axis of the design. This results in a size reduction by almost 75% w.r.t. the initial full-size

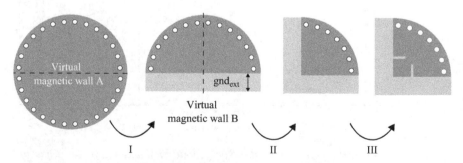

Figure 10.14   Design evolution of the wearable quarter-mode SIW antenna

resonant cavity. To achieve further miniaturization, step 3 introduces slits into the top plane of the top metal patch to effectively lengthen the current patch of the resonant mode, lowering its operating frequency and, hence, the dimensions.

10.4.3.3 Design validation

Simulations and measurement results are used to determine the quality of the design and the fabricated prototype. For the realization of the antenna, all-textile materials are once again used to permit a seamless integration into garments. The substrate material of the antenna is a closed-cell rubber fabric with permittivity = 1.495 and loss tangent = 0.016 and height 3.94 mm. The fabrication procedure is performed as described in Section 10.3.2. Optimization is performed by relying on full-wave electromagnetic solver software, CST microwave Studio. The final layout and its dimensions are visualized in Figure 10.15.

Figures of merit to characterize the performance of the design are the 3D radiation pattern (antenna's radiation performance) and the reflection coefficient (impedance matching) w.r.t. 50 Ω. Once again, the performance is determined in two distinctly different situations: no human body present (free space) and in very close proximity of the user (on-body). Both scenarios are considered because it is impossible to determine the actual distance between body and an integrated antenna into a garment, in the very different situation that might occur: sitting, walking, running, etc.

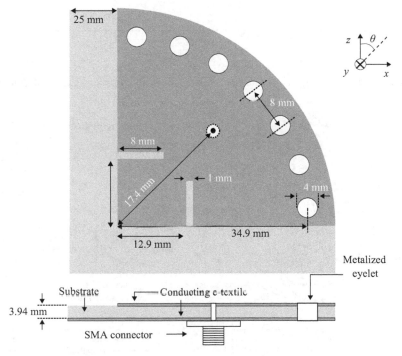

Figure 10.15   Geometry and dimensions of the wearable quarter-mode SIW textile antenna

Free-space performance

The measurements in free space are shown in Figures 10.16 and 10.17a. It can be seen that there is good agreement between simulated and measured reflection coefficients and radiation patterns. The simulated fractional bandwidth is 4.4%, which compares well to the measured value of 4.8%. Radiation patterns are also as expected, with a simulated and measured free-space gain/efficiency of 4.4 dBi/76% and 4.2 dBi/81%.

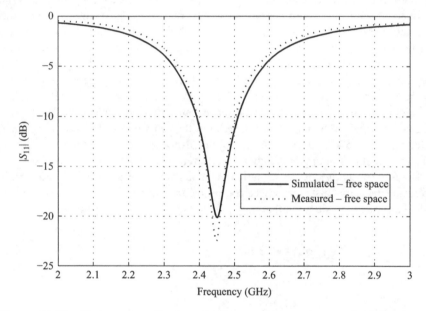

Figure 10.16   Free-space reflection coefficient of the quarter-mode SIW antenna

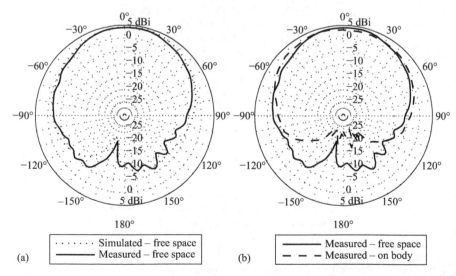

Figure 10.17   Gain pattern of the antenna in (a) free space and (b) on-body

On-body performance

Now, the performance of the antenna, when placed on-body, is compared to the free-space results in order to assess its robustness when deployed in an actual application, like a smart textile. Figures 10.17b and 10.18 show the performance for both cases, proving that the antenna is very robust when positioned near the human body, both in terms of impedance matching and radiation characteristics.

The reflection coefficient shows the same resonance frequency and a measured on-body bandwidth of 5.1%. This is a very stable result for such a compact antenna and is important, because the antenna acts as a stable load impedance matching for the radio front-end to which it is connected. The radiation performance also compares very well, both in terms of pattern and maximal gain, which is now 3.8 dBi. The limited decrease in maximal gain and slight increase in impedance matching is most likely caused by near-field absorption and dissipation of the antenna's electromagnetic fields by the human body. This lowers the Q-factor and reduces the amount of radiated power.

Body absorption not deteriorates the system's performance, but also poses health issues. To determine the amount of radiation to which a person will be subjected when wearing such an on-body antenna, the SAR is computed. In this study, a three-layered body model is adopted to emulate the human body, which is placed at 2-mm distance of the antenna, and 0.5-W input power is applied to the antenna terminals. The simulated SAR value is 0.45 mW/g averaged over 1 g of tissue. This is well below the limit of 1.6 mW/g.

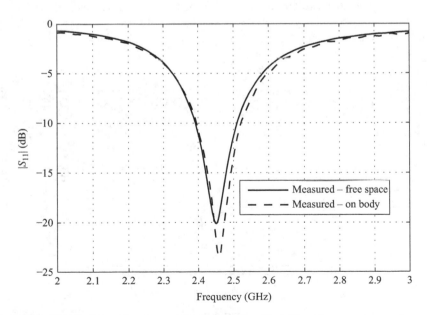

Figure 10.18   Stable reflection coefficient is observed when the quarter-mode antenna is in close proximity to the human body

10.4.4   *Wideband SIW cavity-backed slot antennas*

10.4.4.1   Introduction

Despite the excellent on-body performance of the SIW cavity-backed slot antennas discussed in the previous sections, they all exhibit narrowband behavior due to the excitation of only one resonance frequency [65, 68] or two resonance frequencies in distinct bands [66]. Yet, modern and future wireless communication systems require increasingly higher bandwidths to meet the requirements of various types of future emerging wireless applications, such as uncompressed high-definition (HD) video streaming and accurate submeter localization in indoor confined areas. In literature, several efforts are undertaken to enhance the antenna impedance bandwidth. Gong *et al.* [74] uses a large width-to-length ratio for the radiating slot to enhance impedance bandwidth, whereas Luo *et al.* [75] obtains bandwidth enhancement by merging two simultaneously excited hybrid modes within the desired frequency range. The latter bandwidth enhancement technique is now discussed into more detail, as this technique has already been successfully applied in two wearable textile wideband SIW cavity-backed slot antennas, which will be discussed below.

10.4.4.2   Bandwidth enhancement technique

The basic configuration of an SIW cavity-backed slot antenna that allows exploiting the bandwidth enhancement technique, discussed in Reference 75, is shown in Figure 10.19. The antenna topology consists of a rectangular-shaped SIW cavity, which is split into two subcavities (A and B) by a non-resonant rectangular slot. By carefully selecting the dimensions of the slot and both subcavities, two hybrid mode combinations can be excited at two different, neighboring frequencies. The hybrid mode at the lowest resonance frequency combines a weak TE_{110} and a strong TE_{120} resonances. In both subcavities, this specific combination produces fields which are out-of-phase,

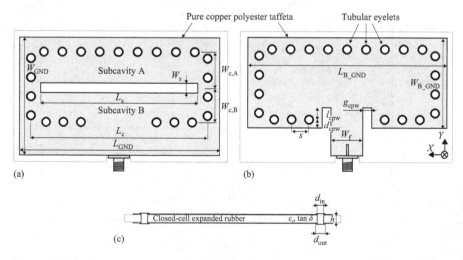

(a)   (b)   (c)

Figure 10.19   Basic geometry of an SIW cavity-backed slot antenna that exploits the bandwidth enhancement technique. (a) Front view, (b) back view, and (c) cross-section

with the dominant field situated in subcavity B. The hybrid mode at the higher resonance frequency is composed of a strong TE_{110} and a weak TE_{120} mode. Then, the fields in both cavity parts are in phase. However, they exhibit a large difference in magnitude, with the dominant field situated in subcavity A. Hence, at both resonance frequencies, the large electric field across the non-resonant slot will cause it to radiate.

Then, bandwidth enhancement is obtained by exciting both resonance frequencies close enough to each other and merging both hybrid modes within the intended frequency range of operation. Therefore, following insights can be exploited [75]. Increasing the cavity size L_c enlarges the resonating area of both hybrid modes, causing the corresponding resonance frequencies to shift to lower frequencies. Cavity dimensions $W_{c,A}$ and $W_{c,B}$ depict the widths of subcavities A and B, respectively. Increasing $W_{c,A}$ enlarges the dominant field resonating area of the highest frequency hybrid mode and the weak field resonating area of the lowest frequency hybrid mode. Hence, the highest resonance frequency decreases correspondingly, whereas the lowest resonance frequency only exhibits a slight decrease. By increasing $W_{c,B}$, a similar effect can be observed. The length of the non-resonant slot L_s should be selected large enough to divide the rectangular cavity into the two subcavities. However, the antenna's center frequency and impedance hardly changes when L_s is slightly varied. The slot width W_s mainly determines impedance matching, whereas the center frequency remains quasi-unaffected when varying W_s.

A 50-Ω GCPW feed line, located in the feed plane at the center point of the largest SIW cavity wall (Figure 10.19b), is adopted as the feeding element. For measurement convenience, an SMA connector is connected to the end of the microstrip line, which results from extending the inner lead of the GCPW feed line.

10.4.4.3 Antenna materials and fabrication procedure

Both designs are manufactured by means of the fabrication procedure discussed in Section 10.3.2. Both designs adopt the same materials as the designs described in Sections 10.4.2. and 10.4.3, being a 3.94-mm-thick closed-cell expanded rubber protective foam as antenna substrate, pure copper-coated nylon taffeta electro-textile as conductive layers, and flat-flange copper tube eyelets to implement the antenna cavity.

10.4.4.4 Wideband 2.45-GHz ISM band and 4G LTE textile SIW cavity-backed slot antenna

The bandwidth enhancement technique is then effectively exploited in Reference 27 to realize a wideband textile SIW cavity-backed slot antenna with an impedance bandwidth of 409 MHz or 15.1%, thereby covering the 2.45-GHz ISM band (2.4–2.4835 GHz) as well as both up- and downlink of the 4G long-term Evolution (LTE) band 7 (2.50–2.57 GHz and 2.62–2.69 GHz, respectively). Such an approach could allow the wearer to connect to a 4G LTE wireless metropolitan area network in the absence of wireless local area network access points at the location of operation.

The geometrical configuration and a realized prototype of the SIW cavity-backed slot antenna are depicted in Figures 10.19 and 10.20, respectively. The dimensions are carefully selected by means of full-wave electromagnetic simulation in CST Microwave Studio to obtain a return loss larger than 10 dB from 2.36 to 2.73 GHz. The final dimensions are shown in Table 10.1 and guarantee coverage of both

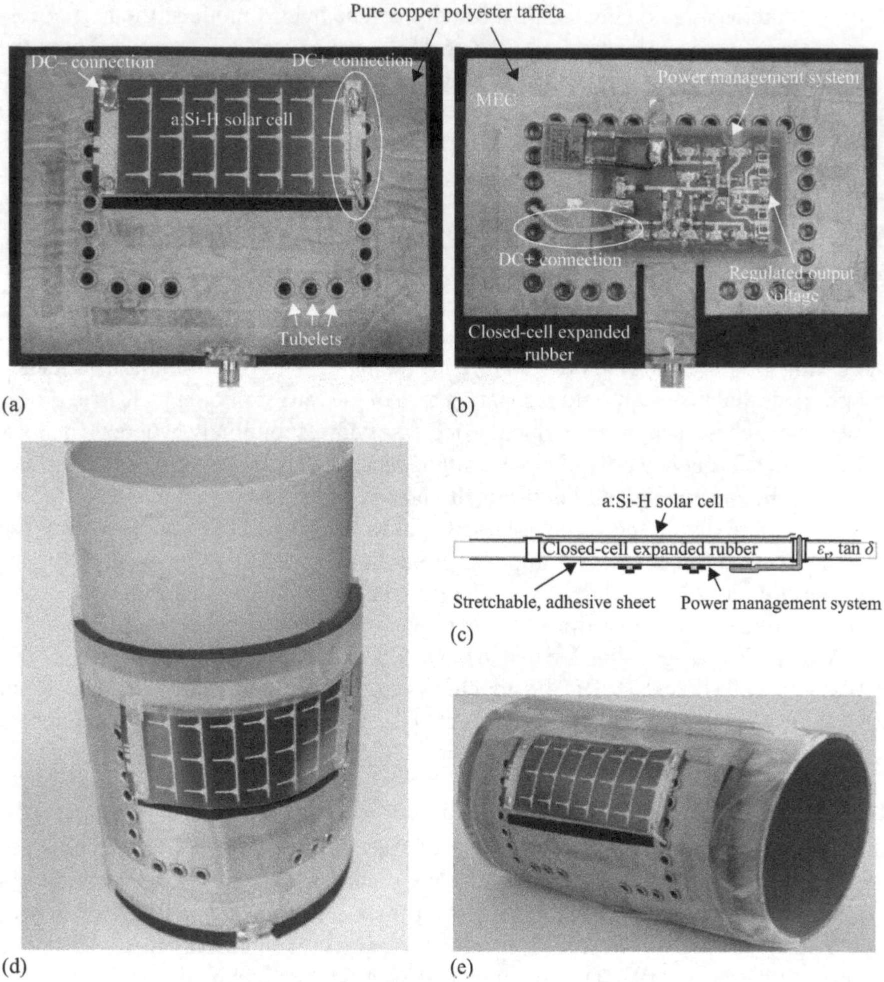

(a)   (b)

(c)

(d)   (e)

*Figure 10.20   The 2.45-GHz ISM band and 4G LTE textile SIW cavity-backed slot
antenna, with integrated flexible solar harvesting system. (a) Front
view, (b) bottom view, (c) cross-sectional view, (d) bent along the
H-plane (bending radius = 5 cm), and (e) bent along the E-plane
(bending radius = 5 cm)*

desired bands with margins of at least 40 MHz to account for variations in material
parameters, inaccuracies of the fabrication process and frequency detuning caused
by bending, crumpling, or proximity of the human body.

The antenna's performance was validated under different operating conditions.
First, the antenna was measured standalone in an anechoic chamber. Figure 10.21
shows that the simulated and measured reflection coefficients are in good agreement.
Good impedance matching with respect to $Z_0 = 50\,\Omega$ from 2.356 to 2.74 GHz, yielding a bandwidth of 409 MHz or 15.1%, can be observed. The measured and simulated
radiation patterns at 2.45 GHz of the antenna in the E and H-plane are shown in

Table 10.1   Dimensions of the wideband 2.45-GHz ISM band and 4G LTE textile SIW cavity-backed slot antenna

Parameter	Value (mm)	Parameter	Value (mm)
L_c	84.5	W_{GND}	93.0
$W_{c,A}$	26.2	L_{GND}	124.5
$W_{c,B}$	26.8	$W_{B,GND}$	81.0
W_S	3.9	$L_{B,GND}$	124.5
L_S	76.5	W_f	15.5
s	8.0	g_{cpw}	2.6
d_{in}	4.0	l_{cpw}	8.0
d_{out}	6.0	d_{cpw}	7.0

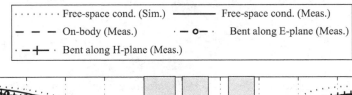

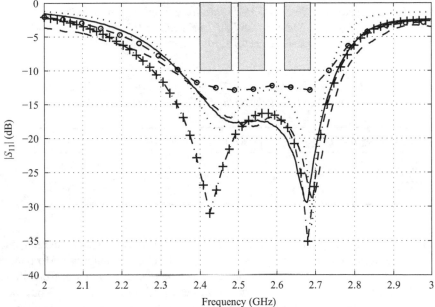

Figure 10.21   Reflection coefficient of the wideband 2.45-GHz ISM band and 4G LTE textile SIW cavity-backed slot antenna in different conditions

Figure 10.22. Lemey *et al.* [27] states that very similar radiation patterns are obtained at 2.53 and 2.65 GHz. Table 10.2 shows the measured antenna gain along broad side, the FTBR and the radiation efficiency at 2.45, 2.53, and 2.65 GHz. Second, the antenna is deployed on the chest of a test person wearing a t-shirt. Because of the presence of the human body, the input return loss depicted in Figure 10.21 exhibits a slight increase in bandwidth to 449 MHz. The on-body radiation pattern in Figure 10.22(b)

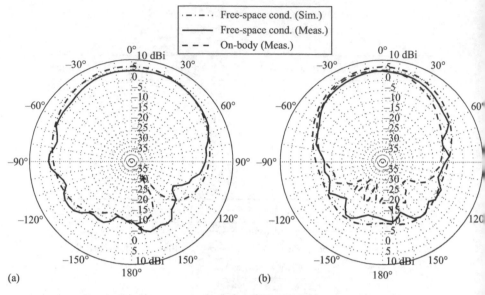

Figure 10.22   *Measured and simulated radiation patterns of the wideband 2.45-GHz ISM band and 4G LTE textile SIW cavity-backed slot antenna at 2.45 GHz. (a) E-plane and (b) H-plane*

Table 10.2   *Measured characteristics of the wideband 2.45-GHz ISM band and 4G LTE textile SIW cavity-backed slot antenna*

Frequency (GHz)	Gain (dBi)	FTBR (dB)	η_{rad} (%)
2.45	4.7	17.1	89
2.53	4.9	17.9	93
2.65	5.1	16.3	92

demonstrates a similar gain along broadside and slightly lower backside radiation, due to additional losses caused by body absorption. Finally, to assess antenna performance when deployed on a human arm, the antenna was bent with a radius of 5 cm along H- and E-plane. Figure 10.21 clearly demonstrates that the antenna remains matched over the frequency ranges of interest, demonstrating its robustness and capability to be deployed in bent conditions.

10.4.4.5   Ultra-wideband (UWB) textile SIW cavity-backed slot antenna

Ultra-wideband (UWB) communication is an emerging wireless technology that offers attractive features over conventional narrowband systems, such as high data rate transmission rates with low power spectral densities [76, 77], high-accuracy

Table 10.3   Dimensions of the UWB textile SIW cavity-backed slot antenna

Parameter	Value (mm)	Parameter	Value (mm)
L_c	75.0	W_{GND}	49.5
$W_{c,A}$	14.34	L_{GND}	85.0
$W_{c,B}$	14.17	$W_{B,GND}$	38.0
W_S	4.2	$L_{B,GND}$	85.0
L_S	67.0	W_f	13.4
s	8.0	g_{cpw}	3.7
d_{in}	4.0	l_{cpw}	4.8
d_{out}	6.0	d_{cpw}	3.5

localization of persons in indoor locations or confined areas [27, 78], and lower sensitivity for narrowband interference. Furthermore, UWB systems exhibit low-power consumption [79], low cost and small size, making the technology very suitable for integration into SFIT systems. The definition of a UWB signal, formulated by the Federal Communications Commission (FCC), requires that the antenna exhibits a −10 dB bandwidth in excess of 500 MHz and/or a fractional bandwidth larger than 20% [77]. Obviously, the ultra-wide bandwidth implies substantially more stringent antenna requirements and different propagation aspects compared to narrow band systems, making the antenna design more challenging [80]. First, matching and efficient radiation should be ensured over an ultra-wide bandwidth. Second, the UWB antenna's radiation pattern and group delay should both be as constant as possible over the entire bandwidth to minimize pulse distortion [80]. Finally, the antenna should maintain these characteristics when worn by a user, even under harsh operating conditions.

Lemey *et al.* [27] demonstrate that the bandwidth enhancement technique can be exploited to design a wearable textile UWB antenna that fulfills all of these stringent requirements. More specifically, a wearable UWB textile antenna based on the geometry depicted in Figure 10.19 was presented for application in the next-generation RF identification (RFID) systems, operating in the low-duty cycle restricted [3.4–4.8] GHz UWB band. The antenna dimensions were carefully selected to obtain a return loss characteristic with maximum bandwidth, exceeding 10 dB well below 3.4 GHz and exceeding 13 dB in the region where both hybrid modes merge, in combination with a stable and broad radiation pattern over that entire impedance bandwidth. In addition, dimensions that resulted in large peaks in the group delay were omitted. Furthermore, special care was taken to prevent out-of-band backside radiation at higher frequencies during the optimization process, as discussed more in depth in Reference 27. The antenna dimensions, after an extensive optimization process by means of the transient solver of CST Microwave studio, are given in Table 10.3.

Figure 10.23 shows the simulated group delay as a function of frequency. Therefore, a second, identical antenna was placed at a distance of 0.4 m along the positive *z*-direction in such a way that both antennas face each other. The group delay was

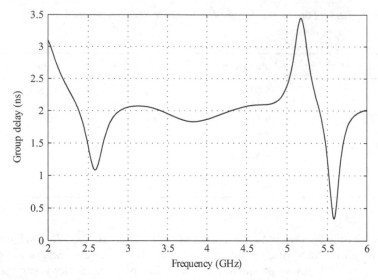

Figure 10.23   Simulated group delay of the textile UWB SIW cavity-backed slot antenna

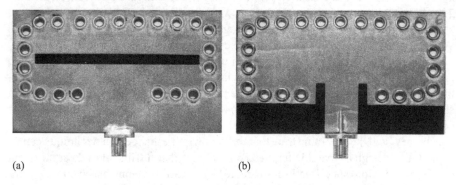

(a)   (b)

Figure 10.24   Prototype of the textile UWB SIW cavity-backed slot antenna. (a) Front view and (b) back view

then obtained by taking the negative derivative of the corresponding S_{21}'s phase. As shown in Figure 10.23, the group delay hardly changes within the 3.4–4.8 GHz band.

A prototype, depicted in Figure 10.24, was realized and measured to validate its excellent UWB performance under free-space conditions and under more realistic (on-body) conditions. First, the antenna performance was verified under free-space conditions in an anechoic room by means of Agilent's N5242A PNA-X Network Analyzer. The simulated and measured reflection coefficients $|S_{11}|$ are depicted in Figure 10.25 and agree well. Both demonstrate good impedance matching to $Z_0 = 50\,\Omega$ from 3.33 to 4.66 GHz, yielding a $-10\,dB$ impedance bandwidth of 1.33 GHz and a fractional bandwidth of 33%. Over the entire 3.4–4.8 GHz band, the return loss

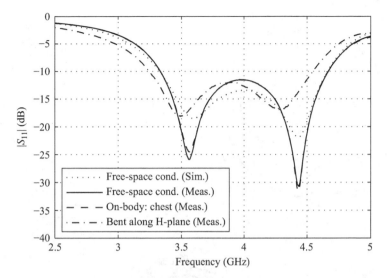

Figure 10.25   Reflection coefficient of the textile UWB SIW cavity-backed slot antenna under different conditions

remains larger than 6 dB. Figure 10.26 depicts the measured radiation patterns in the *E*- and *H*-plane of the antenna at 3.4, 3.85, and 4.3 GHz. These measurements prove that this topology guarantees a radiation pattern that hardly changes over the desired bandwidth, minimizing direction-specific distortion of UWB wave forms. This behavior is also demonstrated in Table 10.4, depicting the measured antenna gain along broadside, FTBR, and 3 dB-beamwidth in the *E*- and *H*-plane at the three frequencies under study. Then, the antenna's performance was extensively tested when deployed onto several body parts of a test person and under deformation, to validate antenna performance in more realistic situations. More specifically, in Reference 27, the antenna was deployed on the arm, leg, chest, and abdomen of an average height and weight male test person wearing a T-shirt and jeans. Only slight deviations were observed when deploying the antenna on different locations of the human body, compared to the free-space conditions. Figure 10.25 shows the antenna's $|S_{11}|$ when deployed on the chest to demonstrate the antenna's stable performance in vicinity of the human body. Finally, the antenna's return loss was also measured after bending the antenna with a radius of 5 cm along the *H*-plane. Then, both resonance frequencies are slightly shifted toward lower frequencies. Yet, the impedance bandwidth only reduces to 1.30 GHz, demonstrating the antenna's robustness when bent.

10.4.5   Textile microwave components

Besides the antennas presented in the previous sections, a variety of SIW components on textile have been implemented and tested, to prove the feasibility and performance of this class of structures for wearable applications.

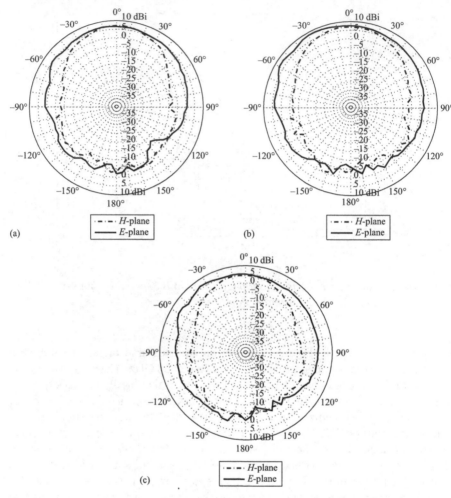

Figure 10.26   *Measured radiation pattern of the textile UWB SIW cavity-backed*
slot antenna at 3.4 GHz (a), 3.85 GHz (b), and 4.3 GHz (c)

Table 10.4   *Measured characteristics of the textile UWB SIW*
cavity-backed slot antenna

Frequency (GHz)	Gain (dBi)	FTBR (dB)	3-dB beamwidth (°)	
			E-plane	*H*-plane
3.40	6.8	10.5	117	46
3.85	6.5	10.5	123	54
4.30	4.6	7.1	133	46

10.4.5.1   SIW interconnect

A straight SIW interconnection on textile was designed for operation in the ISM band around 2.45 GHz [62]. Consequently, the cutoff frequency of the fundamental mode was set to $f_0 = 1.62$ GHz. To this aim, the width of the SIW is $w = 79$ mm, the diameter of metal vias $d = 4$ mm, and their longitudinal spacing $s = 8$ mm (Figure 10.27a).

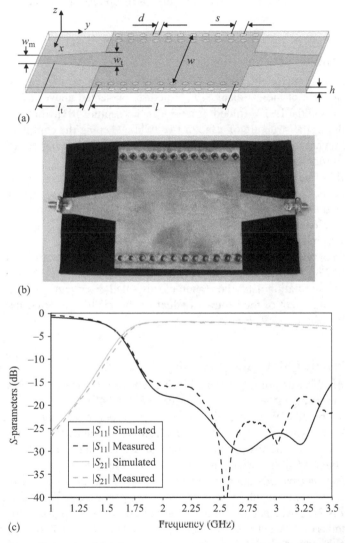

(a)

(b)

(c)

Figure 10.27   Straight SIW interconnect on textile: (a) geometry of the SIW, with transitions to input/output microstrip lines; (b) photograph of the prototype; (c) simulated and measured scattering parameters vs. frequency (©2015 IEEE. Reprinted with permission from Reference 62)

The substrate consists of a foam layer with a thickness of 3.94 mm. Input/output transitions from SIW to microstrip lines were adopted for measurement purposes (with dimensions $l_t = 32.7$ mm, $w_t = 23$ mm, and $w_m = 13$ mm). A prototype of the SIW interconnect on textile is displayed in Figure 10.27b.

Figure 10.27c shows the scattering parameters of the SIW structure, comparing simulation results and measurement data. The measured cut-off frequency of the fundamental SIW mode is 1.65 GHz, and the insertion loss is 2 dB at the frequency of 2.45 GHz.

10.4.5.2   Folded SIW interconnect

A substrate integrated folded waveguide (SIFW) was implemented with the aim to reduce the size of the classical SIW structure, presented above [62]. The SIFW consists of a standard SIW folded around a metal septum [81]: this allows reducing the width of the structure of a factor two, while keeping the complete shielding. The only drawback of this structure is the increased manufacturing complexity, as a double-layer topology is required.

The SIFW interconnect was designed to achieve the same cut-off frequency of the fundamental mode $f_0 = 1.62$ GHz, as the standard SIW interconnect described above. The width of the waveguide resulted to be $w = 41.2$ mm, corresponding to a reduction of 47.8% with respect to the standard SIW structure (Figure 10.28a). A photograph of the prototype is displayed in Figure 10.28b.

The comparison between simulated and measured scattering parameters is shown in Figure 10.27c. The measured cut-off frequency is shifted upward by 45 MHz with respect to the simulation data: the reason of this small discrepancy is attributed to a possible misalignment of the central conductive sheet. The measured insertion loss of the SIFW resulted 1.49 dB at 2.45 GHz.

10.4.5.3   Folded SIW filter

The folded SIW structure was adopted to implement a compact two-pole bandpass SIW filter, operating in the frequency band centered at 2.45 GHz. The filter is based on a dual-mode folded SIW cavity, with three insets cut out in the central metal septum (Figure 10.29a). This filter type allows controlling the bandwidth by changing the length and width of the insets, and the design of the filter is based on the modal analysis of the folded SIW cavity, as discussed in Reference 82.

The filter was manufactured and experimentally tested. Figure 10.29b shows the comparison of simulated and measured scattering parameters. The 3 dB bandwidth of the filter is 725 MHz, and the insertion loss at 2.45 GHz is 2.3 dB. In this filter, the first two modes of the cavity determine the passband of the filter, and the suppression of some of the upper modes due to the input/output striplines guarantees the wide rejection band between 3 and 6 GHz.

This filter topology provides good selectivity, excellent out-of-band performance, along with a compact size: in fact, the area of this filter is approximately four times smaller than the area of a classical second-order cavity SIW filter.

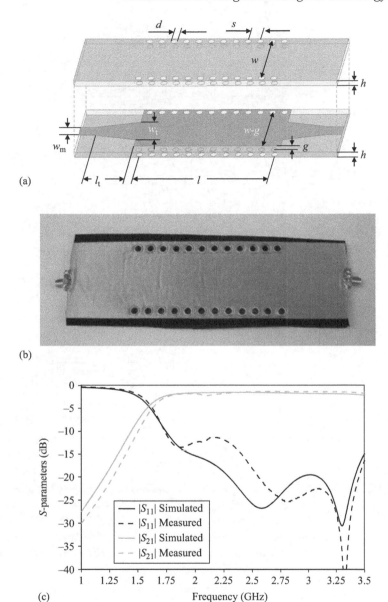

(a)

(b)

(c)

Figure 10.28 *Folded SIW interconnect on textile (dimensions in mm: $l_t = 28.7$,*
$w_i - 19.6$, $w_m - 8.5$, $h - 3.94$, $s - 8$, $d - 4$, $g - 4$, $w - 41.2$, $l - 96$):
(a) geometry of the SIFW, with transitions to input/output striplines,
(b) photograph of the prototype, and (c) simulated and measured
scattering parameters vs. frequency (©2015 IEEE. Reprinted with
permission from Reference 62)

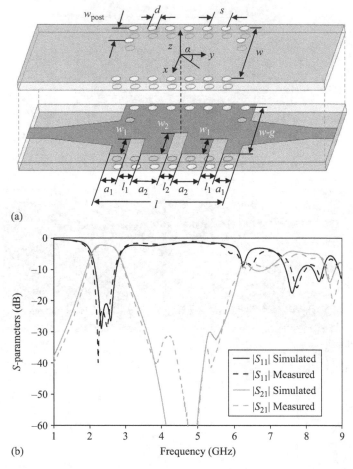

(a)

(b)

Figure 10.29   Folded SIW filter on textile (dimensions in mm: w = 41.2, g = 4,
d = 4, s = 8. l = 56.3, w_{post} = 9.9, l_1 = 6.3, l_2 = 5.9, w_1 = 12,
w_2 = 17, a_1 = 7.2, a_2 = 11.7): (a) geometry of the SIFW filter, with
transitions to input/output striplines and (b) simulated and measured
scattering parameters vs. frequency (©2015 IEEE. Reprinted with
permission from Reference 62)

10.5   Textile SIW antennas as hybrid energy-harvesting platforms

10.5.1   Introduction

A major issue of the current generation smart fabric and interactive textile (SFIT) systems concerns the limited system autonomy, and the corresponding inconvenience of
frequent charging [83] As Huang *et al.* [84] identify the wireless communication

subsystem as one of the major power consumers within a SFIT system, the textile antenna's dimensions should remain of the order of a wavelength to yield large radiation efficiency in proximity of the human body [85]. Moreover, by deploying multiple highly efficient textile antennas on well-considered locations of the wearer's body, shadowing caused by the human body may be avoided, different services in distinct frequency bands can be provided and/or system autonomy can be further extended [86]. System autonomy can even be extended further, as Lemey *et al.* [70] pinpoint SFIT systems as prime candidates to be partly, or even solely, powered by energy-harvesting techniques. Extracting energy at the location of operation from the user's activities [70, 87], or from ambient sources [70, 88], could even make batteries superfluous and could prevent life-threatening situations, due to batteries running out of power during interventions. Obviously, exploiting multiple different energy sources is preferable as such a hybrid approach improves the continuity of energy scavenging and the amount of energy harvesting [89]. As already a large surface in the SFIT system is consumed by the textile (multi-)antenna system, a smart approach is requisite for the integration of multiple energy harvesters and the corresponding power management module, in order to harvest sufficient levels of energy, on the one hand, maintaining the user's comfort on the other hand, and without degrading the textile antenna's performance.

10.5.2 Exploiting the textile antenna as integration platform

Lemey *et al.* [70] describe how a textile antenna's functionality may be further extended by exploiting its surface as an energy-harvesting and power management platform. Their approach is illustrated in Figure 10.30. It enables a compact, highly integrated and unobtrusive design. The selection of an appropriate antenna topology is critical. Obviously, a topology should be selected that facilitates the integration of

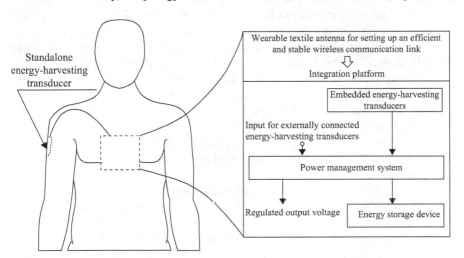

Figure 10.30　*Concept of exploiting the textile antenna as energy harvesting platform*

additional electronic hardware without degradation of radiation performance, in the meantime guaranteeing the wearer's comfort. A plethora of antenna topologies were discussed, identifying textile SIW cavity-backed slot antennas as ideal candidates. First, the cavity-backed slot topology, in combination with the SIW technology, guarantees an excellent isolation from its environment, making it not only very suitable for on-body deployment but also for the integration of additional energy-harvesting or computational hardware. Second, SIW technology allows adopting antenna materials that enable a flexible, low-profile, and conformal design. A third and very important reason to adopt an SIW cavity-backed slot textile antenna concerns the ease by which additional hardware can be integrated, in the meantime minimizing the amount and length of the interconnections by exploiting the topology's properties. Sections 10.5.3 and 10.5.4 elaborate on the latter reason, by discussing two different designs in which energy-harvesting and power management hardware are integrated onto the textile SIW cavity-backed slot antenna, discussed in Sections 10.4.4 and 10.4.1, respectively. Furthermore, both sections also briefly repeat the guidelines to ensure good antenna performance after integration. By following these guidelines, other textile SIW cavity-backed slot antennas, such as the ones discussed in Sections 10.4.2 and 10.4.3, could also be exploited as an integration platform.

10.5.3   Wideband SIW textile antenna with integrated solar harvester

The integration procedure described in Reference 70 and briefly discussed in Section 10.5.2 was for the first time applied in Reference 27. More specifically, in Reference 27, the textile SIW cavity-backed slot antenna, discussed in Section 10.4.4, is exploited as an integration platform for a flexible solar energy-harvesting system, as depicted in Figure 10.20. The flexible solar energy-harvesting system is designed to charge an energy storage device in outdoor environments and to provide a regulated output voltage for adequately powering a wireless communication system.

Therefore, one ultra-thin and ultra-flexible hydrogenated amorphous silicon (a-Si:H) solar cell is integrated on the slot plane of the antenna, whereas a flexible power management system (PMS), including a micro-energy cell (MEC) as energy storage device, is integrated onto the antenna's feed plane, to prevent antenna radiation from coupling into the circuitry. The PMS relies on the MAX17710, an energy-harvesting charger and protector IC by Maxim Integrated, to control the charging process of the MEC and to generate a regulated output voltage. A more elaborate description about the functionality and design procedure of the PMS can be found in Reference 27.

Mechanical flexibility of the design is maintained by implementing the PMS on a flexible polyimide layer and using small electronic components that are distributed over a larger area than necessary. Furthermore, a non-conductive, stretchable sheet is used to fix the solar cell to the slot plane, and the PMS and MEC to the feed plane. The solar cell is positioned in such a way that the radiating slot remains uncovered, whereas the MEC and the PMS are oriented to prevent the GCPW feed line from being covered. The antenna cavity is then exploited as a common DC ground for the PMS, MEC, and flexible solar cell, to minimize the amount and length of interconnections, yielding

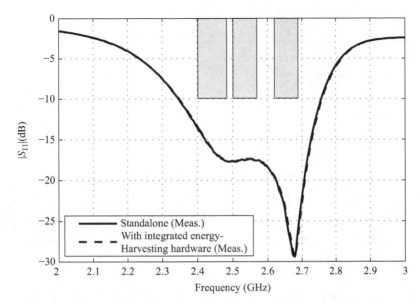

Figure 10.31   *Reflection coefficient of the 2.45-GHz ISM band and 4G LTE textile*
SIW cavity-backed slot antenna, under free-space conditions, to
validate the integration procedure

increased robustness and a higher comfort for the wearer. Furthermore, the positive
connection of the solar cell is routed through one of the hollow tube eyelets, as depicted
in Figure 10.20, further reducing the length of interconnections. Figure 10.20 also
depicts the antenna when bent along the *H*-plane (d) and along the *E*-plane (e), demonstrating its bending capability after integration of the solar energy-harvesting system.

The integration procedure is validated under free-space conditions by comparing
the antenna's standalone performance with the antenna's performance after integration of the solar harvesting system. Figures 10.31 and 10.32 demonstrate that the
integration of the solar cell, PMS, and MEC only has a minor influence on antenna
performance. Moreover, Lemey *et al.* [27] showed that the integration of the flexible
solar harvester allows scavenging up to 53 mW in real-life outdoor environments,
without requiring additional surface of the garment.

10.5.4   SIW cavity-backed slot antenna with integrated hybrid
energy-harvesting hardware

In Reference 70, the integration procedure is applied to integrate a flexible hybrid
energy-harvesting system and PMS onto the textile SIW cavity-backed slot antenna
described in Section 10.4.1. Now, the integration of the flexible hybrid energy-
harvesting system and PMS enables combining the energy scavenged from indoor
artificial light, outdoor solar light, and the heat emanated by the human body to charge
a MEC and to provide a regulated output voltage. Such a hybrid approach is preferable over the single energy source-harvesting technique discussed in Section 10.5.3,

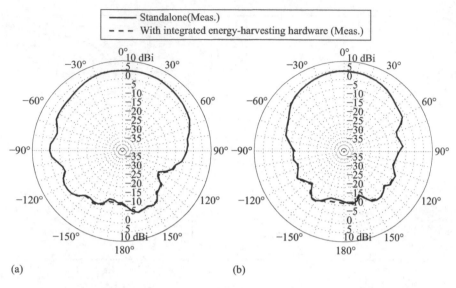

Figure 10.32   *Measured radiation pattern of the 2.45-GHz ISM band and 4G LTE textile SIW cavity-backed slot antenna at 2.45 GHz, under free-space conditions, to validate the integration procedure. (a) E-plane and (b) H-plane*

as it yields a higher continuity of energy scavenging and a larger amount of total harvested energy.

Figure 10.33 shows the integration of two different ultra-thin, flexible a-Si:H solar cells at the antenna's slot plane and a flexible power management module, including a 170-μm-thick MEC, at its feed plane. The flexible PMS consists of an ultra-low voltage step up converter and a central PMS (CPMS). The latter is designed to charge the MEC from two high-voltage DC sources by means of linear harvesting and one low-voltage DC source via an incorporated boost converter, in the meantime protecting the MEC and providing a regulated output voltage. By connecting the smaller solar cell to the boost converter input and the larger solar cell to one of the linear harvesting inputs of the CPMS, energy scavenging from artificial and natural solar light is enabled. Furthermore, by connecting the output of the ultra-low voltage step up converter to the other linear harvesting input of the CPMS, thermal body energy harvesting via an externally connected TEG is enabled. Therefore, the TEG's hot side needs to be tightly attached to the skin of the wearer, whereas its cold side needs to be exposed to the ambient air. Lemey *et al.* [70] describe a procedure to select an appropriate TEG.

A similar approach as in Section 10.5.3 was applied to maintain mechanical flexibility and antenna performance after integration of the energy-harvesting and power management hardware. First, both circuits are etched on a flexible polyimide substrate and small electronic components, distributed over a larger surface than necessary, were used. Second, non-conductive adhesive sheets were applied to glue both solar cells and both circuits to the antenna's slot and feed plane, respectively.

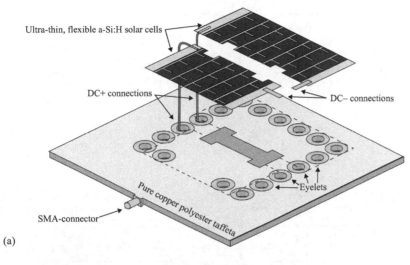

(a)

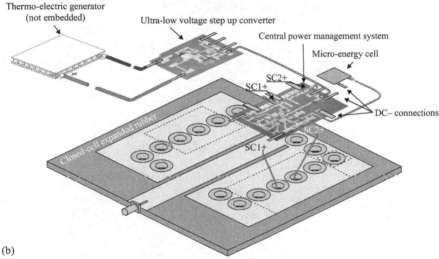

(b)

Figure 10.33   *The wearable textile SIW cavity-backed slot antenna, discussed in*
                Section 10.4.1, with integrated flexible a-Si:H solar cells, PMS, and
                energy storage device. The externally connected TEG is also
                depicted. (a) Slot plane and (b) feed plane

Judicious alignment prevents the slot and feed line from being covered. Therefore,
unlike in Section 10.5.3, the solar cells are patterned to fit the antenna's slot. Finally,
the amount and length of interconnecting wires were minimized by routing the positive
connections of both solar cells through the hollow eyelets to the PMS and exploiting
the antenna cavity as a common DC ground for the solar cells, MEC, and PMS.

    Figures 10.34 and 10.35 demonstrate that the integration of the additional energy-
harvesting and power management hardware, according to Figure 10.33, only has a

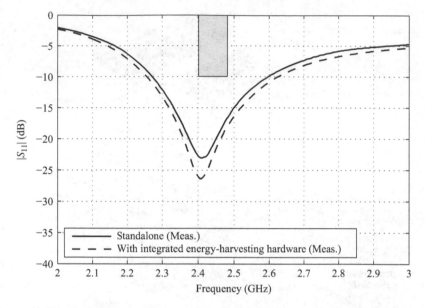

Figure 10.34   Reflection coefficient of the textile SIW cavity-backed slot antenna, under free-space conditions, to validate the integration procedure

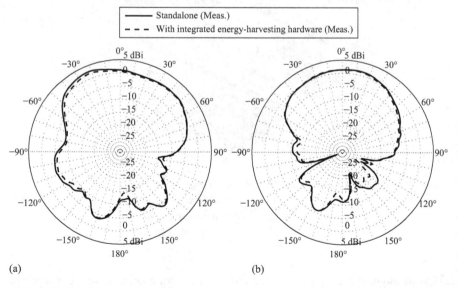

Figure 10.35   Measured radiation pattern of the textile SIW cavity-backed slot antenna at 2.45 GHz, under free-space conditions, to validate the integration procedure. (a) E-plane and (b) H-plane

minor influence on the antenna's reflection coefficient and radiation pattern, respectively. Furthermore, a comprehensive field test was carried out in a realistic indoor environment to emphasize the importance of a hybrid energy-harvesting approach in realistic scenarios [89]. First, the field test clearly demonstrated that the periods during which no energy can be harvested are significantly reduced. Second, it was also shown that, during a significant percentage of the time, energy could be harvested from multiple energy sources. Both effects contribute to higher system autonomy and/or reduced battery size.

References

[1] L. Atzori, A. Iera, and G. Morabito, "The internet of things: a survey," *Computer Networks*, vol. 54, no. 15, pp. 2787–2805, 2010.

[2] D. Miorandi, S. Sicari, F. De Pellegrini, and I. Chlamtac, "Internet of things: vision, applications and research challenges," *Ad Hoc Networks*, vol. 10, no. 7, pp. 1497–1516, 2012.

[3] M. Zorzi, A. Gluhak, S. Lange, and A. Bassi, "From today's INTRAnet of things to a future INTERnet of things: a wireless- and mobility-related view," *Wireless Communications, IEEE*, vol. 17, no. 6, pp. 44–51, Dec. 2010.

[4] G. D. Abowd and E. D. Mynatt, "Charting past, present, and future research in ubiquitous computing," *ACM Transactions on Computer-Human Interaction (TOCHI)*, vol. 7, no. 1, pp. 29–58, 2000.

[5] M. Weiser, "The computer for the 21st century," *Scientific American*, vol. 265, no. 3, pp. 94–104, 1991.

[6] M. Di Rienzo, F. Rizzo, G. Parati, G. Brambilla, M. Ferratini, and P. Castiglioni, "MagIC system: a new textile-based wearable device for biological signal monitoring. Applicability in daily life and clinical setting," in *27th Annual International Conference of the Engineering in Medicine and Biology Society, 2005. IEEE-EMBS 2005*, pp. 7167–7169.

[7] A. Fanelli, M. Ferrario, L. Piccini, *et al.*, "Prototype of a wearable system for remote fetal monitoring during pregnancy," in *Engineering in Medicine and Biology Society (EMBC), 2010*, pp. 5815–5818.

[8] Y. Hao and R. Foster, "Wireless body sensor networks for health-monitoring applications," *Physiological Measurement*, vol. 29, no. 11, p. R27, 2008.

[9] N. Oliver and F. Flores-Mangas, "HealthGear: a real-time wearable system for monitoring and analyzing physiological signals," in *International Workshop on Wearable and Implantable Body Sensor Networks, 2006. BSN 2006*, 2006, 4 pp.

[10] A. Pantelopoulos and N. G. Bourbakis, "A survey on wearable sensor-based systems for health monitoring and prognosis," *IEEE Transactions on Systems, Man, and Cybernetics, Part C: Applications and Reviews*, vol. 40, no. 1, pp. 1–12, Jan. 2010.

[11] M. G. Tsipouras, A. T. Tzallas, E. C. Karvounis, *et al.*, "A wearable system for long-term ubiquitous monitoring of common motor symptoms in patients

with Parkinson's disease," in *2014 IEEE-EMBS International Conference on Biomedical and Health Informatics (BHI)*, 2014, pp. 173–176.

[12] A. Thielens, S. Agneessens, L. Verloock, *et al.*, "On-body calibration and processing for a combination of two radio-frequency personal exposimeters," *Radiation Protection Dosimetry*, vol. 163, no. 1, pp. 58–69, 2015.

[13] A. Thielens, H. De Clercq, S. Agneessens, *et al.*, "Personal distributed exposimeter for radio frequency exposure assessment in real environments," *Bioelectromagnetics*, vol. 34, no. 7, pp. 563–567, 2013.

[14] A. Thielens, P. Vanveerdeghem, S. Agneessens, *et al.*, "Whole-body averaged specific absorption rate estimation using a personal, distributed exposimeter," *IEEE Antennas and Wireless Propagation Letters,* vol. 14, pp. 1534–1537, 2015.

[15] T. Faetti and R. Paradiso, "A novel wearable system for elderly monitoring," *Advances in Science and Technology*, vol. 85, pp. 17–22, 2013.

[16] S. Patel, H. Park, P. Bonato, L. Chan, and M. Rodgers, "A review of wearable sensors and systems with application in rehabilitation," *Journal of Neuroengineering and Rehabilitation*, vol. 9, no. 1, p. 21, 2012.

[17] T. Tamura, T. Yoshimura, M. Sekine, M. Uchida, and O. Tanaka, "A wearable airbag to prevent fall injuries," *IEEE Transactions on Information Technology in Biomedicine,* vol. 13, no. 6, pp. 910–914, 2009.

[18] M. Klann, T. Riedel, H. Gellersen, *et al.*, "Lifenet: an ad-hoc sensor network and wearable system to provide firefighters with navigation support," in *Adjunct Proc. Ubicomp*, 2007.

[19] R. Paradiso, G. Loriga, and N. Taccini, "A wearable health care system based on knitted integrated sensors," *IEEE Transactions on Information Technology in Biomedicine,* vol. 9, no. 3, pp. 337–344, 2005.

[20] S. Agneessens, P. Van Torre, F. Declercq, *et al.*, "Design of a wearable, low-cost, through-wall Doppler radar system," *International Journal of Antennas and Propagation*, vol. 2012, 2012.

[21] A. A. Serra, P. Nepa, and G. Manara, "A wearable two-antenna system on a life jacket for Cospas-Sarsat personal locator beacons," *IEEE Transactions on Antennas and Propagation,* vol. 60, no. 2, pp. 1035–1042, 2012.

[22] J. Chardonnens, J. Favre, B. Le Callennec, F. Cuendet, G. Gremion, and K. Aminian, "Automatic measurement of key ski jumping phases and temporal events with a wearable system," *Journal of Sports Sciences*, vol. 30, no. 1, pp. 53–61, 2012.

[23] C. Katsis, Y. Goletsis, G. Rigas, and D. Fotiadis, "A wearable system for the affective monitoring of car racing drivers during simulated conditions," *Transportation Research Part C: Emerging Technologies*, vol. 19, no. 3, pp. 541–551, 2011.

[24] G. Badiali, V. Ferrari, F. Cutolo, *et al.*, "Augmented reality as an aid in maxillofacial surgery: validation of a wearable system allowing maxillary repositioning," *Journal of Cranio-Maxillofacial Surgery*, vol. 42, no. 8, pp. 1970–1976, 2014.

[25] F. Declercq, A. Georgiadis, and H. Rogier, "Wearable aperture-coupled shorted solar patch antenna for remote tracking and monitoring applications,"

in *Proceedings of the 5th European Conference on Antennas and Propagation (EUCAP)*, 2011, pp. 2992–2996.

[26]   A. Z. Kausar, A. W. Reza, M. U. Saleh, and H. Ramiah, "Energizing wireless sensor networks by energy harvesting systems: scopes, challenges and approaches," *Renewable and Sustainable Energy Reviews*, vol. 38, pp. 973–989, 2014.

[27]   S. Lemey, F. Declercq, and H. Rogier, "Dual-band substrate integrated waveguide textile antenna with integrated solar harvester," *IEEE Antennas and Wireless Propagation Letters*, vol. 13, pp. 269–272, 2014.

[28]   G. Orecchini, L. Yang, M. Tentzeris, and L. Roselli, "Smart Shoe: an autonomous inkjet-printed RFID system scavenging walking energy," in *2011 IEEE International Symposium on Antennas and Propagation (APSURSI)*, 2011, pp. 1417–1420.

[29]   L. Swallow, J. Luo, E. Siores, I. Patel, and D. Dodds, "A piezoelectric fibre composite based energy harvesting device for potential wearable applications," *Smart Materials and Structures*, vol. 17, no. 2, p. 025017, 2008.

[30]   W. Wu, S. Bai, M. Yuan, Y. Qin, Z. L. Wang, and T. Jing, "Lead zirconate titanate nanowire textile nanogenerator for wearable energy-harvesting and self-powered devices," *ACS Nano*, vol. 6, no. 7, pp. 6231–6235, 2012.

[31]   B. Yang and K.-S. Yun, "Piezoelectric shell structures as wearable energy harvesters for effective power generation at low-frequency movement," *Sensors and Actuators A: Physical*, vol. 188, pp. 427–433, 2012.

[32]   M. M. Jackson, C. Zeagler, G. Valentin, *et al.*, "FIDO-facilitating interactions for dogs with occupations: wearable dog-activated interfaces," in *Proceedings of the 2013 International Symposium on Wearable Computers*, 2013, pp. 81–88.

[33]   P. D. McGreevy, M. Sundin, M. Karlsteen, *et al.*, "Problems at the human–horse interface and prospects for smart textile solutions," *Journal of Veterinary Behavior: Clinical Applications and Research*, vol. 9, no. 1, pp. 34–42, 2014.

[34]   J. Cheng, P. Lukowicz, N. Henze, *et al.*, "Smart textiles: from niche to mainstream," *IEEE Pervasive Computing*, vol. 12, no. 3, pp. 81–84, 2013.

[35]   K. Cherenack and L. van Pieterson, "Smart textiles: challenges and opportunities," *Journal of Applied Physics*, vol. 112, no. 9, p. 091301, 2012.

[36]   T. Kellomaki and L. Ukkonen, "Design approaches for bodyworn RFID tags," in *2010 3rd International Symposium on Applied Sciences in Biomedical and Communication Technologies (ISABEL)*, 2010, pp. 1–5.

[37]   K. Koski, "Characterization and Design Methodologies for Wearable Passive UHF RFID Tag Antennas for Wireless Body-Centric Systems," Tampere University of Technology, 2015.

[38]   K. Koski, T. Björninen, L. Sydänheimo, L. Ukkonen, and Y. Rahmat-Samii, "A new approach and analysis of modeling the human body in RFID-enabled body-centric wireless systems," *International Journal of Antennas and Propagation*, vol. 2014, 2014.

[39]   D. Brosteaux, F. Axisa, J. Vanfleteren, N. Carchon, and M. Gonzalez, "Elastic interconnects for stretchable electronic circuits using mid (moulded

interconnect device) technology," in *MRS Proceedings*, 2006, vol. 926, pp. 0926–CC08.

[40]   M. Gonzalez, F. Axisa, M. V. Bulcke, D. Brosteaux, B. Vandevelde, and J. Vanfleteren, "Design of metal interconnects for stretchable electronic circuits," *Microelectronics Reliability*, vol. 48, no. 6, pp. 825–832, 2008.

[41]   B. Huyghe, H. Rogier, J. Vanfleteren, and F. Axisa, "Design and manufacturing of stretchable high-frequency interconnects," *IEEE Transactions on Advanced Packaging*, vol. 31, no. 4, pp. 802–808, 2008.

[42]   M.-C. Gosselin, G. Vermeeren, S. Kuhn, *et al.*, "Estimation formulas for the specific absorption rate in humans exposed to base-station antennas," *IEEE Transactions on Electromagnetic Compatibility*, vol. 53, no. 4, pp. 909–922, 2011.

[43]   W. Joseph, G. Vermeeren, L. Verloock, M. M. Heredia, and L. Martens, "Characterization of personal RF electromagnetic field exposure and actual absorption for the general public," *Health Physics*, vol. 95, no. 3, pp. 317–330, 2008.

[44]   G. Vermeeren, M.-C. Gosselin, S. Kühn, *et al.*, "The influence of the reflective environment on the absorption of a human male exposed to representative base station antennas from 300 MHz to 5 GHz," *Physics in Medicine and Biology*, vol. 55, no. 18, p. 5541, 2010.

[45]   P. S. Hall, Y. Hao, Y. I. Nechayev, *et al.*, "Antennas and propagation for on-body communication systems," *IEEE Antennas and Propagation Magazine*, vol. 49, no. 3, pp. 41–58, Jun. 2007.

[46]   P. S. Hall and Y. Hao, "Antennas and propagation for body centric communications," in *Antennas and Propagation, 2006. EuCAP 2006. First European Conference on*, 2006, pp. 1–7.

[47]   E. Reusens, W. Joseph, B. Latré, *et al.*, "Characterization of on-body communication channel and energy efficient topology design for wireless body area networks," *IEEE Transactions on Information Technology in Biomedicine*, vol. 13, no. 6, pp. 933–945, 2009.

[48]   E. Reusens, W. Joseph, G. Vermeeren, D. Kurup, and L. Martens, "Real human body measurements, model, and simulations of a 2.45 GHz wireless body area network communication channel," in *5th International Summer School and Symposium on Medical Devices and Biosensors, 2008. ISSS-MDBS 2008*, 2008, pp. 149–152.

[49]   E. Reusens, W. Joseph, G. Vermeeren, and L. Martens, "On-body measurements and characterization of wireless communication channel for arm and torso of human," in *4th International Workshop on Wearable and Implantable Body Sensor Networks (BSN 2007)*, 2007, pp. 264–269.

[50]   P. Van Torre, L. Vallozzi, L. Jacobs, H. Rogier, M. Moeneclaey, and J. Verhaevert, "Characterization of measured indoor off-body MIMO channels with correlated fading, correlated shadowing and constant path loss," *IEEE Transactions on Wireless Communications*, vol. 11, no. 2, pp. 712–721, 2012.

[51] A. Tronquo, H. Rogier, C. Hertleer, and L. Van Langenhove, "Robust planar textile antenna for wireless body LANs operating in 2.45 GHz ISM band," *Electronics Letters*, vol. 42, no. 3, pp. 142–143, 2006.

[52] F. Declercq and H. Rogier, "Active integrated wearable textile antenna with optimized noise characteristics," *IEEE Transactions on Antennas and Propagation*, vol. 58, no. 9, pp. 3050–3054, 2010.

[53] A. Dierck, H. Rogier, and F. Declercq, "A wearable active antenna for global positioning system and satellite phone," *IEEE Transactions on Antennas and Propagation*, vol. 61, no. 2, pp. 532–538, 2013.

[54] T. Kellomäki, J. Virkki, S. Merilampi, and L. Ukkonen, "Towards washable wearable antennas: a comparison of coating materials for screen-printed textile-based UHF RFID tags," *International Journal of Antennas and Propagation*, vol. 2012, 2012.

[55] M. L. Scarpello, I. Kazani, C. Hertleer, H. Rogier, and D. Vande Ginste, "Stability and efficiency of screen-printed wearable and washable antennas," *IEEE Antennas and Wireless Propagation Letters*, vol. 11, pp. 838–841, 2012.

[56] W. G. Whittow, A. Chauraya, J. C. Vardaxoglou, *et al.*, "Inkjet-printed microstrip patch antennas realized on textile for wearable applications," *IEEE Antennas and Wireless Propagation Letters*, vol. 13, pp. 71–74, 2014.

[57] A. Kiourti and J. L. Volakis, "High-geometrical-accuracy embroidery process for textile antennas with fine details," *IEEE Antennas and Wireless Propagation Letters*, vol. 14, no. 99, pp. 1–1, 2014.

[58] J. L. Volakis, L. Zhang, Z. Wang, and Y. Bayram, "Embroidered flexible RF electronics," in *2012 IEEE International Workshop on Antenna Technology (iWAT)*, 2012, pp. 8–11.

[59] F. Declercq, H. Rogier, and C. Hertleer, "Permittivity and loss tangent characterization for garment antennas based on a new matrix-pencil two-line method," *IEEE Transactions on Antennas and Propagation*, vol. 56, no. 8, pp. 2548–2554, Aug. 2008.

[60] F. Declercq, I. Couckuyt, H. Rogier, and T. Dhaene, "Environmental high frequency characterization of fabrics based on a novel surrogate modelling antenna technique," *IEEE Transactions on Antennas and Propagation*, vol. 61, no. 10, pp. 5200–5213, Oct. 2013.

[61] F. Declercq and H. Rogier, "Characterization of electromagnetic properties of textile materials for the use in wearable antennas," in *Antennas and Propagation Society International Symposium, 2009. APSURSI'09. IEEE*, 2009, pp. 1–4.

[62] R. Moro, S. Agneessens, H. Rogier, A. Dierck, and M. Bozzi, "Textile microwave components in substrate integrated waveguide technology," *IEEE Transactions on Microwave Theory and Techniques*, vol. 63, no. 2, pp. 422–432, Feb. 2015.

[63] M. Bozzi, A. Georgiadis, and K. Wu, "Review of substrate-integrated waveguide circuits and antennas," *Microwaves, Antennas & Propagation, IET*, vol. 5, no. 8, pp. 909–920, 2011.

[64]   T. Djerafi and K. Wu, "Substrate integrated waveguide(SIW) techniques: the state-of-the-art developments and future trends," *Journal of University of Electronic Science and Technology of China*, vol. 42, no. 2, pp. 171–192, 2013.

[65]   R. Moro, S. Agneessens, H. Rogier, and M. Bozzi, "Wearable textile antenna in substrate integrated waveguide technology," *IET Electronics Letters*, vol. 48, no. 16, pp. 985–987, 2012.

[66]   S. Agneessens and H. Rogier, "Compact half diamond dual-band textile HMSIW on-body antenna," *IEEE Transactions on Antennas and Propagation*, vol. 62, no. 5, pp. 2374–2381, May 2014.

[67]   T. Kaufmann and C. Fumeaux, "Wearable textile half-mode substrate-integrated cavity antenna using embroidered vias," *IEEE Antennas and Wireless Propagation Letters*, vol. 12, pp. 805–808, 2013.

[68]   S. Agneessens, S. Lemey, T. Vervust, and H. Rogier, "Wearable, small, and robust: the circular quarter-mode textile antenna," *IEEE Antennas and Wireless Propagation Letters*, vol. 14, no. , pp. 1482–1485, 2015.

[69]   R. Moro, M. Bozzi, S. Agneessens, and H. Rogier, "Compact cavity-backed antenna on textile in substrate integrated waveguide (SIW) technology," in *European Microwave Conference (EuMC), 2013*, 2013, pp. 1007–1010.

[70]   S. Lemey, F. Declercq, and H. Rogier, "Textile antennas as hybrid energy-harvesting platforms," *Proceedings of the IEEE*, vol. 102, no. 11, pp. 1833–1857, Nov. 2014.

[71]   T. Kaufmann, Z. Xu, and C. Fumeaux, "Wearable substrate-integrated waveguide with embroidered vias," in *2014 8th European Conference on Antennas and Propagation (EuCAP)*, 2014, pp. 1746–1750.

[72]   F. Giuppi, A. Georgiadis, A. Collado, and M. Bozzi, "A compact, single-layer substrate integrated waveguide (SIW) cavity-backed active antenna oscillator," *IEEE Antennas and Wireless Propagation Letters*, vol. 11, pp. 431–433, 2012.

[73]   F. Giuppi, A. Georgiadis, A. Collado, M. Bozzi, and L. Perregrini, "Tunable SIW cavity backed active antenna oscillator," *Electronics Letters*, vol. 46, no. 15, pp. 1053–1055, 2010.

[74]   K. Gong, Z. N. Chen, X. Qing, P. Chen, and W. Hong, "Substrate integrated waveguide cavity-backed wide slot antenna for 60-GHz bands," *IEEE Transactions on Antennas and Propagation*, vol. 60, no. 12, pp. 6023–6026, Dec. 2012.

[75]   G. Q. Luo, Z. F. Hu, W. J. Li, X. H. Zhang, L. L. Sun, and J. F. Zheng, "Bandwidth-enhanced low-profile cavity-backed slot antenna by using hybrid SIW cavity modes," *IEEE Transactions on Antennas and Propagation*, vol. 60, no. 4, pp. 1698–1704, Apr. 2012.

[76]   N. Chahat, M. Zhadobov, R. Sauleau, and K. Ito, "A compact UWB antenna for on-body applications," *IEEE Transactions on Antennas and Propagation*, vol. 59, no. 4, pp. 1123–1131, Apr. 2011.

[77]   F. C. Commission, "Revision of part 15 of the Commission's Rules Regarding Ultra-WidebandTransmission Systems, First Rep. Order (ET Docket 98-153)."

[78]   D. Dardari, R. D'Errico, C. Roblin, A. Sibille, and M. Z. Win, "Ultrawide bandwidth RFID: the next generation?" *Proceedings of the IEEE*, vol. 98, no. 9, pp. 1570–1582, Sep. 2010.

[79] Z. Li, W. Dehaene, and G. Gielen, "A 3-tier UWB-based indoor localization system for ultra-low-power sensor networks," *IEEE Transactions on Wireless Communications*, vol. 8, no. 6, pp. 2813–2818, Jun. 2009.

[80] W. Q. Malik, C. J. Stevens, and D. J. Edwards, "Ultrawideband antenna distortion compensation," *IEEE Transactions on Antennas and Propagation*, vol. 56, no. 7, pp. 1900–1907, Jul. 2008.

[81] N. Grigoropoulos, B. Sanz-Izquierdo, P. R. Young, *et al.* "Substrate integrated folded waveguides (SIFW) and filters," *IEEE Microwave and Wireless Components Letters*, vol. 15, no. 12, pp. 829–831, 2005.

[82] R. Moro, S. Moscato, M. Bozzi, and L. Perregrini, "Substrate integrated folded waveguide filter with out-of-band rejection controlled by resonant-mode suppression," *IEEE Microwave and Wireless Components Letters*, vol. 25, no. 4, pp. 214–216, 2015.

[83] M. A. Hanson, H. C. Powell Jr, A. T. Barth, *et al.*, "Body area sensor networks: challenges and opportunities," *Computer*, vol. 42, no. 1, pp. 58–65, 2009.

[84] L. Huang, V. Pop, R. de Francisco, *et al.*, "Ultra low power wireless and energy harvesting technologies – an ideal combination," in *Proceedings of IEEE International Conference on Communication Systems*, pp. 295–300.

[85] P. Vanveerdeghem, B. Jooris, P. Becue, *et al.*, "Reducing power consumption in bodycentric ZigBee communication links by means of wearable textile antennas," in *Second International Workshop on Measurement-Based Experimental Research, Methodology and Tools*, 2013.

[86] P. Van Torre, L. Vallozzi, A. Dierck, H. Rogier, and M. Moeneclaey, "Power-efficient body-centric communications," in *URSI Benelux Forum*, pp. 8–10.

[87] T. Starner, "Human-powered wearable computing," *IBM Systems Journal*, vol. 35, pp. 618–629, 1996.

[88] B. Markus and H. Jurgen, "Flexible solar cells for clothing," *Materials Today*, vol. 9, no. 6, pp. 42–50, 2006.

[89] S. Lemey and H. Rogier, "Substrate integrated waveguide textile antennas as energy harvesting platforms," International Workshop on Antenna Technology (iWAT), Seoul, 2015, pp. 23–26, 2015. doi: 10.1109/IWAT.2015.7365349

Chapter 11

Ultra wideband body-centric networks for localisation and motion capture applications

Richa Bharadwaj, Qammer H. Abbasi**, John Batchelor[†],
Srijittra Swaisaenyakorn[††], and A. Alomainy[‡]*

Localisation and motion tracking using body-worn antennas are emerging as an important research area based on ultra wideband (UWB) technology. Motion tracking itself is motivated by a variety of applications such as training of athletes, patient monitoring in health care domain, localisation of people in home or office environment, and the human body is an integral part of such applications. Hence, it is important to study the effect of human body on UWB localisation and the accuracy achieved while localizing the antennas present on the body. The choice of sensors, such as compact, efficient and low-cost UWB antennas, makes human localisation and activity monitoring a promising new application made possible by advances in UWB technology. In this chapter, UWB three-dimensional (3D) human body localisation is studied using body-worn antennas placed on different locations on the human body through numerical and experimental investigations. Detailed analysis is performed based on the measurement data in terms of propagation phenomenon for each antenna location and how the presence of human body affects ranging and localisation accuracy. The objective of the work is to achieve high-accuracy localisation of the human body using time of arrival positioning techniques and also evaluate the results with the optical motion capture system which is used as a standard reference.

11.1 Introduction

UWB (3.1–10.6 GHz) is an emerging wireless communications technology that can transmit data at around 100 Mb/s (up to 1000 Mb/s). UWB signals have a fractional

*Department of Electronics and Communication Engineering Thapar University, Patiala, Punjab, India richa@thapar.edu

**Department of Electrical and Computer Engineering Texas A&M University at Qatar, Doha, Qatar qammer.abbasi@qatar.tamu.edu

[†]School of Engineering and Digital Arts The University of Kent, Canterbury, Kent, UK J.C.Batchelor@kent.ac.uk

[††]School of Engineering and Digital Arts The University of Kent, Canterbury, Kent, UK S.Swaisaenyakorn-2082@kent.ac.uk

[‡]School of Electronic Engineering and Computer Science, Queen Mary University of London, UK a.alomainy@qmul.ac.uk

bandwidth of larger than 20% or an absolute bandwidth of at least 500 MHz [1, 2]. Impulse radio ultra wideband technology (IR-UWB) has several advantages such as immunity to multipath interference, low cost, high data rate, portable, ease of implementation and low energy consumption making it suitable for body-centric and wearable networks applications [3]. The inherent characteristics of UWB radio technology makes it a natural choice for high precision indoor positioning and in scenarios where "everybody and everything" is connected by different types of communication links: human to human, human to machine, machine to human and machine to machine. In the future, the need for even higher data rates is expected to develop jointly with a flourishing increase in large numbers of wireless devices embedded in common appliances, sensors, beacons as well as identification tags, spontaneously interacting in ambient intelligence networks.

In recent years, localisation and tracking of human subjects have received significant interests for several attractive applications in wireless body sensor networks [4–8]. In particular, wearable wireless systems provide detection and monitoring applications to access human activity for sports, health care, military applications and day-to-day life [9]. The human body is an integral part of various indoor body area networks/personal area networks applications such as motion tracking, patient monitoring, training of athletes; hence, it is important to study the effect of human body on UWB localisation and the accuracy achieved while localising the antennas/ sensors present on different locations of the body [4]. Along with the indoor propagation environment, the human body is also a complex medium from the radio propagation perspective and hence, is essential to understand and characterise the effect of the human body on the antenna characteristics, the radio propagation channel parameters and the overall communication system performance in the presence of various objects/obstacles in an indoor environment.

UWB has gained widespread use in commercial and industrial applications and also in research endeavours with various applications in indoor localisation and tracking. Commercial localisation systems such as Time Domain and Ubisense, which utilise time of arrival (TOA), and angle of arrival localisation techniques have specified 3D real-time accuracy of 10–15 cm with indoor operating ranges of over 50 m [10, 11]. 3D motion tracking products based upon miniature (Microelectromechanical systems (MEMS)) inertial sensors and UWB technology enables 5–8 cm positioning accuracy in an area of $20 \times 20\,\mathrm{m}^2$ [12]. On the research side of UWB positioning applications, major advances have been made in the fields of high-accuracy 3D positioning for surgical navigation, low power, integrated UWB Complementary metal-oxide semiconductor (CMOS) solutions [13]. An accuracy of high centimetre range (1–5 cm) is being achieved using impulse-based UWB systems as stated in the open literature by [14, 15] and sub-millimetre range accuracy is possible using carrier-based UWB systems as proposed in the literature [16].

11.2 Indoor propagation channel and multipath environment

The indoor radio propagation channel depends on various factors which include building structures, layout of rooms, objects and the type of construction materials used due to which there is reduction in signal strength and attenuation of the signal. Due to the varied nature of the propagation channel and presence of various objects, large-scale and small-scale propagation losses occur which are best described with the help of statistical models [17]. A complete channel model comprises of the path loss model and the multipath model. The path loss is defined as the reduction in power density of an electromagnetic wave as it propagates through free space. The multipath model describes how the signal energy is dispersed over the multipath components.

In a typical complex indoor environment, a signal, as it travels through the wireless channel, undergoes many kinds of propagation effects such as reflection, diffraction and scattering (Figure 11.1), apart from line of sight (LOS) communication due to the presence of buildings, walls, doors, furniture, human subjects and other such obstructions [4]. The physics of the above phenomena may also be used to describe small-scale fading and multipath propagation [3, 4, 18]. Multipath results when the transmitted signal arrives at the receiver by more than one path. The multipath signal components combine at the receiver to form a distorted version of the transmitted waveform. Therefore, there would be multipath interference, causing multipath fading. The multipath components can combine constructively or destructively depending on phase variations of the component signals. The destructive combination of the multipath components can result in a severely attenuated received signal. Multipath fading degrades the performance of the wireless communication systems because, in the propagation environment, the signal arriving at a receiver experiences the effects of various propagation-dependent mechanisms. Therefore, accurate channel characterisation is required to provide a reliable simulation model. The amplitude of the fading can follow different distributions, such as Rician, Rayleigh, Nakagami, Log-normal, Gamma, Normal and Weibull [9].

The UWB signal has a very wide bandwidth (in the order of gigahertz); hence, the multipath components tend to form clusters of rays caused by building structures

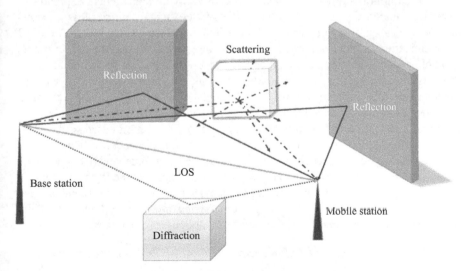

Figure 11.1   *Multipath phenomena in an indoor environment: reflection, diffraction and scattering. Line of sight (LOS) propagation between the base station and mobile station [4]*

and other large reflectors. The rays within a cluster are attributable to reflections from objects in close proximity to the transmitter and receiver [19]. The Saleh–Valenzuela (S–V) channel model is widely used in research for UWB systems as it takes into consideration the clustering phenomenon of the multipath components. In the S–V model, multipath components arrive at the receiver in clusters. Cluster arrivals are Poisson distributed and so are the subsequent arrivals in each cluster. The impulse response of the model [19] can be presented as:

$$h(t) = \sum_{l=0}^{L} \sum_{k=0}^{K} \beta_{k,l} \delta(t - T_l - \tau_{k,l}) \tag{11.1}$$

where L is the cluster of the multipath channel, K is the ray number within each cluster, T_l is the excess delay of the lth cluster and $\tau_{k,l}$ is the delay of the kth path within the lth cluster. The model is independent of the used antennas and includes the frequency dependence of the path loss as well as many generalisations of the S–V model, like mixed Poisson times of arrival and delay-dependent cluster decay constants.

11.3　IR-UWB technology

UWB technology in the range 3.1–10.6 GHz (as per the Federal Communications Commission (FCC) regulations) has been topic of exciting research and development to explore for consumer and commercial applications, especially for short distance and personnel networking including tracking and localisation [2]. Due to the carrier-less characteristics, no sinusoidal carrier is required to raise the signal to a certain frequency band, UWB systems are also referred to as carrier-free or impulse radio (IR-UWB) communication systems [3, 20]. IR-UWB employs a baseband signal with pulse duration in the order of sub-nanoseconds and with signal energy spread over several gigahertz. One of the key benefits of IR-UWB is its carrier-less transmission, which can substantially reduce the development costs [21, 22]. In an IR-UWB communications system, a number of UWB pulses are transmitted per information symbol, and information is usually conveyed by the timings or the polarities of the pulses (Figure 11.2). For positioning systems, the main purpose is to estimate position-related parameters of this IR-UWB signal, such as its TOA. This technology has a low transmitting power and, because of narrowness of the transmitted pulses, has a fine time resolution which provides high precision for location-based applications. The implementation of this technique is simple as no mixer is required, which means low-cost transmitters and receivers. Direct Sequence Ultra Wideband (DS-UWB) and Time Hopping Ultra Wideband (TH-UWB) are two variants of the IR technique [20]. These IR techniques are different multiple access techniques that spread signals over a very wide bandwidth. Because of spreading signals over a very large bandwidth, the IR technique can combat interference from other users or sources.

11.3.1　*Advantages and disadvantages of IR-UWB technology*

UWB offers several inherent properties which will make short-range communication cost-effective, high-speed, simple, miniaturised and portable devices available for

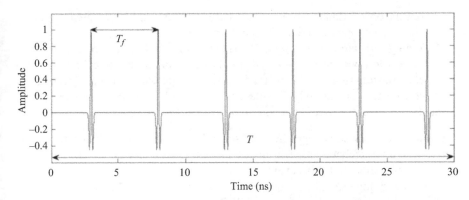

Figure 11.2 *UWB signal consisting of short duration pulses with a low duty cycle, where T is the signal duration and T_f represents the pulse repetition interval or the frame interval [4]*

positioning, communications and radar applications [3, 4, 20–22]. The unique benefits of UWB technology are derived from the wide broadband nature allowing more smart solutions for wireless sensor networks (WSNs). Some of the advantages of UWB technology are listed below:

- *Accurate position estimation:* Due to extremely narrow pulses, IR-UWB systems have emerged as a promising solution for high-resolution indoor positioning and ranging applications.
- *Penetration through obstacles:* Unlike narrowband technology, UWB systems can penetrate effectively through different materials. UWB can carry signals through many obstacles (even lossy and opaque objects) that usually reflect signals at more limited bandwidths and at higher power.
- *Secure:* UWB has low probability of detection and interception. UWB provides high secure and high reliable communication solutions. Due to the low energy density, the UWB signal is noise like, which makes unintended detection quite difficult.
- *Robustness to multipath:* The phenomenon known as multipath is unavoidable in wireless communications channels. The very short duration of UWB pulses makes it less sensitive to multipath and NLOS situations, hence is very suitable for positioning applications in indoor and outdoor environments.
- *High-speed data rate transmission:* UWB has an ultra-wide frequency bandwidth; it can achieve huge capacity as high as hundreds of megabit per second or even several gigabit per second within distance of 1–10 m.
- *Less power consumption:* The transmission of short nanosecond pulses, rather than continuous waveforms, allows pulse generators, amplifiers and receivers to not work continuously, but to be turned on for only a few nanoseconds in each repetition period.
- *Low-cost transceiver designs:* Due to the carrierless nature of UWB signals, transmission of signals requires fewer radio frequency components than

carrier-based transmission. For this reason UWB transceiver architecture is significantly simpler and thus cheaper to build.

- *Suitable for human localisation:* UWB technology is safer for the user because of the very low peak power (-41.3 dBm). Hence, it is very suitable for body-worn antenna/sensors localisation and tracking as the signals are harmless (at such low power) for the human body. Besides, since in UWB systems short nano pulses are transmitted, the user is not continuously exposed to the radiation.

However, with every advantage, there are some disadvantages and challenges to surmount before the technology performs up to its full potential [4, 21, 22]. UWB is robust to multipaths, but this advantage comes at the significant receiver hardware cost due to very high sampling requirement. As UWB operates below noise floor, it is difficult to recognise and synchronise UWB signal at the receiver end. UWB transmitter power limitation poses significant challenges when designing UWB systems to achieve the performance desired at an adequate transmission range and to design UWB waveforms that efficiently utilise the bandwidth and power allowed by the FCC spectral mask [10]. The ultrashort duration of UWB pulses leads to a large number of resolvable multipath components at the receiver. Each resolvable pulse undergoes different channel fading, which makes multipath energy capture a challenging problem in UWB system design.

11.3.2 Body-centric UWB localisation applications

UWB is an excellent signalling choice for high-accuracy localisation in short to medium distances due to its high time resolution and inexpensive circuitry. It is also considered to be the unique signalling choice for short-range, high data rate communications such as in WSNs [4, 20–23]. The UWB technology is an excellent match that makes these exciting applications possible with high centimetre range accuracies for short-range body-centric applications. Some of the important applications are listed below:

- *Sports:* Sensors for athletes' performance monitoring and enhancement to improve outcomes in major events. Accurate tracking of limb motion for sports person using small, lightweight and cost-effective body-worn antennas/sensors based on UWB technology.
- *Health care:* UWB can be used as the communication link in a sensor network making it suitable for wireless body area networking for fitness and medical purposes. The UWB sensors are lightweight and wireless, hence provide freedom to the patient from the tangle of wired sensors. UWB tags/sensors can be used in many medical situations to determine pulse rate, temperature, medical imaging and surgical techniques along with accurate tracking and localisation of patients, providing significant improvement in patient care, increase in efficiency and reduction in operating costs.
- *Military:* Locating people in high-security areas and tracking the positions of the military personnel. Location of various military equipment and vehicles in cluttered environments.

- **Home and office environments:** Locating inhabitants in home and office environments and accurate tracking of objects and persons in indoor environments. Home security, remote operation and control of home appliances such as TV, ovens, lamps and office equipment.
- **Entertainment and digital arts:** Accurate tracking and motion capture of various postures and movements made by artists, dancers and actors.

11.4 UWB body-centric localisation scheme

The high time resolution of the UWB signals makes TOA-based approach most suitable for ranging and localisation [3, 20, 24] for WSNs. The accuracy of prediction of TOA is one of the most important parameter for positioning systems. The TOA between the mobile (Tx) and base stations (Rx) is estimated by classifying the channel impulse response (CIR), statistical information derived from the radio propagation environment and peak detection techniques [24]. First, the magnitude and phase of each frequency components are measured on the analyser. An inverse fast Fourier transform (IFFT) is then applied to obtain the impulse response of the measured channels. For N propagation paths between the transmitter and receiver, with the amplitude, phase and delay of the kth path being α_k, φ_k and τ_k, respectively. The CIR [25] is given by:

$$h(t) = \sum_{k=1}^{N} \alpha_k e^{j\varphi_k} \delta(t - \tau_k) \tag{11.2}$$

Figure 11.3 show a flow chart of the proposed body-centric localisation algorithm which is based on TOA ranging and localisation techniques [7]. First steps include computation of the CIR and received signal waveforms using IFFT and convolution techniques. Further channel classification is performed using various features that can distinguish between LOS and NLOS situations. Based on these results, TOA estimation is performed using different peak detection techniques for LOS scenarios and threshold-based algorithms for NLOS scenarios. Later, 3D localisation results are computed based on the range estimates using least square data fusion techniques.

11.4.1 NLOS identification

11.4.1.1 Received signal amplitude

Sine-modulated Gaussian pulse is used as UWB source pulse and the frequency range is from 3 GHz to 10 GHz [7]. In NLOS conditions, signals are considerably more attenuated, weak and have smaller energy and amplitude due to propagation through obstacles or obstructions. The strength of NLOS signals varies greatly depending on the type of obstruction present.

11.4.1.2 RMS delay spread

The Root Mean Square (RMS) delay spread is defined as:

$$\tau_{\text{rms}} = \sqrt{\frac{\sum_k (\tau_k - \tau_m)^2 . |h(\tau_k; d)|^2}{\sum_k |h(\tau_k; d)|^2}} \tag{11.3}$$

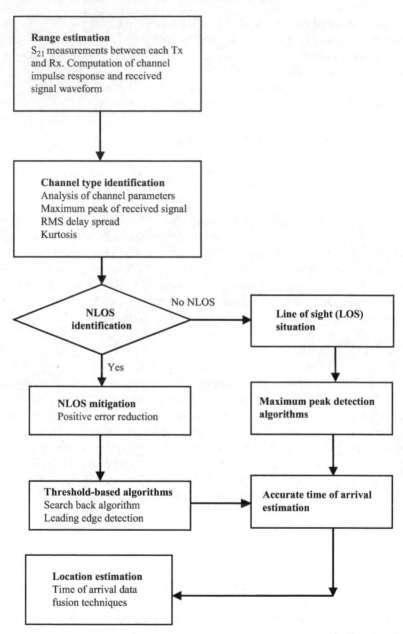

Figure 11.3   Proposed localisation scheme for UWB localisation in realistic environment with multipath and NLOS situations. ©2015 IEEE. Reprinted with permission from Reference 7

where τ_m is the mean excess delay, and τ_k are the multipath delays relative to the first-arriving multipath component and d is the separating distance between the Tx and Rx [26]. This parameter helps to distinguish between LOS and NLOS links as the RMS values for NLOS scenarios are much larger.

11.4.1.3   Kurtosis

The NLOS scenarios have dense multipath components and also more number of weaker components before the maximum peak [27]. Hence, overall NLOS scenarios have wider distribution of multipath components in comparison to LOS scenarios which have a clearly distinguishable maximum peak with multipath components of very low magnitude. Kurtosis index is a simple and efficient parameter to precisely differentiate between the indoor channel conditions and thereby identify the best and worst mobile station–base station (MS–BS) link in the indoor environments. The kurtosis is a statistical parameter that indicates the fourth-order moment of the received signal amplitude [18, 19]. Kurtosis κ is mathematically defined as follows:

$$\kappa(x) = \frac{1}{\sigma^4} \frac{\sum_i (x_i - \bar{x})^4}{N} \tag{11.4}$$

where σ is the standard deviation of the variable x and $\bar{x}$ is the mean value of x. N is the number of samples of x. The kurtosis index κ is supposed to be high in case of LOS conditions while it has low values for signals received under NLOS conditions.

11.4.2   Non-line of sight mitigation

For LOS case, the TOA is determined through peak detection algorithm in which the dominant peak of the CIR will give an estimate of the TOA [7, 28, 29]. Figure 11.4 shows the CIR strongest peak detection technique for LOS situation. It can also be observed that the LOS situation has very less multipath; hence, the peak is easily detectable. However, in urban or indoor environments the accuracy in estimation of TOA depends on the multipath components, non-line of sight (NLOS) situations, low signal-to-noise ratios, sampling precision of the analyser and the environment in which measurements are taking place. In NLOS situations, the direct signal between the transmitter and the receiver is blocked due to the presence of obstructions; hence, the transmitted signal can only reach the receiver through a reflected, diffracted or scattered path. The strongest path in such scenarios does not give the direct path estimate, leading to large ranging errors. Situations, in which the first path is not the strongest path, can be still detectable by appropriate algorithm and understanding of the environment in which measurements are taking place. Due to the large bandwidth of a UWB signal, multipath components are usually resolvable but can be challenging in dense multipath environments. Threshold-based algorithms such as search back technique and leading edge detection methods [29, 30] are used to reduce the error obtained in range estimation for NLOS situations (Figure 11.4). Threshold-based TOA estimation is specifically attractive due to its low complexity and computational burden, which are crucial for many low-cost battery powered

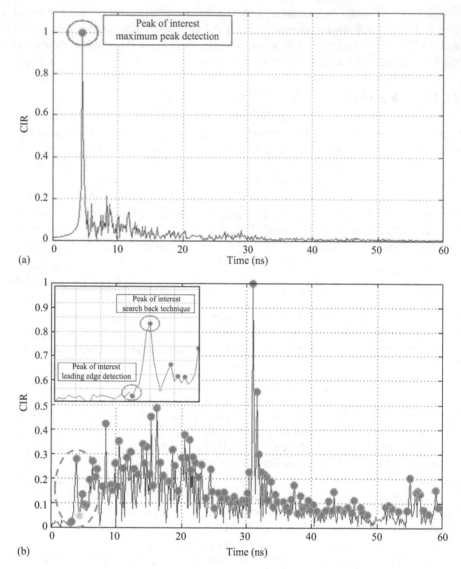

Figure 11.4   Maximum peak detection technique for (a) line of sight situations.
Maximum peak detection technique for (b) non-line of sight situations:
threshold-based and leading-edge technique. Zoomed region of the
channel impulse response showing the detection of peaks [4]

devices, wireless antenna networks and radio frequency identification. These search algorithms compare individual signal samples with a certain threshold in order to identify the first-arriving signal and obtain the range information. Moreover the sensitivity of the system is quite dependent on the threshold value which can be selected based on the noise level or signal peak.

11.4.3 TOA data fusion method

The TOA data fusion method is based on combining estimates of the TOA of the MS signal that arrives at four different BSs [7, 31]. Since the wireless signal travels at the speed of light ($c = 3 \times 10^8$ m/s), the distance between the MS and BSi is given by:

$$r_i = (t_i - t_0)\, c \tag{11.5}$$

where t_0 is the time instant at which the MS begins transmission and t_i is the TOA of the MS signal at BSi. The distances r_1, r_2, r_3, r_4 can be used to estimate (x_m, y_m, z_m) by solving the following set of equations:

$$r_1^2 = x_m^2 + y_m^2 + z_m^2 \tag{11.6}$$

$$r_2^2 = (x_2 - x_m)^2 + (y_2 - y_m)^2 + (z_2 - z_m)^2 \tag{11.7}$$

$$r_3^2 = (x_3 - x_m)^2 + (y_3 - y_m)^2 + (z_3 - z_m)^2 \tag{11.8}$$

$$r_4^2 = (x_4 - x_m)^2 + (y_4 - y_m)^2 + (z_4 - z_m)^2 \tag{11.9}$$

These equations can be solved by using least square solution. Subtracting (11.6) from (11.7)–(11.9) and rearranging the terms in the following matrix form:

$$
\begin{bmatrix} x_2 & y_2 & z_2 \\ x_3 & y_3 & z_3 \\ x_4 & y_4 & z_4 \end{bmatrix}
\begin{bmatrix} x_m \\ y_m \\ z_m \end{bmatrix} =
\frac{1}{2}
\begin{bmatrix} K_2^2 - r_2^2 + r_1^2 \\ K_3^2 - r_3^2 + r_1^2 \\ K_4^2 - r_4^2 + r_1^2 \end{bmatrix}
\tag{11.10}
$$

where

$$K_i^2 = x_i^2 + y_i^2 + z_i^2$$

It can be rewritten as:

$$Hx = b \tag{11.11}$$

$$
H = \begin{bmatrix} x_2 & y_2 & z_2 \\ x_3 & y_3 & z_3 \\ x_4 & y_4 & z_4 \end{bmatrix}; \;
x = \begin{bmatrix} x_m \\ y_m \\ z_m \end{bmatrix}; \;
b = \frac{1}{2} \begin{bmatrix} K_2^2 - r_2^2 + r_1^2 \\ K_3^2 - r_3^2 + r_1^2 \\ K_4^2 - r_4^2 + r_1^2 \end{bmatrix}
$$

The target coordinates can be found by rearranging the matrix equation through the following equation:

$$\hat{x} = (H^T H)^{-1} H^T b \tag{11.12}$$

11.5 BS configurations for UWB localisation

The configuration of the BSs has a direct effect on the accuracy of localisation estimation of the MS. A trade-off between the number of BSs and accuracy has to be considered in order to have a compact, simple and efficient UWB 3D positioning system.

Related work reported in the open literature is based on 3D localisation using a minimum of four BSs or more for localising an object [16, 32]. It is observed that different BS configurations and number of BSs used have an effect on the accuracy obtained in positioning [4–6]. Three-dimensional localisation is achieved by using two configurations using three BSs only and one with four BSs occupying less coverage area, which is more compact than the conventional four BSs or more requirement. Three novel BS models are proposed in Reference 33 for 3D localisation namely the Y-shape configuration, L-shape configuration and mirror-based BS configuration and are validated theoretically and experimentally using a compact tapered slot UWB antenna [34]. The above methods reduce the number of BSs required for obtaining localisation in 3D and the amount of area and complexity required for setting up the system. In addition, Cartesian and directional information is obtained for the proposed methods giving good accuracy in centimetre range. All the proposed configurations can be used for various indoor applications for localisation of objects or body-worn antennas/sensors on human. The position of the BS antennas is chosen in such a way that localisation information can be obtained accurately in three dimensions. The antenna at BS1 is considered as reference zero coordinate which acts as reference to find the position of the target placed inside the volume of the cuboid. In order to find the direction in which target is moving with respect to the reference base station "BS1" at (x_1, y_1, z_1), the estimated target location (x_m, y_m, z_m) in Cartesian coordinate system can be is defined as (R, θ, φ) in the spherical coordinate system.

11.5.1   Cuboid-shape configuration

The cuboid-shape configuration is shown in Figure 11.5(a). Four BSs are used to estimate the location of the unknown target. This configuration is most commonly used for localisation as given in the literature [4, 33]. Each BS is placed at the vertices of the cuboid with two at the bottom and two at the upper vertices of the cuboid. Eight BSs can also be used to get more accurate results but it would make the localisation system more complex to install and more expensive. TOA data fusion is used to obtain the location of the target in 3D. The unknown coordinates at $P(x_m, y_m, z_m)$ found through the TOA data fusion method is converted to spherical coordinates.

11.5.2   Y-shape configuration

The Y-shape configuration is shown in Figure 11.5(b) [4, 33]. The four BSs are arranged in shape of letter Y. The distance between the BSs can be adjusted according to the area under observation and for obtaining precise measurement. The proposed configuration occupies less space and gives accurate results which are comparable with the conventional cuboid-shape BS configuration. The TOA data fusion method is used for estimation of the unknown coordinates.

11.5.3   Geometric dilution of precision

The positioning precision depends significantly on the geometry of the BSs distribution. Geometric dilution of precision (GDOP) [35] indicates the effectiveness of a

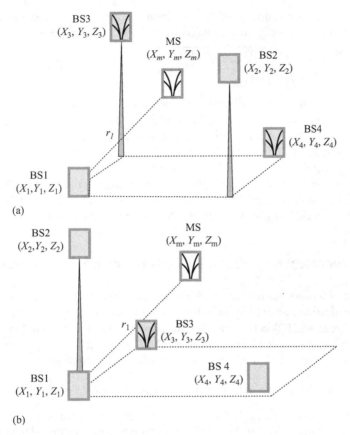

*Figure 11.5   The configuration of the base stations. (a) Cuboid shape and
(b) Y shape with BS1 at origin O. The target antenna (MS) whose
coordinates have to be estimated is placed in the centre of the cuboid*

geometric configuration. It is an indicator of 3D positioning accuracy as consequence
of relative position of the BSs with respect to a MS. A very close bunching or poorly
spaced satellites give poor GDOP value whereas well-distributed satellites yield good
GDOP. GDOP can be defined as:

$$\text{GDOP} = \sqrt{\text{tr}(H^T H)^{-1}} \tag{11.13}$$

where $H = \begin{bmatrix} a_{x1} & a_{y1} & a_{z1} & 1 \\ a_{x2} & a_{y2} & a_{z2} & 1 \\ a_{x3} & a_{y3} & a_{z3} & 1 \\ a_{x4} & a_{y4} & a_{z4} & 1 \end{bmatrix}$

and $a_{xi} = \dfrac{(x_i - x_m)}{r_i}, \quad a_{yi} = \dfrac{(y_i - y_m)}{r_i}, \quad a_{zi} = \dfrac{(z_i - z_m)}{r_i}$

where x_i, y_i, z_i correspond to the BS positions, x_m, y_m, z_m are the estimated MS positions and r_i is the estimated range between the MS and each BS:

$$HDOP = \sqrt{((H^T H)^{-1})_{1,1} + ((H^T H)^{-1})_{2,2}} \qquad (11.14)$$

$$VDOP = \sqrt{((H^T H)^{-1})_{3,3}} \qquad (11.15)$$

where VDOP is the vertical dilution of precision which is measure of localisation accuracy in 1D (height (z-axis)) and HDOP is the horizontal dilution of precision which is the measure of localisation accuracy in 2D (x- and y-axis). The average GDOP, HDOP, and VDOP values for cuboid-shape configuration is 2.6, 1.15, and 2.3 respectively and for the Y-shape configuration, the average values are 3.7, 1.6, and 3.22 respectively. Overall the configurations show GDOP in the range of DOP values (2–5) stating good accuracy. Best GDOP results are obtained for the MS placed in the centre of the cuboid with BS placed at the corners of the cuboid.

11.6 Numerical investigation of UWB localisation accuracy

The choice of sensors, such as compact, efficient and low-cost UWB antennas, makes human localisation and activity monitoring a promising new application made possible by advances in UWB technology. A human body model simulated in Computer Simulation Technology (CST) microwave studio © is used to localise antennas placed on various locations of the body by applying different BS configurations.

11.6.1 Numerical analysis of body-worn antennas

A compact, omni-directional and low-cost UWB tapered slot antennas (TSAs) [34] are used in this study (Figure 11.6). The antenna has excellent impedance matching with return loss below -10 dB and radiation performance in the UWB range with relatively constant gain across the whole frequency band. Figure 11.6(b) shows the simulated return loss of the TSA antenna.

The frequency range used in the numerical analysis is 3–4 GHz and sine-modulated Gaussian pulse is used for signal transmission [36]. Sub-band of the UWB bandwidth is considered keeping in mind the constraints and time consumption in simulating the full UWB bandwidth range using CST simulation with the inclusion of the human body and BS antennas. Single-layer human body model [10] consisting of a muscle tissue and having a dielectric permittivity value of 51.44 at 3.5 GHz (centre frequency) with 14 antennas is simulated using CST Microwave Studio [37]. The height of the human body model is 1.72 m and has an average build with Body mass index (BMI) of 23.7. The TSA antennas are used as beacons, which are placed approximately 5 mm away from the human body surface. Three antennas are placed on each arm and leg and two on the torso as shown in Figure 11.7. Five different positions of the arm are also simulated with each having three antennas placed near the shoulder, elbow and wrist. The interval between the simulated arm positions is 20°. Various techniques and algorithms like peak detection technique, TOA positioning algorithms are used to find the location of each antenna as mentioned in Section 11.4.

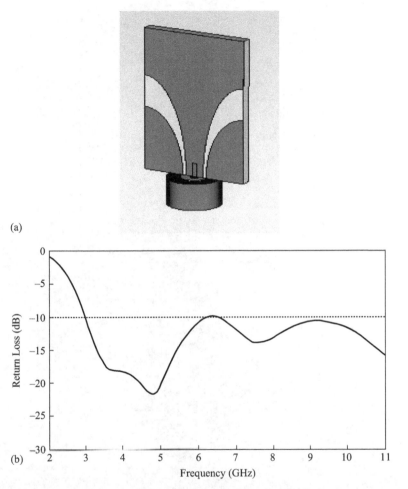

(a)

(b)

*Figure 11.6   Tapered slot co-planar ultra wideband (UWB) antenna: (a) CST model
[[36], reproduced by courtesy of The Electromagnetics Academy]. (b)
Simulated return loss of the antenna for 3–10 GHz frequency range [4]*

11.6.2   Analysis of body-worn antenna localisation

It is observed that the overall error in estimation of the position reduces from antennas
located near the shoulder towards the antenna placed near the wrist (Figure 11.8). A
similar trend of decrease in error is observed for the antennas placed near the thigh
to the ankle (Figure 11.8). The average percentage decrease in error is from 1 to 3 cm
for arms and 2–4 cm for the legs. This is due to the fact that the antenna is near torso
for the case of shoulder and thighs; hence, there is more interference which leads to
less accurate results. Around 57% of the estimated values have localisation error of
less than 3 cm for TOA data fusion method [4, 36, 38].

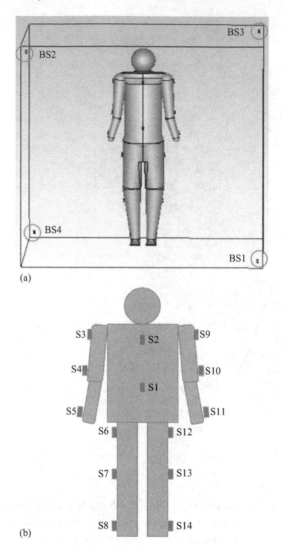

Figure 11.7   (a) CST human body model with the base station positions
represented in circles. (b) Schematic of the human body model
depicting the position of the 14 body-worn antennas. [[36],
reproduced by courtesy of The Electromagnetics Academy]

Figure 11.9 shows the estimated and actual positions of the antennas placed on the arm for five different positions. It is observed that high-accuracy results are obtained for all the three joint locations considered (wrist, elbow and shoulder) when compared with the actual coordinates obtained from CST microwave. This shows the potential of UWB to be used to track limb movements effectively. The estimated and actual positions of the antennas for whole body localisation are shown in Figure 11.9 with an average localisation of 2.5–4.5 cm [36, 38]. The average localisation accuracy

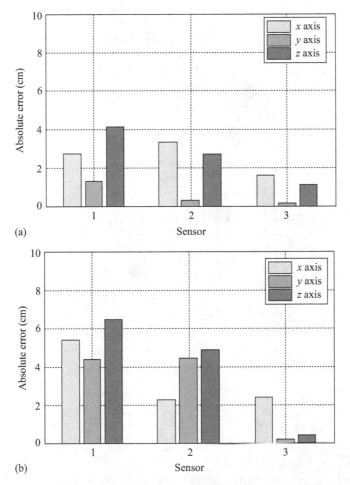

Figure 11.8   Localisation of (a) left arm [[36], reproduced by courtesy of The Electromagnetics Academy] and (b) bight arm. It is observed that the overall error reduces from antenna 1 (near the shoulder) towards the antenna 3 (near the wrist). Also, the error obtained for the x-, y- and z-coordinates is showing different range of error obtained [4]

for directional azimuth and elevation angles estimation is approximately 1°–2°. The variation in accuracy can be attributed to the level achieved in estimation of the TOA, i.e. range estimation values, the type of source pulse/bandwidth used and the presence of human body, which acts like an obstacle causing delay and distortion of the received pulse. Good accuracy within the centimetre range is observed demonstrating the potential of using UWB technology for accurate motion tracking and 3D

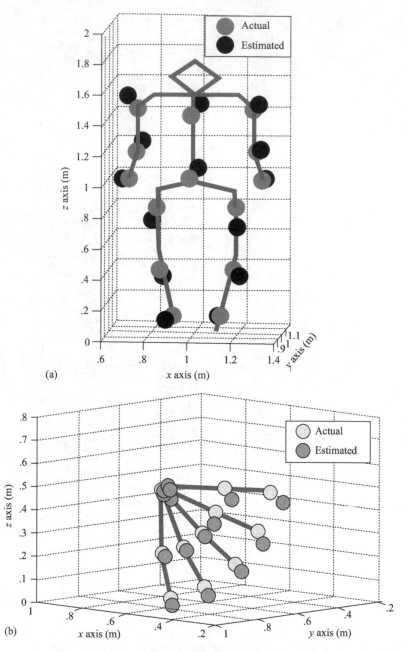

*Figure 11.9   (a) Antenna locations on the body comparing actual and estimated
positions for TOA. ©2012 IEEE. Reprinted with permission from
Reference 38. (b) 3D tracking of the arm movements with tapered slot
UWB antennas placed on the arm [4]*

body-worn sensor localisation in certain applications such as filming, entertainment, leisure sport exercise, medical and tracking of people in home/office environment.

11.6.3 Effect of the presence of obstacles near BSs

Numerical simulations have been performed using different kinds of obstacles near the BSs and compared with the situation when no obstacle is present to study the effect on localisation accuracy. The position of the antennas and BS antennas are kept the same as that of the case with no obstacles present. Blocks of glass and wood have been considered as objects that act like obstacles during the localisation simulation as shown in Figure 11.10. It is observed that there is distortion in the received pulse leading to variations in the estimated range value of the BS and body-worn antenna (MS). More variations and distortions of the received signal are observed for the glass object as it has higher permittivity of 4.28 in comparison to wood, which has a permittivity of 1.2. Considering an example of antenna 6 which is placed near the thigh (Figure 11.11), it is observed that different BSs give different received signal output depending on whether LOS or NLOS scenario occurs between the MS–BS link [36]. From Figure 11.10, as the obstacle is closer to BS3, more reflections and also increased wave distortion levels (NLOS case) are created. The delay is also dependent on the size, material type of the obstacles and the overall propagation phenomenon. It is observed that the major role of permittivity of the obstacle material is prominent when the MS and BSs are in NLOS situation [4]. BS2 is in LOS range of antenna 6; hence, similar received signal waveforms are obtained for the three different cases (no obstacle, wood and glass) studied for BS2. As expected, higher error values are obtained in estimation of antenna positions when located on the human model due to the presence of obstacles. Table 11.1 summarises the error obtained for the three different cases: no object present, wood and glass block as obstacles.

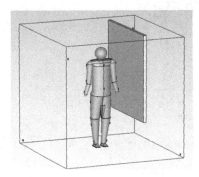 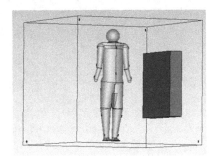

Figure 11.10   *Presence of obstacles (glass: (150 × 5 × 150) cm³ with permittivity 4.28 and wood: (100 × 20 × 100) cm³ with permittivity 1.4) during localisation and dimensions of each obstacle. [[36], reproduced by courtesy of The Electromagnetics Academy]*

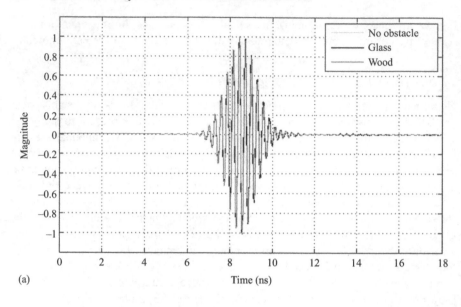

(a)

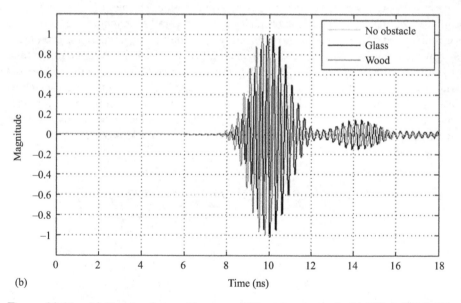

(b)

Figure 11.11   *(a) Received signal between BS2 and antenna 6, (b) BS3 is in NLOS range of antenna 6 for the three different cases. [[36], reproduced by courtesy of The Electromagnetics Academy]*

Table 11.1   Estimated localisation error for the presence of no obstacle scenario with the results of wood and glass as obstacle. [[36], reproduced by courtesy of The Electromagnetics Academy]

	Average estimated error		
	x-axis (cm)	*y*-axis (cm)	*z*-axis (cm)
None	2.77	2.57	4.48
Wood	3.95	4.04	4.79
Glass	4.50	6.5	6.7

11.7   Body-worn antennas localisation in realistic indoor environment

A number of issues and challenges prevail for accurate localisation of human subject in an indoor environment before the system can be deployed for commercial applications. These include effect of human body in the localisation area, portable and cost-effective localisation system, multi-user interference, multipath effects and mitigation of NLOS propagation. This work presents detailed study and analysis of UWB human body localisation using compact, wearable antennas placed on the upper body using analytical and numerical techniques [7]. Very limited work is presented in the open literature in the field of localisation of body-worn antennas using UWB technology. The main objective of the work is to achieve high-accuracy localisation of the body-worn antennas using channel information from statistical analysis and TOA positioning techniques; hence, study the effects of the indoor environment and presence of the human subject in the localisation area on the UWB indoor channel are investigated.

11.7.1   Measurement set-up

The body-worn antennas localisation was performed over the frequency band of 3–10 GHz using UWB compact and cost-effective TSAs [34] which act as the body-sworn antenna and BSs. Measurements were performed in the Body-Centric Wireless Sensor Laboratory at Queen Mary, University of London [7] in order to take into account the effects of the indoor environment on the radio propagation channel. Real human test subject of height 1.68 m and average built is chosen for localizing antennas on the body. Eight antenna locations were chosen with six at the joints of the arm (wrist, elbow and shoulder) and two on the torso (chest and waist). The distance between the human body surface and the antenna is around 5 mm.

  The antennas were connected to an Agilent four-port programmable vector network analyser (PNA-X), model number N5244A, by low loss coaxial cables to

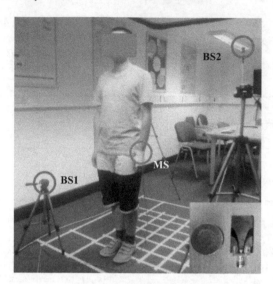

Figure 11.12   Measurement set-up with tapered slot UWB antenna in inset showing body-worn sensor on the left wrist and base stations localisation used for localisation of the body-worn antennas in an area of 1.5 × 1.5 m²

measure the transmission response (S_{21}) between the transmitting (MS) and receiving (BS) antenna. The PNA is set to 6400 data samples, which are sufficient to capture all the necessary impulse response information required to provide appropriate statistical data set. The BSs are positioned in cuboid-shape configuration [33] as shown in Figure 11.12 to obtain high-accuracy positioning in three dimensions with BS1 as reference zero coordinate. The MS is moved in 49 different positions with spacing of 15 cm each in an area sized 1.5 × 1.5 m² for each antenna location. Cartesian coordinates and directional information related to the position of the target are obtained through TOA data fusion and peak detection algorithms. Measurement set-up is shown in Figure 11.12.

11.7.2   NLOS identification and mitigation

Different channel conditions such as LOS, partial non-line of sight (PNLOS) and NLOS have been observed from the measured data for each antenna location on the body (Figure 11.13) [28, 29]. From the measurement data, it can be observed that the complex propagation phenomenon gives rise to number of error sources such as multipath and NLOS situations significantly degrade the ranging performance, which is due to the large positive bias (high multipath situations and delay in propagation) occurrence in such channels. Figure 11.14 depicts two waveforms received in the LOS and NLOS condition supporting the observations. Hence, it is necessary to identify and mitigate the presence of NLOS effects [29] in order to obtain accurate localisation results.

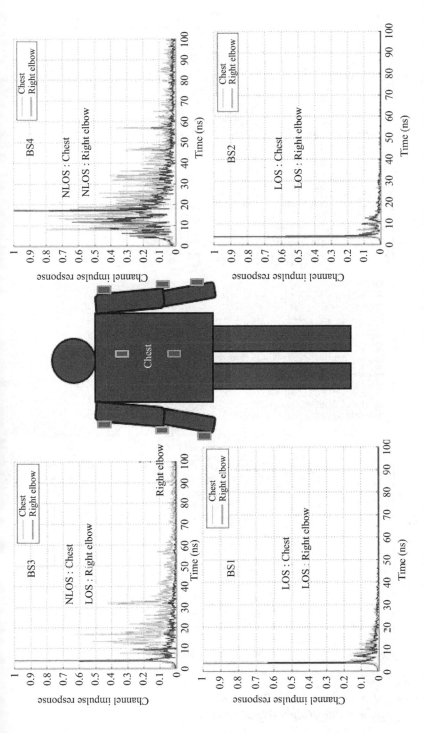

Figure 11.13 *Different kinds of channel impulse responses observed: line of sight direct path, line of sight with high multipath and detectable non-line of sight path. ©2015 IEEE. Reprinted with permission from Reference 7*

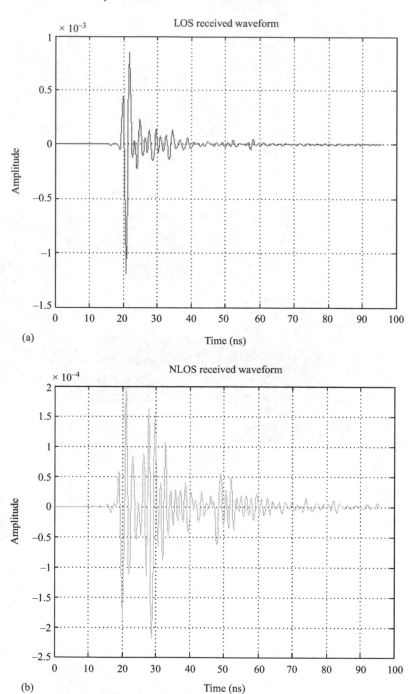

Figure 11.14   Measured received UWB signal: (a) line of sight (LOS) and
(b) non-line of sight (NLOS) signals [4]

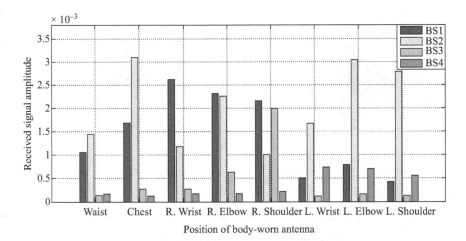

Figure 11.15 *A graphical comparison of amplitude of the received signals for the different antenna locations on the body with respect to base station locations. ©2015 IEEE. Reprinted with permission from Reference 7*

11.7.2.1 Amplitude of received signal

Figure 11.15 shows the strength of the received signal versus position of the BS averaged over the 49 grid locations for each mobile antenna location. It is interesting to observe that the relation between the Rx signal and the antenna position is a complex integrated effect of, the distance of the antenna from the BS, height of the BSs, position of the wearable antenna and the extent of NLOS situation present between the antenna and BS [7]. Results and analysis based on the received signal amplitude demonstrate that the peak amplitude magnitude on average is 11 times less for NLOS situations (BS3 and BS4) in comparison to LOS situations (BS1 and BS2). BS4 in general shows minimum signal amplitude (with least for chest location) as it is in NLOS situation with most of the antenna locations and is at a lower height (0.6 m) leading to reduction in signal strength and high multipath propagation. Highest amplitude is observed for the antenna placed on the chest for BS2 due to the height of the antenna location, visibility and orientation of the antenna with respect to BS2.

11.7.2.2 RMS delay spread

Highest multipath is generally observed for the situation when antennas are placed on the torso region (i.e. waist and chest) especially when there is NLOS situation between the antenna and the BS. The reason for the increase in multipath is due to the fact that the antennas are present on the torso region which causes interference [7]. In the measurements conducted, BS3 and BS4 are at the back of the human subject; hence, the body itself acts like an obstruction causing delay and high multipath propagation. An increase of 3.5 times in the RMS delay spread is observed when the BS–MS link is in NLOS in comparison to LOS situation.

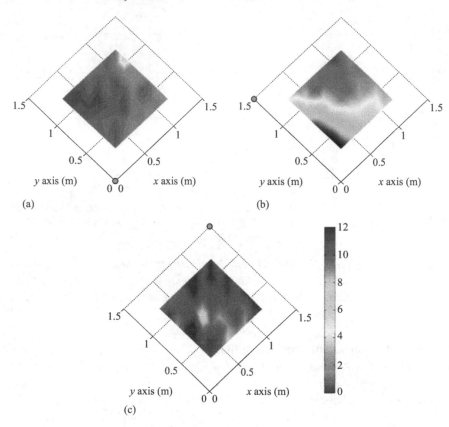

*Figure 11.16   Surface graphs for RMS delay spread for various receiver positions
with respect to the body-worn antenna locations: (a) BS1 – right
elbow, (b) BS3 – right elbow and (c) BS4 – chest. ©2015 IEEE.
Reprinted with permission from Reference 7*

Highest multipath (10.11 ns) is observed when the antenna is placed in the centre
of the chest for BS4. Lowest multipath (1 ns) is observed for BS1 – right wrist link
and for BS2 – left elbow and shoulder link, which is due the direct path propagation
between the BS and antenna, the distance between the Tx and Rx antennas, and also the
antennas are facing each other bringing the antennas in LOS situation. Figure 11.16
shows different cases of the wearable antenna locations and the RMS delay spread
surface distribution over the area of localisation. It can be observed for Figure 11.16(a)
that BS1 (elbow) values are in range of 1–4 ns; for Figure 11.16(b) BS3 (elbow)
distinctly variable ranges of values are observed which is dependent on the position
of the wearable antenna with respect to BS3; for Figure 11.16(c) BS4 (chest) values
are in range of 8–12 ns. The darkest region occupying least area at the bottom of the
surface graph (Figure 11.16(c)) shows that the antenna is in NLOS situation as the
RMS values are high and as it goes further towards the other end, the antenna gets in
direct view of the BS3; hence in LOS situation and low RMS values are obtained.

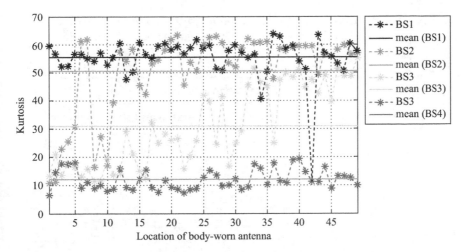

Figure 11.17   Kurtosis index values and their mean values calculated over the 49 grid locations for right elbow. ©2015 IEEE. Reprinted with permission from Reference 7

11.7.2.3   Kurtosis parameter analysis

For the torso region, the BS3 and BS4 are in NLOS situation, showing lower level of kurtosis index in the range of 10–20 and BS1 and BS2 are in LOS situation showing higher values of kurtosis index in the range of 40–60. The benefit of using kurtosis as a feature of differentiation between different channel scenarios is that it can classify the situations more distinctly, i.e. differentiate between partial LOS/NLOS and LOS/NLOS situations [7, 27]. The mid-range values of kurtosis are in the range of 30–40, still showing more chances of LOS situations in comparison to NLOS situations. BS1 shows higher values of kurtosis for the right arm and BS2 for the left arm as the BS is in direct LOS situation with the respective antenna positions. A detailed description of kurtosis values for the right elbow is shown in Figure 11.17, displaying values for each antenna position on the grid with respect to the BSs. BS1 and BS2 are in LOS situation and BS4 in NLOS. For antenna at BS3, there is an increase magnitude from antenna location 1 to antenna location 49 on the grid, which depicts the relation between distance of the antenna from the BS, position of the antenna location with respect to the BS bringing it from NLOS (location 1) to LOS (location 49) situation.

11.7.2.4   NLOS mitigation using threshold-based techniques

By using NLOS mitigation techniques such as leading edge detection techniques, and channel information regarding the localisation environment, the NLOS ranging error can be reduced to a level which is comparable with the LOS ranging error, hence enhancing the performance of the UWB localisation system [28]. After applying TOA-based range estimation techniques and positive bias error correction for NLOS

Table 11.2   *Average localisation accuracy for the body-worn antennas*
placed on the upper half of the human body. ©2015 IEEE.
Reprinted with permission from Reference 7

	Average localisation error (cm)		
Antenna position	*x*-axis (cm)	*y*-axis (cm)	*z*-axis (cm)
Left arm			
Left shoulder	2.16	1.26	2.34
Left elbow	1.14	1.13	1.93
Left wrist	1.52	1.21	2.45
Right arm			
Right shoulder	2.27	1.23	2.51
Right elbow	1.09	1.14	1.76
Right wrist	1.48	1.29	2.68
Torso			
Chest	2.31	2.25	3.21
Waist	2.10	2.18	3.06

situations, higher accuracy results are obtained. In LOS conditions, a ranging error below 3 cm occurs in more than 90% of the measurements. On the other hand, in NLOS conditions ranging error below 3 cm occurs in less than 20% of the measurements and increases to 70% after applying NLOS mitigation techniques [7].

11.7.3   Accuracy and error range analysis

11.7.3.1   Antenna localisation accuracy

Table 11.2 lists the accuracy achieved by applying TOA least square localisation algorithm [7] in 3D for the eight antenna locations chosen on the upper body when compared with actual coordinates that have been calculated geometrically. Highest accuracy results have been observed for the antenna placed on the elbows (right and left arms). Accuracy results below a centimetre have been observed in more than 50% of the antenna locations with an average accuracy of 1–2 cm [7]. Figure 11.18 shows the localisation results for the right elbow when compared to actual positions. This can be attributed to the fact that the position of the elbow falls in the centre of the volume of the cuboid in *z*-direction, hence along with *x*- and *y*-axis best accuracy results are obtained in the *z*-direction also.

The antenna on the elbow is in LOS situations for three out of four BSs and the antenna on the arm has better visibility and in comparison to antennas placed on the torso region. Lower accuracy is observed for the antennas on the chest and waist in the range of 2–3 cm. For the shoulder and wrist antenna locations, an average accuracy of 1.5–2.5 cm has been observed. Overall, very high accuracy in the range of 1–3 cm has been observed which makes UWB technology very suitable for indoor tracking and localizing applications [7].

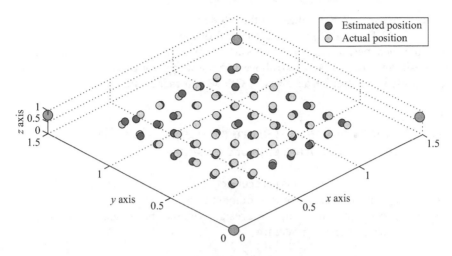

Figure 11.18   Comparison of actual and estimated positions of the wearable antennas: right elbow. ©2015 IEEE. Reprinted with permission from Reference 7

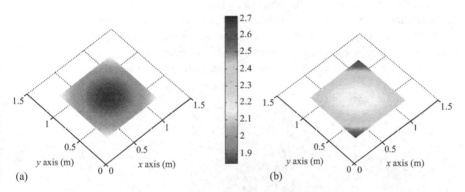

Figure 11.19   Sliced surface graph for GDOP values for different heights (a) 0.4 m and (b) 0.8 m of the sensor locations in the volume of 1.5 × 1.5 × 0.8 m³

11.7.3.2   Geometrical dilution of precision analysis

Low DOP values are obtained for all the antenna locations in the range of values (1–2) which is considered as excellent accuracy values [7, 33]. The GDOP is computed for the different positions of the antenna locations for each BS configuration in an area of 1.5 × 1.5 m². The average GDOP, HDOP and VDOP values for cuboid-shape configuration used are 2.15, 1.22 and 1.64, respectively. It is observed that the VDOP values are slightly higher in comparison to HDOP values. This is because the distance of the BSs in the z-direction is only 0.8 m where as in the $x - y$-plane the distance is 1.5 m which gives better coverage area in the $x - y$-plane. Figure 11.19 shows surface graphs for GDOP values for the whole localisation area for different heights ((a) elbow (0.4 m) and (b) shoulder (0.8 m)) of the antennas placed on the body. In the experiment performed, the elbow is at a height of 0.4 m from BS1; hence, it gives best positioning results. As we go above and below the centre of the cuboid, there will be decrease in localisation accuracy as GDOP values increase.

11.8   Localisation of body-worn antennas using UWB and optical motion capture system

The main objective of the work is to accurately determine and track a person's position and motion activity in an indoor environment using UWB technology based on TOA positioning techniques and evaluate the results with the optical motion capture system which is used as a standard reference.

11.8.1   Measurement set-up for upper body localisation

Human body localisation is performed at the motion capture studio housed at Kent University, UK [39]. A human test subject of height 1.8 m and average build is chosen for localising the antennas on the body in an area of 2×2 m^2 as shown in Figure 11.20 in the frequency range 3–10 GHz. Twelve antenna locations are chosen with six at the joints of the arms and six on the torso. The subject is made to sit on a chair which is placed in the centre of the localisation area of. A vector network analyser is used to capture S_{21} parameters for each antenna location on the body and receiver antenna.

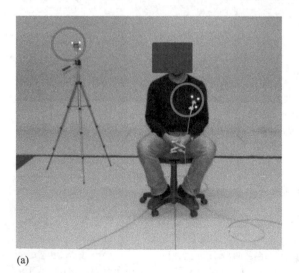

(a)   (b)

Figure 11.20   *(a) Human subject sitting with UWB antenna placed on the chest and the base station antenna on the tripod stand shown in circles. (b) TSA antenna placed on the tripod stand with reflective markers*

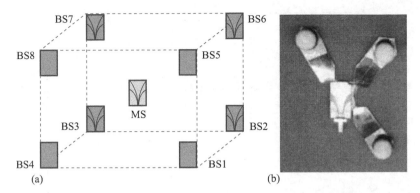

Figure 11.21   *Eight BS configurations with antennas placed at the vertices in a*
2 × 2 × 0.45 m³ volume with MS in the centre. Four BS
configurations: (a) BS (1,6,3,8) for cuboid shape and BS (1,2,4,5) for
Y-shape configuration. The subject is facing BS (1,4,5,8) showing
higher probability of LOS situation between MS and BS link. (b) TSA
antennas placed on plastic base with reflective markers. ©2014
IEEE. Reprinted with permission from Reference 39

TSAs are used as the body-worn antenna and also as the receiver antennas which are placed in three different configurations [33, 39] (Figure 11.21(a)) (cuboid shape, eight BSs; cuboid shape, four BSs; Y shape, four BSs).

The TSA antennas [11] are mounted on plastic frames (Figure 11.21(b)) with three markers, each to allow estimation of the position of the antennas in 3D space through VICON optical motion capture system with passive infrared reflecting markers [40]. The optical motion capture system is used to compare localisation results and also to obtain exact coordinates of the BSs (receiver antennas). Eight-camera system was chosen to capture movement without obscuring of the markers in a 3 × 3 m² space with an accuracy of 1 cm. By using data from the coordination of at least two spatially separated cameras, the position of the marker can be determined. The distance between the human body surface and the antenna is around 5 mm. TOA ranging and positioning techniques are applied to obtain coordinates of the antenna with respect to BS1.

11.8.2   Localisation results and analysis

From the measurement data, it is observed that different situations of LOS and NLOS are obtained depending upon the BS and body-worn antenna location. Figure 11.22 shows the CIR obtained for base stations 1 and 2 for the body-worn antenna placed on the chest (antenna 12). More multipath/high RMS delay spread is generally observed for the situation when antennas are placed on the chest (such as antenna 12) in comparison to the antennas placed on the arm (such as antenna 1). The reason for the increase in multipath is due to the fact that the antennas are present on the torso region which causes interference. Low RMS delay spread is observed for MS–BS link such as BS1,5 for antenna 12 as it is in LOS situation with the two BSs and high

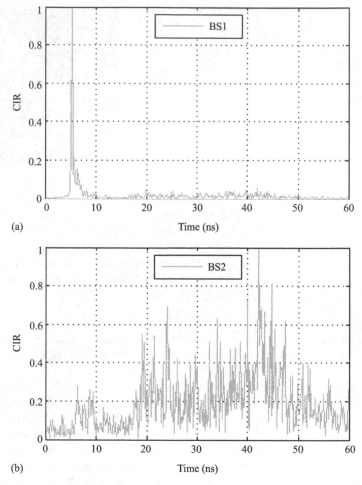

Figure 11.22   Channel impulse response for different base stations 1 and 2 for body-worn antenna location 12. (a) BS1 is in line of sight and (b) BS2 is in non-line of sight with the body-worn antenna

RMS delay spread is observed for BS2,3,4,6,7,8 as NLOS situation is formed. For the body-worn antenna 1, low RMS delay spread is observed for MS–BS3,4,7,8 and higher values for MS–BS1,2,5,6.

Some of the factors affecting localisation accuracy are precise estimation of TOA between the MS and BS, operating bandwidth, antenna efficiency and presence of the human body which acts like an obstacle causing delay and interference. The accuracy of the positioning system also depends on the number of BSs used and the distribution of the BSs. The antennas should have sufficient spacing between each other in order to keep minimum interference and coupling between the antennas. Around 0.5 m distance between the antennas which is five times the wavelength at

Table 11.3   *Average localisation error for the body-worn antennas. ©2014*
IEEE. Reprinted with permission from Reference 39

	Average localisation accuracy	
	x-axis (cm)	**y-axis (cm)**
	Eight base stations	
Cuboid shape	1.74	2.32
	Four base stations	
Cuboid shape	2.51	2.82
Y shape	3.72	3.79

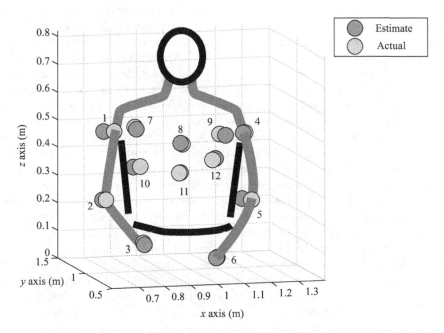

Figure 11.23   *Estimated and actual positions of the antennas placed on the body*
for eight base station configurations. ©2014 IEEE. Reprinted with
permission from Reference 39

the lower frequency in the band (3 GHz) is sufficient. High localisation accuracy is achieved as shown in Table 11.3 for various BS configurations studied.

The estimated and actual positions of the antennas are shown in Figure 11.23 for eight BS configurations showing high-accuracy position estimation (1–2 cm). The localisation error obtained is similar to that of the motion capture system with eight cameras. As the number of BSs is increased, the area under localisation is better covered and also because of the usage of trilateration technique to estimate unknown locations, additional antennas will enhance the least square solution accuracy. For the configurations considering four BSs, the error is increased by 0.5–1 cm for the cuboid-shape configuration [39]. For the Y-shape configuration, which is more compact and easily set-up, 1–1.5 cm increase in average error is obtained. This configuration has the substantial advantage of using fewer BSs and requiring less coverage area for localisation in comparison to the two other configurations applied.

11.9 Summary

Human body 3D localisation applying compact and cost-effective wearable antennas placed at different locations on the body has been presented taking into account various numerical simulation scenarios and realistic indoor environments. The overall average accuracy in 3D localisation is around 2.5–4.5 cm for the case of simulations which are conducted at a lower UWB bandwidth range. Based on the measured data for realistic environments, channel classification has been performed in detail using features such as RMS delay spread, signal amplitude and kurtosis which assist in identifying LOS and NLOS situations. Simple and effective techniques for identifying and mitigating NLOS situations have been stated and a UWB localisation scheme for human body tracking has been proposed. GDOP analysis has been performed for validating the compact BS configuration used with results showing GDOP values in the range of 1–2, which are considered as high-accuracy values. Results demonstrated that the UWB system performance is highly affected by the position of the antennas on the body with respect to the BS location and also by the type of channel between each BS and MS link, thus showing the importance of considering these parameters while making an optimal UWB localisation system for body-centric wireless communications. High 3D accuracy is obtained in centimetre range (1–3 cm) suitable for tracking, patient monitoring, training of athletes and motion capture applications with higher accuracy obtained for the limbs in comparison to the torso region, as there will be more interference in the signal propagation by the torso. Best results are obtained when the antenna is placed on the elbow (0.5–1.5 cm) for the BS considered; hence, this location can be best suited for UWB indoor tracking of human subjects. Average localisation accuracy as small as 1–2 cm has been achieved, which is comparable to common commercial optical systems. The work carried out gives an insight regarding the propagation phenomenon when antennas are placed on the human body and the factors affecting the accuracy achieved while localizing the antennas present on the human body.

References

[1]   Z. Sahinoglu, S. Gezici, and I. Guvenc, *Ultra-Wideband Positioning Systems: Theoretical Limits, Ranging Algorithms, and Protocols*, Cambridge University Press, New York, 2008.

[2]   "First Report and Order, Revision of part 15 of the commission's rule regarding ultra-wideband transmission system FCC 02-48," *Federal Communications Commission*, 2002.

[3] M. Z. Win and R. A. Scholtz, "Impulse radio: how it works," *IEEE Communications Letter*, vol. 2, no. 2, pp. 36–38, February 1998.

[4] R. Bharadwaj, "Investigation of 3D positioning accuracy enhancement using impulse ultra wideband radio," Ph.D. Thesis, Queen Mary, University of London, May 2015.

[5] W. Mekonnen, E. Slottke, H. Luecken, C. Steiner, and A. Wittneben, "Constrained maximum likelihood positioning for UWB-based human motion tracking," *International Conference on Indoor Positioning and Indoor Navigation, IPIN 2010*, Zurich, Switzerland, pp. 1–10, September 2010.

[6] D. Zhang, F. Xia, Z. Yang, L. Yao, and W. Zhao, "Localisation technologies for indoor human tracking," in *Proceedings of the IEEE International Conference on Future Information Technology (FutureTech'10)*, Busan, Korea, pp. 1–6, May 2010.

[7] R. Bharadwaj, C. G. Parini, and A. Alomainy, "Experimental investigation of 3D human body localisation using wearable ultra wideband antennas," *IEEE Transactions on Antennas and Propagation*, vol. 63, no. 11, pp. 5035–5044, November 2015.

[8] J. Hamie, B. Denis, and M. Maman, "On-body localization experiments using real IR-UWB devices," in *2014 IEEE International Conference on Ultra-WideBand (ICUWB)*, Paris, pp. 362–367, 1–3 September 2014.

[9] P. S. Hall and Y. Hao, *Antennas and Propagation for Body-Centric Wireless Communications*, Artech House, Norwood, MA, 2006.

[10] Hardware Datasheet, Ubisense, Cambridge, UK, 2007. Available from http://www.ubisense.net/media/pdf/Ubisense%20System%20Overview%20V1.1.pdf.

[11] Time domain. Available from http://www.timedomain.com/datasheets/TD_DS_P410_RCM_FA.pdf.

[12] Xsens Motion Grid. Available from http://www.xsens.com/en/general/motion-grid.

[13] M. R. Mahfouz, A. E. Fathy, M. J. Kuhn, and Y. Wang, "Recent trends and advances in UWB positioning," in *IEEE MTT-S International Microwave Workshop on Wireless Sensing, Local Positioning, and RFID (IMWS 2009)*, Cavtat, pp. 1–4, 24–25 September 2009.

[14] Z. N. Low, J. H. Cheong, C. L. Law, W. T. Ng, and Y. J. Lee, "Pulse detection algorithm for line-of-sight (LOS) UWB ranging applications," *IEEE Antennas and Wireless Propagation Letters*, vol. 4, pp. 63–67, 2005.

[15] R. Zetik, J. Sachs, and R. Thoma, "UWB localisation – active and passive approach [ultra wideband radar]," in *Proceedings of the 21st IEEE Instrumentation and Measurement Technology Conference (IMTC 04)*, Como, vol. 2, pp. 1005–1009, 18–20 May 2004.

[16] Z. Cemin, M. J. Kuhn, B. C. Merkl, A. E. Fathy, and M. R. Mahfouz, "Real-time noncoherent UWB positioning radar with milimeter range accuracy: theory and experiment," *IEEE Transactions on Microwave Theory and Techniques*, vol. 58, no. 1, pp. 9–20, January 2010.

[17] T. S. Rappaport, *Wireless Communications Principles and Practice*, Prentice Hall Inc., Englewood Cliffs, NJ, 1996.

[18] H. Hashemi, "The indoor radio propagation channel," *Proceedings of IEEE*, vol. 81, no. 7, pp. 943–968, 1993.

[19] A. A. Saleh and R. A. Valenzuela, "A statistical model for indoor multipath propagation," *IEEE Journal on Selected Areas in Communications*, vol. 5, no. 2, pp. 128–137, February 1987.

[20] S. Gezici and H. V. Poor, "Position estimation via ultra-wide-band signals," *Proceedings of the IEEE*, vol. 97, no. 2, pp. 386–403, February 2009.

[21] D. Adalja, "A comparative analysis on indoor positioning: techniques and systems," *International Journal of Engineering Research and Applications*, vol. 3, pp. 1790–1796, 2013.

[22] C. C. Chong, F. Watanabe, and H. Inamura, "Potential of UWB technology for the next generation wireless communications," in *Proceedings of the Ninth IEEE International Symposium on Spread Spectrum Technique and Applications*, Manaus-Amazon, pp. 422–429, August 2006.

[23] Y. Rahayu, T. A. Rahman, R. Ngah, and P. S. Hall, "Ultra wideband technology and its applications," in *Fifth IFIP International Conference on Wireless and Optical Communications Networks (WOCN'08)*, Surabaya, pp. 1–5, 5–7 May 2008.

[24] R. Bharadwaj, Q. H. Abbasi, A. Alomainy, and C. Parini, "Ultra wideband sub-band time of arrival estimation for location detection," in *2011 Loughborough Antennas and Propagation Conference (LAPC)*, Loughborough, pp 1–4, 14–15 November 2011.

[25] D. Humphrey and M. Hedley, "Prior models for indoor super-resolution time of arrival estimation," in *IEEE 69th Vehicular Technology Conference, VTC Spring 2009*, Barcelona, pp. 1–5, 26–29 April 2009.

[26] N. A. Alsindi, B. Alavi, and K. Pahlavan, "Measurement and modeling of ultrawideband TOA-based ranging in indoor multipath environments," *IEEE Transactions on Vehicular Technology*, vol. 58, no. 3, pp. 1046–1058, March 2009.

[27] S. Marano, W. M. Gifford, H. Wymeersch, and M. Z. Win, "NLOS identification and mitigation for localisation based on UWB expcrimental data," *IEEE Journal on Selected Areas in Communications*, vol. 28, no. 7, pp. 1026–1035, September 2010.

[28] L. Mucchi and P. Marcocci, "A new parameter for UWB indoor channel profile identification," *IEEE Transactions on Wireless Communications*, vol. 8, no. 4, pp. 1597–1602, April 2009.

[29] R. Bharadwaj, C. Parini, and A. Alomainy, "Indoor tracking of human movements using UWB technology for motion capture," in *Eighth European Conference on Antennas and Propagation (EuCAP 2014)*, The Hague, The Netherlands, 6–11 APRIL 2014.

[30] I. Guvenc and Z. Sahinoglu, "Threshold-based TOA estimation for impulse radio UWB systems," in *2005 IEEE International Conference on Ultra-Wideband, 2005 (ICU 2005)*, Zurich, pp. 420–425, 5–8 September 2005.

[31] A. H. Sayed, A. Tarighat, and N. Khajehnouri, "Network-based wireless location: challenges faced in developing techniques for accurate wireless location

information," *IEEE Signal Processing Magazine*, vol. 22, no. 4, pp. 24–40, July 2005.

[32] Y. Ruiqing and L. Huaping, "UWB TDOA localisation system: receiver configuration analysis," in *2010 International Symposium on Signals Systems and Electronics (ISSSE)*, Nanjing, vol. 1, pp. 1–4, 17–20 September 2010.

[33] R. Bharadwaj, A. Alomainy, and C. G. Parini, "Ultra-wideband-based 3-D localisation using compact base-station configurations," *IEEE Antennas and Wireless Propagation Letters*, vol. 13, pp. 221–224, 2014.

[34] A. Alomainy, A. Sani, A. Rahman, J. G. Santas, and Y. Hao, "Transient characteristics of wearable antennas and radio propagation channels for ultrawideband body-centric wireless communications," *IEEE Transactions on Antennas and Propagation*, vol. 57, no. 4, pp. 875–884, April 2009.

[35] B. Li, A. G. Dempster, and J. Wang, "3D DOPs for positioning applications using range measurements," *Wireless Sensor Network*, vol. 3, no. 10, pp. 334–340, 2011.

[36] R. Bharadwaj, A. Alomainy, and C. Parini, "Numerical Investigation of body-worn UWB antenna localisation techniques for motion capture applications," in *Progress in Electromagnetics Research Symposium (PIERS 2013)*, Stockholm, pp. 1183–1187, 12–15 August 2013.

[37] M. Ur Rehman, Y. Gao, X. Chen, and C. G. Parini, "Effects of human body interference on the performance of a GPS antenna," in *The Second European Conference on Antennas and Propagation (EuCap 2007)*, Edinburgh, pp. 1–4, 11–16 November 2007.

[38] R. Bharadwaj, A. Alomainy, and C. Parini, "Localisation of body-worn sensors applying ultra wideband technology," in *2012 IEEE Asia-Pacific Conference on Antennas and Propagation (APCAP)*, Singapore, pp. 106–107, 27–29 August 2012.

[39] R. Bharadwaj, S. Swaisaenyakorn, C. G. Parini, J. Batchelor, and A. Alomainy, "Localisation of wearable ultra wideband antennas for motion capture applications," *IEEE Antennas and Wireless Propagation Letters*, vol. 13, pp. 507–510, 2014.

[40] S. Swaisaenyakron, P. R. Young, and J. C. Batchelor, "Animated human walking movement for body worn antenna study," in *2011 Loughborough Antennas and Propagation Conference (LAPC)*, Loughborough, pp. 1–4, 14–15 November 2011.

Chapter 12

Down scaling to the nano-scale in body-centric nano-networks

Ke Yang*, Nishtha Chopra*, Qammer H. Abbasi†,
Khalid Qaraqe†, and Akram Alomainy*

12.1 Development of nano-communication

As Metin Sitti said, small-scale network has a quite bright future, especially in health-care and bioengineering scenario because the corresponding devices in the network are "unrivalled for accessing into small, highly confined and delicate body sites, where conventional medical devices fall short without an invasive intervention" [1].

Nano-technology has gained a great attention since it was put forward in 1959 [2]. The most fundamental elements to materialize nano-technology are the development of battery-free nano-devices, which can be used to accomplish simple tasks such as computation, sensing, and communication [3]. Furthermore, by combining these basic units, the capacity of these nano-devices could be substantially expanded and much more complex tasks can be targeted, which highlights the concept of nano-networks. The latter opens the door to an immense range of applications ranging from medical technologies to flexible electronics [4]. On the other hand, researches on body area networks (BAN) at microwave frequencies have obtained great achievements, and it is pointed out by Prof. Metin Sitti that the entire network systems would be shrunk into nano-scale with the nano-robots and molecular machine as the elements in the near future [1]. Additionally, combined with the concept of internet of things, the internet of multi-media nano-things has been introduced and detailed in Reference 5, which indicates the significance of the study of nano-communication. As the name indicates, nano-communication encapsulates the communication between devices at the nano-scale applying novel and modified communication and radio propagation principles in comparison to conventional and existing solutions as further explained in this manuscript [3]. Among three scenarios of body-centric communications, namely: in-body, on-body, and off-body [6], where the in-body application related to medical healthcare is the most promising and of great interest.

*School of Electronic Engineering and Computer Science, Queen Mary University of London, Mile End Road, London, UK E1 4NS
†265, Texas A&M University at Qatar, Education City, Doha, Qatar

Table 12.1   Overview of the envisioned applications [1, 4]

	Biomedical [10]	Environmental	Industrial	Military
Health monitor	• Active visual imaging for disease diagnosis [11–15] • Mobile sensing for disease diagnosis [16–19]	Bio-degradation [7]	Product quality control [20]	Nuclear, biological, and chemical defences [21]
Therapy	• Tissue engineering [22–24] • Bio-hybrid implant [25, 26] • Targeted therapy/drug delivery [27–31] • Cell manipulation [11], [32–35] • Minimally invasive surgery [36–38]	Bio-control [39–41]	Intelligent office [9]	Nano-fictionalized equipment [42]

12.2   Applications of nano-communication

There are a great number of potential applications of nano-networks, which can be mainly divided into four groups: biomedical, environmental, industrial, and military [4, 7], which have been summarized and classified in Table 12.1. From it, we can see that nano-networks are born for biomedical fields due to its advantages of size, bio-compatibility, and bio-stability. Nano-devices spreading over the human body can monitor the human physical movement. For example, nano-pressure-sensors distributed in eyes can detect the intraocular pressure (IOP) for the early diagnosis and treatment of glaucoma to prevent vision loss [1]. At the same time, the nano-devices deployed in bones can monitor the bone-growth in young diabetes patients to keep them from osteoporosis [1]. Furthermore, nano-robots inside the biological tissues can detect and then eliminate malicious agents or cells, such as viruses or cancer cells, make the treatment less invasive and real time [8]. Moreover, networked nano-devices will be used for organ, nervous track, or tissue replacements, i.e., bio-hybrid implants. While in Reference 9, the concept of intelligent office was proposed, shown in Figure 12.1. Nano-transceivers are attached to all the elements in the office and even their internal components, which enables them connect to the Internet all the time; therefore, the user can keep track of the location and status of all the belongings in an effortless fashion. At the same time, all the nano-sensors detect the user's movement to make essential activity according to the corresponding user's behaviour/needs.

Figure 12.2 shows an example of health monitor system with nano-network: the nano-machines spreading over the clothes can sense the change of the surrounding and make the corresponding response to protect the user or make the user comfortable; the nano-sensors inside the body can sense the body information to indicate the health

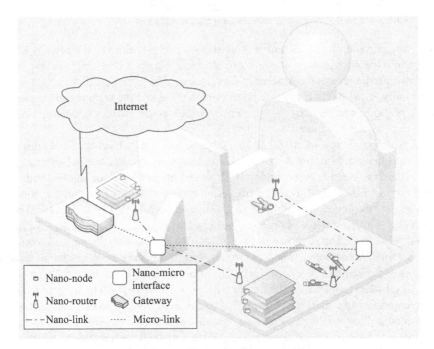

Legend:
- Nano-node
- Nano-router
- Nano-link
- Nano-micro interface
- Gateway
- Micro-link

Figure 12.1 Network architecture of the e-office. ©2010 IEEE. Reprinted with permission from Reference 9

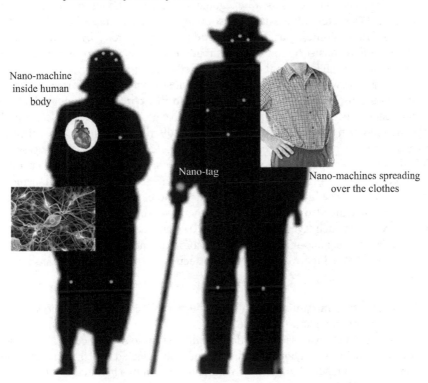

Nano-machine inside human body

Nano-tag

Nano-machines spreading over the clothes

Figure 12.2 Architecture of a health monitor nano-network

level, for example, the nano-sensor around the heart can monitor the activity of the heart while the ones in neuro-system can release the pain if necessary; nano-phones can entertain the user in the process of exercise or make a phone call, but much more importantly, it can be used as the nano-micro interface and the nano social tags inside the users' body can be used as the identity which could make the communication between people much more convenient.

A conceptual network model of IoNT, based on On-Off Keying (OOK) protocol and Time Division Multiple Access (TDMA) framework, was done in Reference 43, where the nano-sensors were randomly deployed at organs of the human body and may be moved by body fluid. The suggested model assumes hexagonal cell-based nano-sensors deployed in cylindrical shape three-dimensional (3-D) hexagonal pole, which is closer to the shape of organs. The network architecture of the IoNT for intra-body disease detection is shown in Figure 12.3(a). As normal, the network contains nano-sensors, nano-routers, nano-micro interface, and gateway. When nano-sensors detect specific symptoms or virus by means of molecules [44] or bacteria behaviours [45], simple data (e.g., 1 for detection or 0 for non-detection) will be transmitted over short ranges to nano-routers to inform the existence of symptoms or virus because of the limited capacity of the nano-sensors. Because of the relatively more computational resources, nano-routers can aggregate the data and send the related information to the nano-micro interface. The gateway (i.e., micro-scale device) enable the remote control of the entire system over the Internet possible. Each cell, the smallest living unit of organs, is considered as a hexagonal shape cell, shown in Figure 12.3(b), resulting a 3-D structure for the nano-sensor networks. A 3-D space for the individual target organ, for example, heart, lungs, and kidney, is constructed by the accumulated unit layers which consist of unit cells. As many nano-sensors as possible are put in the model where each hexagonal cell has one active nano-sensor.

Nano-sensors within each layer construct a cluster, where the information sensed by each nano-sensor can be transmitted to the nano-micro interface through the nano-router of each cluster. Each hexagonal cell may have more than one nano-sensors. To make the load of the energy consumption evenly distributing in the nano-networks, only the nano-sensor with the most energy can be selected as the active nano-sensor while the others will go into the sleep mode which would be used for the next data transmission. The information detected by the nano-sensors would be transmitted within the occurrence layer to the nano-router at the centre cell (annulus A_0) of the same layer which are also served as an active nano-sensor. Then the data collected by the nano-router of each layer are sent to layer 0's nano-router which will forward the collected data to the nano-micro interface to communicate with micro-scale devices, i.e., gateway. From Figure 12.3, it can be seen that there are three transmission methods:

- Direct data transmission: the data would be sent directly from the nano-sensor to the corresponding nano-router [46].
- Multi-hop data transmission: the data would be sent to the adjacent nano-sensors which locate in the neighbour cell and is randomly selected until the corresponding nano-router is reached in the way of annulus by annulus hierarchically.

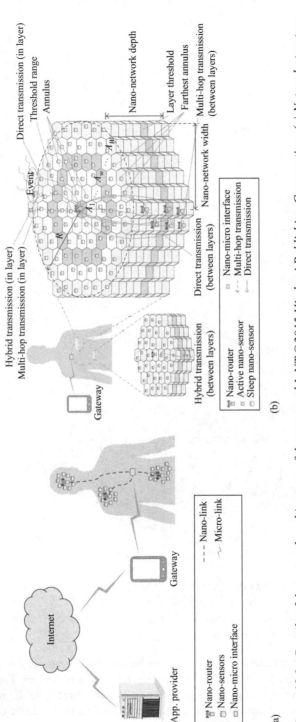

Figure 12.3 *Details of the network architecture of the proposed IoNT. (a) Network structure of the proposed IoNT and (b) Structure of the nano-cell of IoNT. Reprinted with permission from Reference 43*

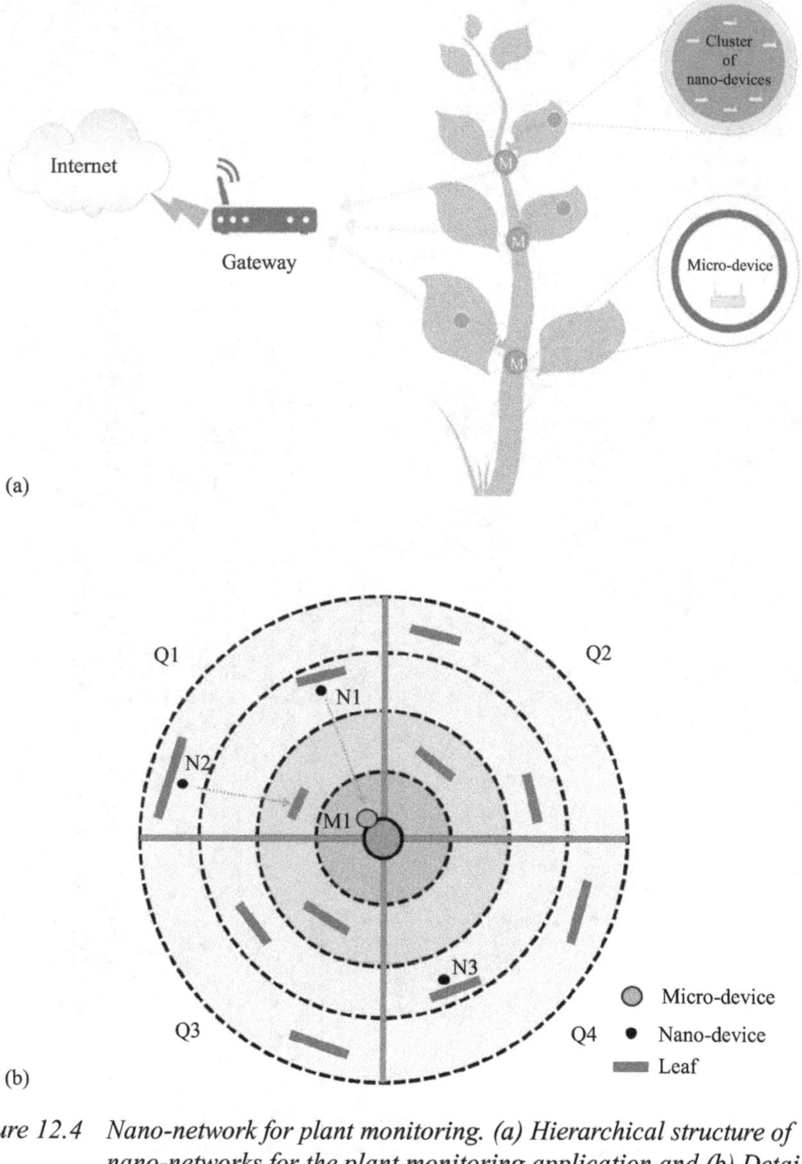

(a)

(b)

*Figure 12.4   Nano-network for plant monitoring. (a) Hierarchical structure of
nano-networks for the plant monitoring application and (b) Details of
the nodes distribution. ©2015 Elsevier. Reprinted with permission
from Reference 47*

- Hybrid data transmission: the data can be sent by combining the multi-hop and
  direct transmission methods. First, a threshold range is defined. Then, within the
  threshold range, every nano-sensor sends the data directly to the nano-router in
  the same layer. Otherwise, the multi-hop transmission should be used when the
  nano-sensor is outside the range.

Similarly, there are also three data transmission methods between layers, shown in Figure 12.3(b).

Figure 12.4 shows an example of using nano-network for agricultural crop-monitor [47]. The chemical compound released by plants can be caught by the nano-sensors, and such information can be sent to the micro-devices to analyse the knowledge of environmental conditions and plant interaction patterns; thus, an enhanced chemical defence systems could be developed, or the underground soil condition can be retrieved by using plants as sensors. A large numbers of nano-devices equipped with chemical nano-sensors and THz radio units are deployed on the plant, which can transfer the data detected to a micro-scale networking device over a short distance, where the transmission frequency can be dynamically selected in order to optimize throughput of the network. The hierarchical structure of this monitoring network is illustrated in Figure 12.4(a) where numerous nano-sensors are situated on the plant through suspension in a spray applied to the plants [47]. The chemical nano-devices deployed in the nano-network comprise a power block and a communication block along with relevant sensors, processing and storage units. The nano-devices are supposed randomly scattered on the plant leaves, and clusters are formed based on their location and proximity to micro-devices which are located at specified points on the stem to manage clusters of nano-devices, shown in Figure 12.4(b). The information can only be sent by single-hopped way which means that there is no Non-Line of Sight (NLoS).

12.3 Available paradigms of nano-communication

To connect the nano-devices, the communications between them need to be completed. According to Reference 4, nano-communication can be divided into two scenarios: (1) communication between a nano-machine and a larger system such as micro/macro-system and (2) communication between two or more nano-devices. Furthermore, the methods of electromagnetic, acoustic, nano-mechanical, or molecular can all be applied to nano-communications [42], and some of them will be discussed in this section.

12.3.1 Molecular communication

As the original idea of nano-communication, molecular communications are considered as the most promising paradigm to achieve the nano-communication, because there are numerous examples in nature for us to learn and study. In molecular communication, an engineered miniature transmitter releases small particles into a propagation medium while the molecules are applied to encode, transmit, and receive information. Usually, in molecular communication, information can be encoded in various ways, such as in the release time of the molecules (timing modulation) [48, 49], in the concentration of the molecules or number of molecules per unit area (concentration modulation) [50, 51], in the number of molecules (amplitude modulation) [52], in the identities of the molecules, *etc*. Molecular communication can be classified into several categories such as walkway-based where molecules propagate along a predefined pathway via molecular motors, flow-based where molecules propagate in a guided fluidic medium, diffusion-based where molecules propagate in a fluidic medium via spontaneous diffusion, *etc*.

As the most general and widespread scheme found in nature, the diffusion-based molecular communication (DMC) is the most widely investigated. Some of the most prominent works include mathematical framework for a physical end-to-end channel model for DMC [53], development of an energy model for DMC [54], modelling of diffusion noise [55], channel codes for reliability enhancement [56], and relaying-based solutions for increasing the range of DMC [57, 58].

However, the flow-based molecular communication is much more accurate which has also been caught attention recently [59, 60]. Srinivas *et al.* [49] first studied the case of molecules with timing modulation in the fluidic medium while flow-based molecular communication with concentration modulation was initially analysed later [51]. In 2014, flow-based molecular communication with amplitude modulation scheme was studied [52]. Meanwhile, a more general flow-based molecular communication model has been proposed recently which can be applied to both molecule shift keying (MoSK) and concentration shift keying (CSK) [61], where the effects of the moving medium on the signal propagation and bit error rate (BER) performance was investigated.

12.3.2   Acoustic communication

Acoustic propagation introduces slight pressure variations in the fluid or solid medium, which satisfy the wave equation. The behaviour of the nano-robots is relevant to their physical properties, surrounding medium, and the working frequency. The feasibility of *in vivo* ultrasonic communication is evaluated by Hogg *et al.* [62], where communication effectiveness, power requirements, and effects on nearby tissue were examined on the basis of discussion on the principles. Based on this knowledge, an isolated robot and an aggregated robot were designed to use in blood vessel. Later, the nano-scale opto-ultrasonic communications in biological tissues was discussed in References 8 and 63, where the generation and propagation model were studied, and in line with Reference 62 the hazards and design challenges were investigated.

12.3.3   EM communication

As the name indicates, electromagnetic methods use the electromagnetic wave as the carrier and its properties like amplitude, phase, *etc.* are used to encode the information.

The possibility of EM communication is first discussed in Reference 7 on the basis of the fact that THz band can be used as the operation frequency range for future EM nano-transceivers because of the emerging new materials like carbon nano-tube (CNT) and graphene [44] while Reference 64 demonstrates the theoretical model of the nano-network whose nodes are made of CNT. Later, the channel model for THz wave propagating in the air with different concentrations of the water vapour was presented in Reference 65, and the corresponding channel capacity was also studied. Based on the characteristics of the channel, a new physical-layer aware medium access control protocol, Time Spread On-Off Keying (TS-OOK), was proposed [5]. Meanwhile, the applications of THz technology in imaging and medical field [66, 67] have also achieved great development, and the biological effects of THz radiation are reviewed in Reference 68 showing minimum effect on the human body and no strong evidence of hazardous side effects.

Later, a more detailed model of THz communication was proposed with the consideration of multi-ray scenario; thus, the propagation models for reflection, scattering, and diffraction were considered in Reference 69. At the same time, the scattering effects of small particles were discussed with the analysis frequency and the impulse responses [70]. Also, the finite-difference time-domain (FDTD) and the ray-tracing (RT) technique were compared to evaluate the reception quality in nano-network with the consideration of two cases: line-of-sight (LOS) and multiple objects dispersed near the LOS [71]; then, the conclusion was drawn that at the THz band RT is as good as FDTD. Meanwhile, a discussion of the use of VHF band was conducted with the study of the BER performance of the nano-receiver made of CNT [72].

12.4 Current study on body-centric nano-networks at THz band

12.4.1 Numerical modelling of THz wave propagation in human tissues

12.4.1.1 Relationship of optical parameters to electromagnetic parameters

At optical frequencies, all the information is usually delivered in terms of refractive index or index of refraction:

$$\tilde{n}(f) = n_r(f) - jn_i(f). \tag{12.1}$$

The real part n_r can be usually obtained directly from the measurement while the imaginary part n_i (which would also be referred to as κ, named extinction coefficient) can be calculated from the absorption coefficient α. The relationship between the two can be given by:

$$n_i(f) = \frac{\alpha(f)\lambda_o}{4\pi}, \tag{12.2}$$

where, $\lambda_o = c/f$ is the wavelength in free-space.

From $\tilde{n}$, we can obtain the relative permittivity constant ε as:

$$\varepsilon(f) = \tilde{n}(f)^2 = n_r(f)^2 - \kappa(f)^2 - j2n_r(f)\kappa(f). \tag{12.3}$$

Thus, we can obtain:

$$\begin{cases} \varepsilon' = n_r(f)^2 - \kappa(f)^2, \\ \varepsilon'' = 2n_r(f)\kappa(f), \end{cases} \tag{12.4}$$

where ε' and ε'' are the real and imaginary parts of the permittivity ε.

The absorption coefficient of some human tissues, i.e. blood, skin, and fat, is shown in Figure 12.5(a) [73, 74] while the EM parameters, i.e., permittivity of

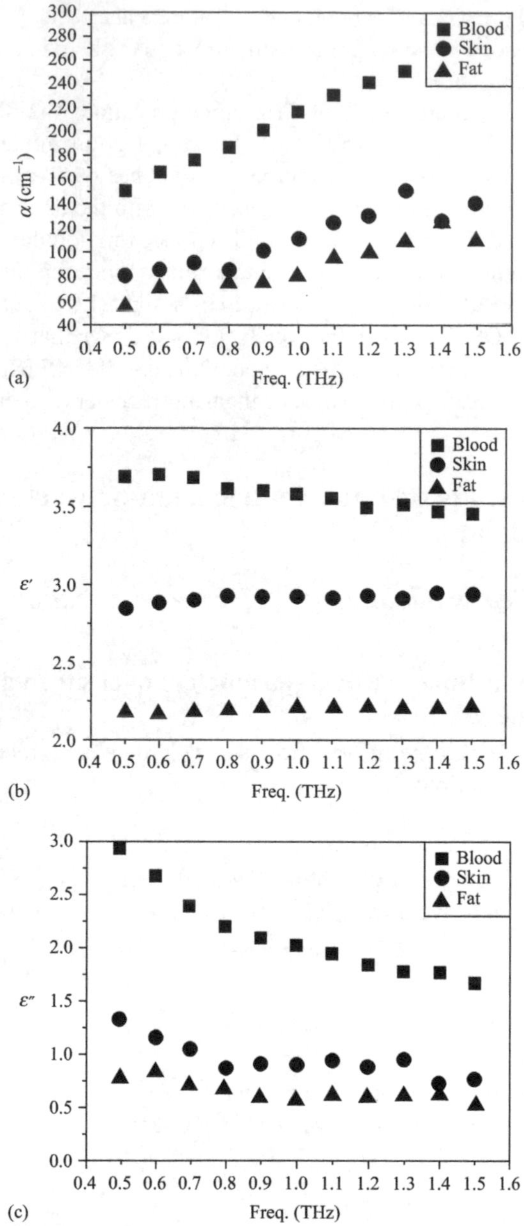

Figure 12.5   Optical and electromagnetic parameters of human tissues (blood, skin, and fat). (a) Measured absorption coefficient α. ©2003 Springer. Reprinted with permission from References 73, 74, (b) Real part of the relative permittivity (calculated from (12.4), and (c) imaginary part of the relative permittivity (calculated from (12.4)). ©2015 IEEE. Reprinted with permission from Reference 6

the corresponding tissues, calculated from (12.4), were shown in Figures 12.5(b) and 12.5(c).[1]

12.4.1.2 Path loss

A modified Friis equation has been proposed by Jornet *et al.* in Reference 65 to calculate the path loss of the THz channel in water vapour, which can be divided into two parts: the spread path loss PL_{spr} and the absorption path loss PL_{abs}. Similarly, the path loss in human tissues can also be divided into two parts:

$$PL_{total}[dB] = PL_{spr}(f,d)[dB] + PL_{abs}(f,d)[dB], \tag{12.5}$$

where f stands for the frequency while d is the path length.

The spread path loss is introduced by the expansion of the wave in the medium, which is defined as:

$$PL_{spr}(f,d) = \left(\frac{4\pi d}{\lambda_g}\right)^2, \tag{12.6}$$

where $\lambda_g = \lambda_o/n_r$ stands for the wavelength in medium with free-space wavelength λ_o, and d is the travelling distance of the wave. In this study, the electromagnetic power is considered to spread spherically with distance.

The absorption path loss accounts for the attenuation caused by the molecular absorption of the medium, where part of the energy of the propagating wave is converted into internal kinetic energy of the excited molecules in the medium. The absorption loss can be obtained from the transmittance of the medium $\tau(f,d)$:

$$PL_{abs} = \frac{1}{\tau(f,d)} = e^{\alpha(f)d}. \tag{12.7}$$

The dependency of the channel path loss for blood, skin, and fat on the distance and frequency is shown in Figure 12.6. It is demonstrated that there are some fluctuations in each individual figure due to the fact that absorption path loss is related to the extinction coefficient, κ, which is not an analytical function along the required frequency band, in addition to the expected increase in path loss values with larger distances and higher frequency components. For different tissues, the path loss varies with blood experiencing the highest losses, followed by the skin due to the water concentration, which contributes a significant absorption path loss. At the level of the millimetres, the path loss of the blood is around 120 dB, while the skin is around 90 dB and the fat is around 70 dB. Compared with the channel attenuation of the molecular communication [50], the future of the EM paradigms is promising because at 1 kHz (here, the frequency is the operation frequency of the RC circuit which depicted the emission and absorption process of the diffusion-based particle communication) and at a distance of 0.05 mm the molecular channel attenuation is above 140 dB which is substantially higher than the case for blood at the distance of 1 mm applying THz EM communication mechanism. In Reference 75, the capacity was also compared between the two paradigms, showing that that the EM communication keeps extremely high

---

[1] The refractive index n_r is 1.97, 1.73, and 1.58 for blood, skin, and fat, respectively.

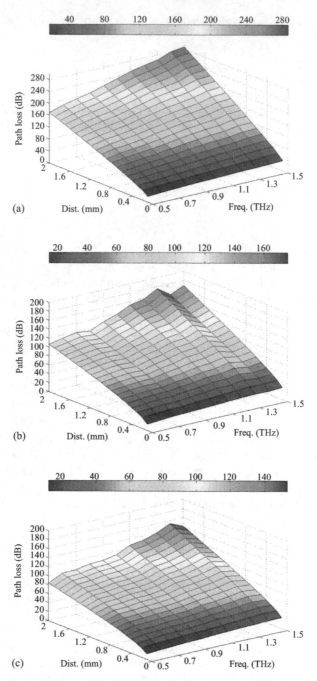

Figure 12.6   Total path loss as a function of the distance and frequency for different human tissues. (a) Blood, (b) skin, and (c) fat. ©2015 IEEE. Reprinted with permission from Reference 6

data rate until the distance is shorter than 10 mm while the molecular communication scheme provides much lower capabilities.

12.4.1.3 Noise

The molecules along the path not only introduce the attenuation of the wave but also introduce the noise because their internal vibration, provoked by the incident wave, would turn into the emission of EM radiation at the same frequency [76], which can be measured by the parameter of the emissivity of the channel, ξ:

$$\xi(f,d) = 1 - \tau(f,d). \tag{12.8}$$

where, $\tau(f,d) = e^{-\alpha(f)d}$ is the transmissivity of the medium, f is the frequency of the EM wave, d stands for the path length.

Thus, the equivalent noise temperature due to molecular absorption can be obtained:

$$T_{mol}(f,d) = T_0\xi(f,d), \tag{12.9}$$

where T_0 is the reference temperature, f is the frequency of the EM wave, d stands for the path length, ξ refers to the emissivity of the channel given by Eq. 12.8. It should be noted that this kind of noise only appears around the frequencies in which the molecular absorption is quite high.

The total noise temperature of the system T_{noise} is composed of the system electronic noise temperature, T_{sys}, and the total antenna noise temperature, T_{ant}, which includes not only the molecular absorption noise temperature, T_{mol}, but also other contributions from several sources, T_{other}, such as the noise created by surrounding nano-devices or the same device:

$$T_{noise} = T_{sys} + T_{ant} = T_{sys} + T_{mol} + T_{other}. \tag{12.10}$$

For a given bandwidth, B, the total system noise power at the receiver can be calculated as follows:

$$P_n(f,d) = \int N(f,d)df = k_B \int T_{noise}(f,d)df, \tag{12.11}$$

where N stands for the noise power spectral density; k_B is the Boltzmann constant; T_{noise} is the equivalent noise temperature.

Because the electronic noise temperature of the system is assumed to be low due to the electron transport properties of graphene [77], the main factor affecting the channel performance will be the molecular absorption noise temperature, which indicates $T_{noise} \approx T_{mol}$.

The molecular absorption noise temperature is shown in Figure 12.7. It can be seen that the noise temperature increases with the rise of the frequency and distance, which will lead to the rise of the noise power. At the level of millimetres, the molecular noise temperature reaches 310 K, the normal human temperature.

12.4.1.4 Channel capacity of human tissues at THz band

In order to evaluate the potential of the Terahertz band, the channel capacity would be used as the performance metric. In the analysis, THz band is considered as a single

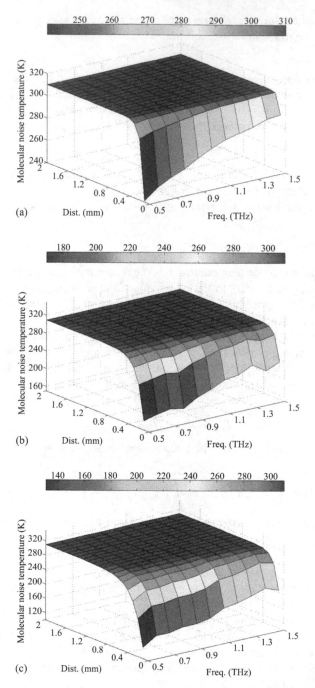

*Figure 12.7   Noise temperature as a function of the distance and frequency for
different human tissues. (a) Blood, (b) skin, and (c) fat. ©2015 IEEE.
Reprinted with permission from Reference 6*

transmission window which is almost 1 THz wide (from 0.5 THz to 1.5 THz) because of the limit of the current database for human tissues at THz band.

According to the *Shannon's theory*, the channel capacity can be obtained as follows [78]:

$$C = B \log_2 \left(1 + \frac{S}{N} \right),$$ (12.12)

where B is the whole bandwidth of the system while $\frac{S}{N}$ stands for the signal-noise-ratio.

From the previous investigation, the channel is frequency-selective and the noise is non-white, thus the whole bandwidth needs to be divided into many narrow sub-bands which can be chosen small enough to make the channel appear non-selective and the noise p.s.d. locally flat [78]. The ith sub-band is centred at f_i with the bandwidth $\Delta f (i = 1, 2, 3 \dots)$. Then, (12.12) can be rewritten as:

$$C(d) = \sum_i \Delta f \log_2 \left[1 + \frac{S(f_i)}{PL(f_i, d)S_N(f_i, d)} \right],$$ (12.13)

where $S(f_i)$ is the power spectral density of the transmitted signal while Δf is fixed to 0.1 THz. And here the frequency band covering from 0.5 THz to 1.5 THz with 1 THz bandwidth is considered.

From (12.12), it can be easily seen that besides the effect of both path loss and noise temperature communication capabilities are also strictly influenced by the way how the transmitted power, P_{tx}, is distributed in the frequency domain. In line with Reference 65, three communication schemes (i.e., *flat, pulse-based,* and *optimal*) are considered in this work and characterized in what follows.

- *Flat communication*

The total power transmission, P_{tx}, is uniformly distributed over the entire operating band. Thus, the corresponding power spectral density is:

$$S_f(f) = \begin{cases} S_0 = P_{tx}/B & \text{if } f_m \leq f \leq f_M \\ 0 & \text{otherwise,} \end{cases}$$ (12.14)

where, obviously, $\int_{f_m}^{f_M} S_f(f) df = P_{tx}$, that is, $S_0 = P_{tx}/B$.

- *Pulse-based communication*

Taking into account capabilities of graphene-based nano-electronic, the pulse generated by a nano-device, i.e., the wave form used to transmit the logical 1, can be modelled with a n-th derivative of a Gaussian-shape, i.e., $\phi(f) = (2\pi f)^{2n} e^{(-2\pi\sigma f)^2}$ [65]. Hence, the power spectral density can be expressed as:

$$S_p(f) = a_0^2 \cdot \phi(f),$$ (12.15)

where σ and a_0^2 are the standard deviation of the Gaussian pulse and a normalizing constant, respectively.

Considering that $\int_{f_m}^{f_M} S_p(f) df = P_{tx}$, a_0^2 is given by:

$$a_0^2 = P_{tx} \Big/ \int_{f_m}^{f_M} \phi(f).$$ (12.16)

• *Optimal communication*
This scheme aims at maximizing the overall channel capacity by optimally adapting
the power allocation as a function of frequency-selective properties of the channel.
The optimal transmission scheme can be obtained solving the following optimization
problem with the consideration of (12.13):

$$
\begin{cases}
\max \left\{ \sum_i \Delta f \log_2 \left[1 + \dfrac{S_o(f_i)}{PL(f_i, d)N(f_i, d)} \right] \right\}, \\
\text{subject to} \quad \sum_i S_o(f_i)\Delta f = P_{tx}.
\end{cases}
\tag{12.17}
$$

As known, the maximum value of a concave function, like the one in (12.17), can
be done by using Lagrange multiplier, λ. Thus, the optimization problem can be
rewritten as:

$$
\max \left\{ \sum_i \Delta f \left(\log_2 \left[1 + \frac{S_o(f_i)}{PL(f_i, d)N(f_i, d)} \right] + \lambda S_o(f_i) \right) - P_{tx} \right\}.
\tag{12.18}
$$

The maximum is found by equating to zero the derivative of the argument of
(12.18) with respect to $S_o(f_i)$ and λ. This produces:

$$
\frac{1}{\ln(2)[S_o(f_i) + PL(f_i, d)N(f_i, d)]} = \lambda \quad \forall i.
\tag{12.19}
$$

That is, the overall channel capacity is maximized when:

$$
S_o(f_i) + PL(f_i, d)N(f_i, d) = \beta,
\tag{12.20}
$$

where β is a constant to be evaluated.

The problem can be solved by using the *water-filling* principle, which adopts
an iterative procedure for finding the most suitable power distribution on available
sub-bands. In details, at the n-th step, β is computed as:

$$
\beta(n) = \frac{1}{L(n)} \left[\frac{P_{tx}}{\Delta f} + \sum_i PL(f_i, d)N(f_i, d) \right],
\tag{12.21}
$$

where $L(n)$ is the number of sub-bands at the n-th step.

In particular, considering the i-th sub-band, the power spectral density is set as
$S_o(f_i) = \beta - A(f_i, d)N(f_i, d)$. If it results $S_o(f_i) \leq 0$, then the corresponding power
spectral density is set to 0 and (12.21) should be computed again (without considering
the identified sub-band) as long as there are no sub-bands with a negative $S_o(f_i)$. At
the end, the procedure optimally distributes the total power transmission over the
available sub-bands by assigning higher power spectral density values to sub-bands
offering better channel conditions (i.e., lower path loss and lower noise power).

Figure 12.8 shows the capacity for different tissues with different power allocations. From the figure, it can be seen that with the increase of the distance the
capacity will keep almost the same at first and then drop significantly at some point
(for different tissues and different power allocation, the point changes). And also we
can see that the optimal one always occupies the top while the performance of the

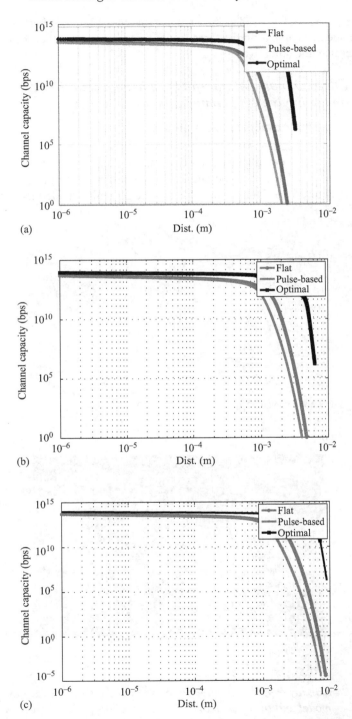

*Figure 12.8   Capacity for different tissues with different power allocations ©2015
IEEE. Reprinted with permission from Reference 6. (a) Blood,
(b) skin, and (c) fat*

pulse-based one is not as good as the other two but the differences between them are not very huge. By comparing the three figure listed, we can see that the capacity of fat is better than the other two.

12.4.2   *Effects of non-flat interfaces in human skin tissues on the in vivo THz communication channel*

12.4.2.1   **Applied numerical skin models**

Human skin structure

The skin is a complex heterogeneous and anisotropic medium, where the small parts, like blood vessel and pigment content, are spatially distributed in depth [79] [80], as shown in Figure 12.9. Human skin [81] can be generally divided into three main

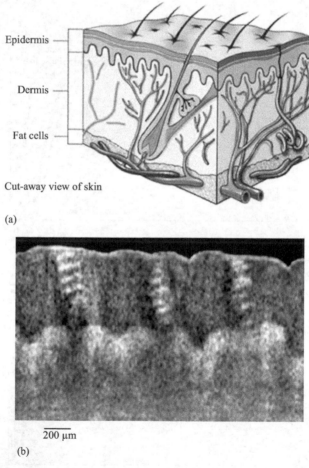

(a)

(b)

Figure 12.9   Skin layer model and the corresponding OCT image. (a) Layered skin model and (b) OCT image of the skin ©2013 IEEE. Reprinted with permission from Reference 84

visible layers from the surface: the epidermis ($\sim$100 μm, blood-free layer), the dermis ($\sim$1–4 mm, vascularized layer) and the subcutaneous fat (from 1 to 6 mm dependent on the location) as shown in Figure 12.9 [82]. The epidermis contains two sub-layers: stratum corneum with only dead squamous cells and the living epidermis layer, where most of the skin pigmentation stay. The stratum corneum is a thin, stratified, and highly specialized layer which is accumulated on the outermost skin surface. The dermis, supporting the epidermis, is thicker and mainly composed of collagen fibres and intertwined elastic fibres enmeshed in a gel-like matrix. The subcutaneous fat layer is composed of the packed cells with considerable fat, where the boundary is not well demarcated; thus the thickness of this layer varies widely for different parts of human body [83].

Numerical skin models
● *Models with rough boundary*
From the skin model and its optical coherent tomography (OCT) image shown in Figure 12.9 [84], it can be observed that the interface between the layers is non-flat [85], and the dermal ridges prolong and penetrate into the epidermis. In this chapter, the influence of rough interface between these two layers was studied, making the model more physiologically plausible. CST Microwave Studio^{TM} software package using Finite Integration Technique was selected to simulate the propagation and interaction of THz wave in the 3-D numerical model of skin, shown in Figure 12.10 where the interfaces between the epidermis and dermis are 3-D sine and 3-D sinc wave, respectively. A cube, L μm wide and H μm high was built to model the skin cube, including a h_{epi} μm upper layer as the epidermis and a h_{derm} μm layer as the dermis. The bisected cross-sections of both models are shown in Figure 12.11 and the corresponding applied functions, depicting the 3-D surface, were shown below:

$$z = \pm A \cos\left(\frac{2\pi x}{S}\right) \cos\left(\frac{2\pi y}{S}\right), \tag{12.22}$$

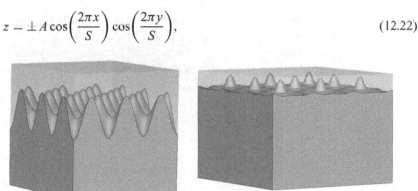

(a)   (b)

Figure 12.10 Numerical models with different function interfaces, developed in CST Microwave Studio. (a) Stratified skin model with 3-D sine function as the interface and (b) stratified skin model with 3-D sinc function as the interface. ©2015 Elsevier. Reprinted with permission from Reference 86

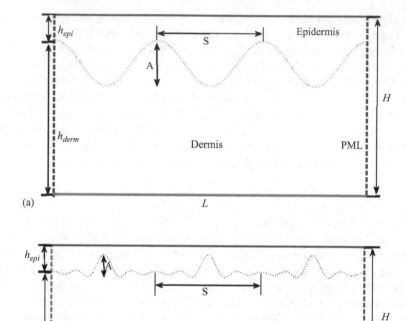

Figure 12.11   Cross-section of the applied models in Figure 12.10 (L is the cube length, H is the cube height, h_{epi} and h_{derm} is the thickness of epidermis and dermis, A is the amplitude of the surface while S is the span of the surface). (a) Stratified skin model with 3-D sine function as the interface in CST and (b) Stratified skin model with 3-D sinc function as the interface in CST. ©2015 Elsevier. Reprinted with permission from Reference 86

where A and S are short for amplitude and span, respectively; "+" stands for the peak scenario while "−" refers to the valley case (in Figure 12.11(a), "−" is chosen):

$$z = \begin{cases} \pm A \dfrac{\sin\left(\sqrt{(x-nS)^2+(y-nS)^2}\right)}{\sqrt{(x-nS)^2+(y-nS)^2}}, & \text{when } -(2n-1)S/2 \le x, y \le (2n+1)S/2; \\ 0, & \text{else;} \end{cases}$$

$$(12.23)$$

Table 12.2   Different parameters of simulated models using sine and sinc functions as the interface (unit: μm)

Model	Sine	Sinc
H	2105	1600
L	2100	2100
h_{epi}	[405–105]	[40–240]
h_{derm}	[1000–1700]	[1260–1460]

Table 12.3   Parameters of the dipole antennas used as transmit and receive elements for the EM propagation power loss study across and within the human skin tissue models

	Dipole on the skin surface	Dipole in the skin
Arm length (μm)	43	36
Radius (μm)	5	5

where A represents the amplitude of the sinc function while S is the span of sinc function shown in Figure 12.11(b); $n = L/S$ is the ratio of skin cube length to sinc function span; "+" stands for the peak scenario while "−" refers to the valley case (Figure 12.11(b) shows the case of "+").

The parameters of the two models in Figure 12.10 are summarized in Table 12.2.[2]

Two dipoles are also modelled as the nano-antennas, working at THz band. One is placed on the skin surface while the other is located within the skin. Two discrete ports were applied to the dipoles and the transmission coefficient, $S21$, was recorded to investigate the power loss. The dimensions of the dipoles are shown in Table 12.3, and they are optimized to ensure that the impedance matching is better than -10 dB. The open boundary, i.e., Perfect Match Layer (PML) was applied to ensure that the reflections of the abruptly-cutting boundaries are neglectable. The nano-second long Gaussian pulse is applied in the Time-Domain simulation while hexahedral mesh type was applied. The permittivities applied to the different parts of skin are shown in Figure 12.12 [87].

● *Skin models with sweat ducts*

For this part of the study, similar detailed skin model as shown in Figure 12.10(a) was applied with the inclusion of sweat ducts, as shown in Figure 12.13. The sweat duct is 265 μm in height and 40 μm in diameter. It includes three turns to be considered as a helical antenna (for helical antenna, number of turns should be at least 3).

---

[2] For sine model the amplitude of the surface A is 350 μm while for sinc model the amplitude of the surface A is 200 μm.

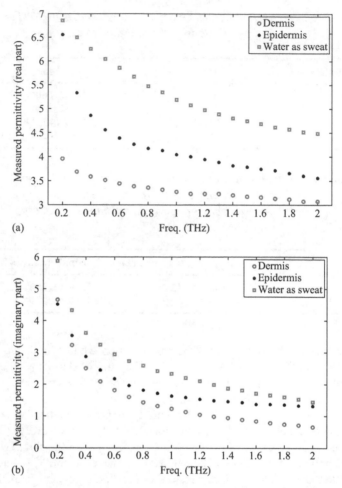

Figure 12.12   Permittivity of different skin layers and sweat at THz frequencies:
(a) Real part of the relative permittivity and (b) imaginary part of the
relative permittivity. ©2007 International Society for Optics and
Photonics. Reprinted with permission from Reference 86

Only when the sweat duct contains sweat, containing 99% water and 1% salt and amino acid [88], it can be regarded as a conductor; and hence the permittivity of water at THz band was used to model sweat duct, as shown in Figure 12.12.

12.4.2.2   Analysis of skin-internal non-flat interfaces on the THz EM channel

• *Effects of different shapes*
Four scenarios were investigated with model parameters as shown in Table 12.4.

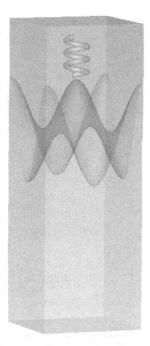

Figure 12.13   Numerical model of skin while including sweat ducts ©2015 Elsevier. Reprinted with permission from Reference 86

Table 12.4   Amplitude and span change for the different interface models (unit: μm)

Sine model		Sinc model	
S = 700	*A* = 350	*S* = 700	*A* = 200
A	*S*	*A*	*S*
175	350	100	300
245	560	120	400
350	700	140	500
420	910	160	600
525	1050	180	700
		200	

The models with different amplitudes and spans (shown in Table 12.4) are simulated. The power losses for individual cases were recorded. Take the sine model as an example, the line array recorded is illustrated in Figure 12.14. Then the mean values and deviations of the power losses are calculated (as shown in Figure 12.15) and compared to the power losses obtained from the common flat one, where it can

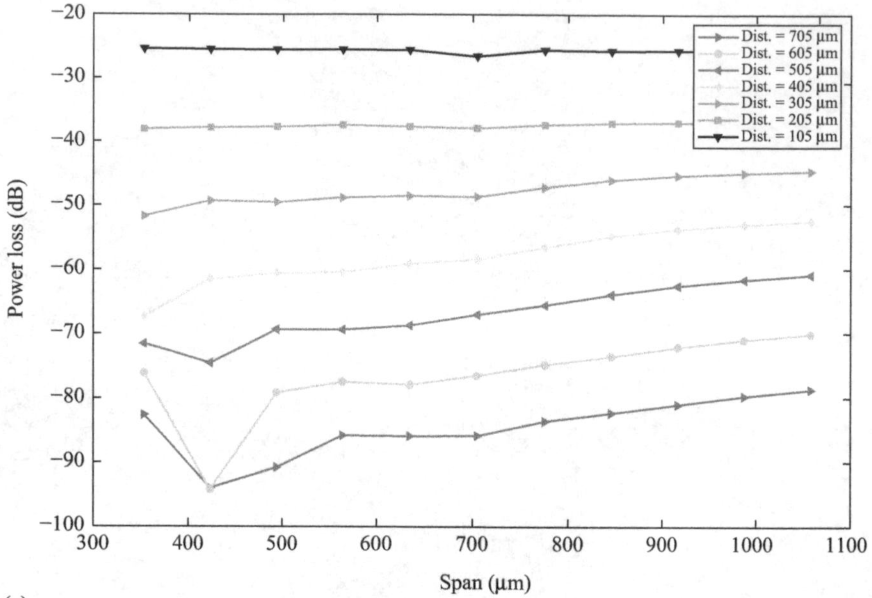

(a)

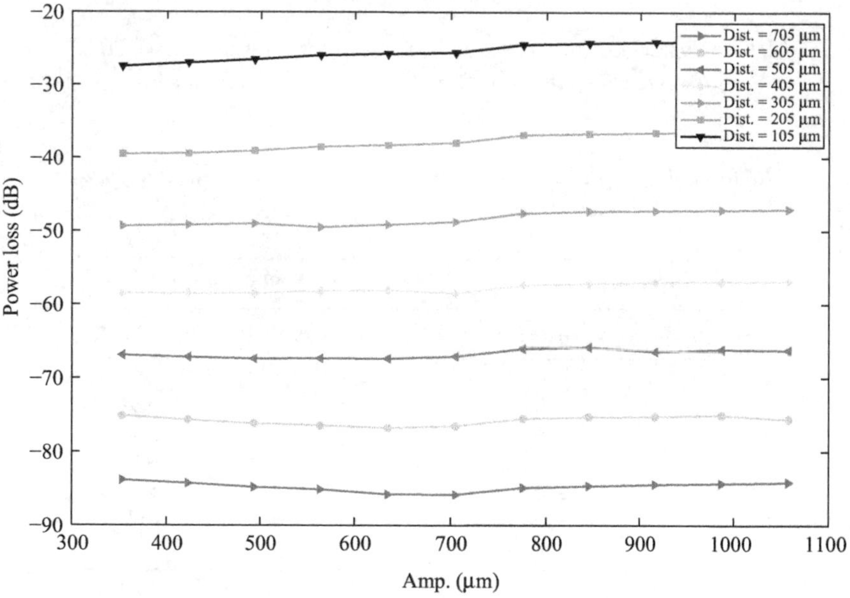

(b)

Figure 12.14   Effects of parameter changes on power loss for sine-wave model listed in Table 12.4. (a) Power loss of different span values when A is set to 350 μm and (b) power loss of different amplitude values when S is set to 700 μm. ©2015 Elsevier. Reprinted with permission from Reference 86

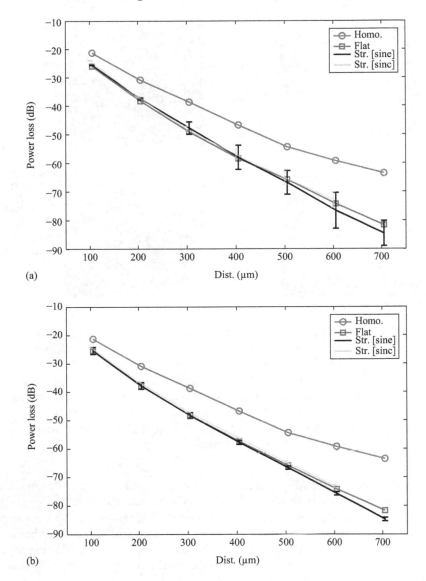

Figure 12.15 *Dependence of the power loss on the distance for different models. (a) Influence of span change on power loss and (b) influence of amplitude change on power loss. © 2015 Elsevier. Reprinted with permission from Reference 86*

be seen that all values decrease due to the distance increment. The power loss of the flat model and the mean values of stratified models with 3-D sine and sinc surfaces are close to each other, pointing out that the influence of the rough interface could be neglected, when the general study is being conducted. Additionally, the deviation of

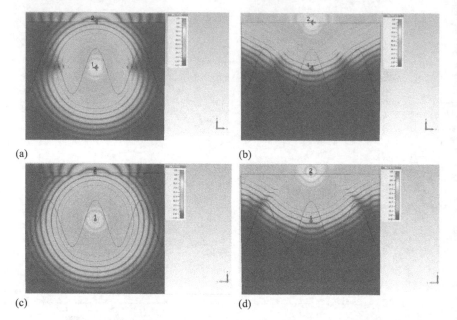

(a)   (b)

(c)   (d)

Figure 12.16   Electric field distribution of the dipoles at 1 THz for the peak
scenario of the sine wave model. (a) Electric field distribution for the
dipole in skin at the plane of xoz (y = 0), (b) electric field
distribution for the dipole in skin at the plane of xoz (y = 0), (c)
Electric field distribution for the dipole on skin surface at the plane
of yoz (x = 0), and (d) Electric field distribution for the dipole on
skin surface at the plane of yoz (x = 0). © 2015 Elsevier. Reprinted
with permission from Reference 86

the stratified model with 3-D sine interface is greater than the one with sinc interface. The biggest difference is around 2 dB, and they both increase with the increasing distance. By comparing the results shown in Figure 12.15(a) and 12.15(b), it can be concluded that the span variation would cause more changes to the power loss because of its larger deviations, reaching around 10 dB (sine model) and 5 dB (sinc model) at the distance of 600 μm, respectively, while the deviation of the amplitude is always less than 1 dB.

- *Effects of the antenna location*

The impact of the location of the antennas is also studied. In one of the scenarios, both antennas are placed along the line going through the peak of the 3-D wave as shown in Figure 12.16(a) while another going through valley of the wave as shown in Figure 12.17(a). The field distributions for both cases of the sine model are shown in Figures 12.16 and 12.17. It can be easily seen that the applied PML boundary has no effect to the performance of the antenna and the interface between

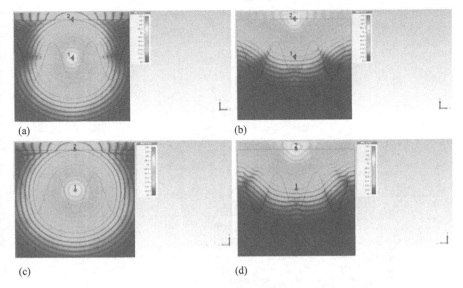

Figure 12.17 *Electric field distribution of the dipoles at 1 THz for the valley*
scenario of the sine wave model. (a) Electric field distribution
for the dipole in skin at the plane of xoz (y = 0), (b) electric field
distribution for the dipole in skin at the plane of xoz (y = 0),
(c) electric field distribution for the dipole on skin surface at the
plane of yoz (x = 0), and (d) electric field distribution for the dipole
on skin surface at the plane of yoz (x = 0). ©2015 Elsevier. Reprinted
with permission from Reference 86

the epidermis and dermis barely affects the wave propagation. The results shown in Figure 12.18 agree well indicating that the effect of the antenna location can be ignored, when the antennas are located above/below the peak/valley boundary for the 3-D sine model. However, when the span and amplitude values go beyond 300 μm, the difference between the power losses is noticeable with a 5 dB difference at 600 μm (as shown in Figure 12.19). The power loss for the scenario that two dipoles locate above/below the peak is larger for the one with the valley scenario because for the peak scenario the second antenna would be in the dermis most of the time.

- *Effects of the sweat duct*

The effects of the sweat duct for the sine wave model were investigated, shown in Figure 12.20. It is observed that with the increase of the distance the power loss increases, as predicted. When PEC was applied to the sweat duct to make it as the ideal conductor, the power loss is 5 dB less than the case of water applied to the

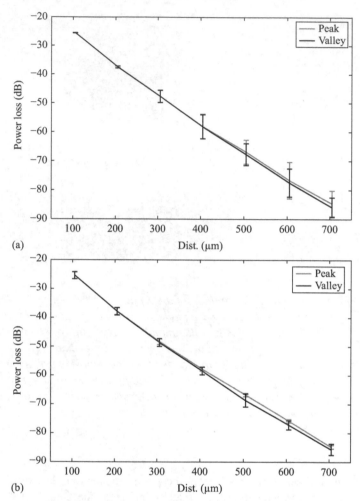

Figure 12.18   *Comparison between the two scenarios: two dipoles located*
above/beneath the peak and valley of the interface for the model
with 3-D sine function as the interface. (a) Effects of span alteration
on the power loss with respect to distance and (b) Effects of
amplitude alteration on the power loss with respect to distance.
©2015 Elsevier. Reprinted with permission from
Reference 86

sweat duct, which shares similar power loss with the model without sweat duct. This indicates that if sweat duct is working as PEC, the power loss would be reduced; however, in most cases, the sweat duct is full of sweat containing 99% water, which has limited influence on the overall communication link quality. Analysis clearly shows the importance of the study of the parameters of the human tissue, especially at THz band.

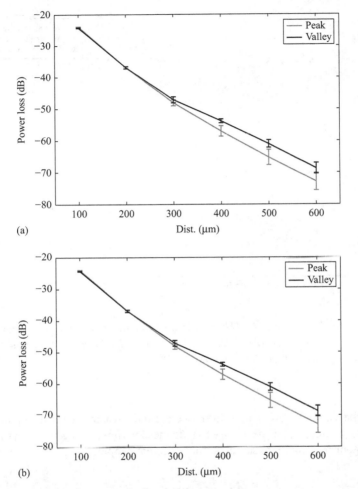

Figure 12.19 *Comparison between the two scenarios: two dipoles located above/beneath the peak and valley of the interface for the model with 3-D sinc function as the interface (a) Effects of span alteration on the power loss with respect to distance and (b) Effects of amplitude alteration on the power loss with respect to distance. © 2015 Elsevier. Reprinted with permission from Reference 86*

12.5 Future work

In line with the work presented, the future research directions which would make potential and natural progression to complete the studies in the thesis are summarized as follows.

Measurement on the parameters of the human tissues

Although several optical parameters are provided, it is still not enough. First, the experiment was conducted long time ago and the number of tissues was limited.

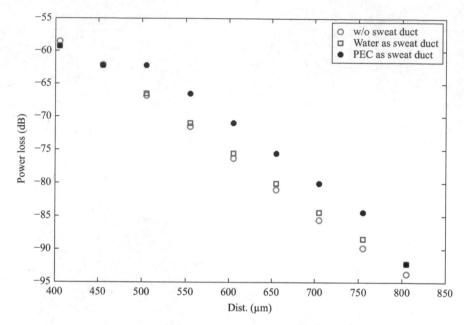

Figure 12.20   Power loss at 1 THz for three scenario: (a) without the sweat duct in epidermis; (b) water is considered as the sweat in sweat duct; (c) the sweat duct is considered as PEC

Second, the samples were mainly from one old male which lowers the credibility of the results. Therefore, it is necessary to redo the experiments to get the parameters of interest by measuring more samples from different people.

Safety issues: heating problems of THz wave

Safety issue is always the main consideration when people talks about nano-network, especially when the nano-devices are applied to the in-body scenario. For THz communication, heating problem was the main concern in the academic field. Therefore, the study of the THz wave heating effects on the human tissue would be conducted to make the standard and requirement for communicating or sensing.

Interaction between the nano-devices and the surrounding environment

From the study of the models of nerve system and skin, it seems dispensable to study the detailed model when the size of the functional devices goes down to milli/nano-scale. The interaction between the environment and the devices should be study to make sure the devices function normal.

Simulating body-centric nano-networks using the state-of-the-art simulator

Although there are lots of communication paradigms for nano-communication, the study on interaction between each two different communications – for example, the EM communication and the molecular communication – is still missing. It is generally believed that by emerging all the communications together the nano-network would be much more flexible and powerful. With the comprehensive simulator proposed in IEEE P1906.1 based on the ns-3 platform in mind, different communication schemes would be studied by comparing to the EM one. And at the same time, the relationship between them and EM method should be further studied.

References

[1]   Metin Sitti, Hakan Ceylan, Wenqi Hu, *et al.* Biomedical applications of unteth-ered mobile milli/microrobots. *Proceedings of the IEEE*, 103(2):205–224, 2015.

[2]   Richard P. Feynman. There's plenty of room at the bottom. *Engineering and Science*, 23(5):22–36, 1960.

[3]   Stephen F. Bush. *Nanoscale Communication Networks*. London: Artech House, 2010.

[4]   Ian F. Akyildiz, Fernando Brunetti, and Cristina Blázquez. Nanonetworks: a new communication paradigm. *Computer Networks*, 52(12):2260–2279, 2008.

[5]   Josep Miquel Jornet, Joan Capdevila Pujol, and Josep Solé Pareta. Phlame: A physical layer aware mac protocol for electromagnetic nanonetworks in the terahertz band. *Nano Communication Networks*, 3(1):74–81, 2012.

[6]   Ke Yang, Alice Pellegrini, Max O. Munoz, Alessio Brizzi, Akram Alomainy, and Yang Hao. Numerical analysis and characterization of THz propaga-tion channel for body-centric nano-communications. *IEEE Transactions on Terahertz Science and Technology*, 5(3):419–426, May 2015.

[7]   Ian F. Akyildiz and Josep Miquel Jornet. Electromagnetic wireless nanosensor networks. *Nano Communication Networks*, 1(1):3–19, 2010.

[8]   G. Enrico Santagati and Tommaso Melodia. Opto-ultrasonic communica-tions for wireless intra-body nanonetworks. *Nano Communication Networks*, 5(1):3–14, 2014.

[9]   Ian F. Akyildiz and Josep Miquel Jornet. The internet of nano-things. *Wireless Communications, IEEE*, 17(6):58–63, 2010.

[10] Robert A. Freitas. Nanotechnology, nanomedicine and nanosurgery. *Interna-tional Journal of Surgery*, 3(4):243–246, 2005.

[11] Zhuan Liao, Rui Gao, Can Xu, and Zhao-Shen Li. Indications and detection, completion, and retention rates of small-bowel capsule endoscopy: a systematic review. *Gastrointestinal Endoscopy*, 71(2):280–286, 2010.

[12] Tetsuya Nakamura and Akira Terano. Capsule endoscopy: past, present, and future. *Journal of Gastroenterology*, 43(2):93–99, 2008.

[13]   Guobing Pan and Litong Wang, Swallowable Wireless Capsule Endoscopy: Progress and Technical Challenges. *Gastroenterology Research and Practice*, vol. 2012, Article ID 841691, 9 pages, 2012. doi:10.1155/2012/841691

[14]   M. Fluckiger and Bradley J. Nelson. Ultrasound emitter localization in heterogeneous media. In *Engineering in Medicine and Biology Society, 2007. EMBS 2007. 29th Annual International Conference of the IEEE*, pages 2867–2870. IEEE, 2007.

[15]   Kang Kim, Laura A. Johnson, Congxian Jia, *et al.* Noninvasive ultrasound elasticity imaging (UEI) of Crohn's disease: animal model. *Ultrasound in Medicine & Biology*, 34(6):902–912, 2008.

[16]   Olgaç Ergeneman, Görkem Dogangil, Michael P., *et al.* A magnetically controlled wireless optical oxygen sensor for intraocular measurements. *Sensors Journal, IEEE*, 8(1):29–37, 2008.

[17]   J. Matthew Dubach, Daniel I. Harjes, and Heather A. Clark. Fluorescent ion-selective nanosensors for intracellular analysis with improved lifetime and size. *Nano Letters*, 7(6):1827–1831, 2007.

[18]   Jianping Li, Tuzhi Peng, and Yuqiang Peng. A cholesterol biosensor based on entrapment of cholesterol oxidase in a silicic sol–gel matrix at a prussian blue modified electrode. *Electroanalysis*, 15(12):1031–1037, 2003.

[19]   Padmavathy Tallury, Astha Malhotra, Logan M. Byrne, and Swadeshmukul Santra. Nanobioimaging and sensing of infectious diseases. *Advanced Drug Delivery Reviews*, 62(4):424–437, 2010.

[20]   Jonathan W. Aylott. Optical nanosensors – an enabling technology for intracellular measurements. *Analyst*, 128(4):309–312, 2003.

[21]   Ph Avouris, G. Dresselhaus, and M.S. Dresselhaus. Carbon nanotubes: synthesis, structure, properties and applications. *Topics in Applied Physics*, pp. 1035–1044, 2000.

[22]   S. Tasoglu, E. Diller, S. Guven, M. Sitti, and U. Demirci. Untethered micro-robotic coding of three-dimensional material composition. *Nature Communications*, 5, Article number: 3124, 2014.

[23]   Ira J. Fox, George Q. Daley, Steven A. Goldman, *et al.* Use of differentiated pluripotent stem cells as replacement therapy for treating disease. *Science*, 345(6199):1247391, 2014.

[24]   Sangwon Kim, Famin Qiu, Samhwan Kim, *et al.* Fabrication and characterization of magnetic microrobots for three-dimensional cell culture and targeted transportation. *Advanced Materials*, 25(41):5863–5868, 2013.

[25]   K. Eric Drexler. *Nanosystems: Molecular Machinery, Manufacturing, and Computation*. New York: John Wiley & Sons, Inc., 1992.

[26]   Robert A. Freitas. What is nanomedicine? *Nanomedicine: Nanotechnology, Biology and Medicine*, 1(1):2–9, 2005.

[27]   Rodrigo Fernández-Pacheco, Clara Marquina, J. Gabriel Valdivia, *et al.* Magnetic nanoparticles for local drug delivery using magnetic implants. *Journal of Magnetism and Magnetic Materials*, 311(1):318–322, 2007.

[28] Robert A. Freitas. Pharmacytes: an ideal vehicle for targeted drug delivery. *Journal of Nanoscience and Nanotechnology*, 6(9–10):2769–2775, 2006.

[29] Brian P. Timko, Tal Dvir, and Daniel S. Kohane. Remotely triggerable drug delivery systems. *Advanced Materials*, 22(44):4925–4943, 2010.

[30] Sehyuk Yim and Metin Sitti. Shape-programmable soft capsule robots for semi-implantable drug delivery. *IEEE Transactions on Robotics*, 28(5):1198–1202, 2012.

[31] Rika Wright Carlsen and Metin Sitti. Bio-hybrid cell-based actuators for microsystems. *Small*, 10(19):3831–3851, 2014.

[32] Ching-Jen Chen, Drs Yousef Haik, and Jhunu Chatterjee. Development of nanotechnology for biomedical applications. In *Emerging Information Technology Conference, 2005*, pp. 1–4. IEEE, 2005.

[33] Edward B. Steager, Mahmut Selman Sakar, Ceridwen Magee, *et al.* Automated biomanipulation of single cells using magnetic microrobots. *The International Journal of Robotics Research*, 32(3):346–359, 2013.

[34] Tomohiro Kawahara, Masakuni Sugita, Masaya Hagiwara, *et al.* On-chip microrobot for investigating the response of aquatic microorganisms to mechanical stimulation. *Lab on a Chip*, 13(6):1070–1078, 2013.

[35] Deok-Ho Kim, Pak Kin Wong, Jungyul Park, Andre Levchenko, and Yu Sun. Microengineered platforms for cell mechanobiology. *Annual Review of Biomedical Engineering*, 11:203–233, 2009.

[36] Kyoung-Chul Kong, Jinhoon Cha, Doyoung Jeon, and Dong-Il Dan Cho. A rotational micro biopsy device for the capsule endoscope. In *2005 IEEE/RSJ International Conference on Intelligent Robots and Systems (IROS 2005)*, pages 1839–1843. IEEE, 2005.

[37] Piero Miloro, Edoardo Sinibaldi, Arianna Menciassi, and Paolo Dario. Removing vascular obstructions: a challenge, yet an opportunity for interventional microdevices. *Biomedical Microdevices*, 14(3):511–532, 2012.

[38] Sehyuk Yim, Evin Gultepe, David H. Gracias, and Metin Sitti. Biopsy using a magnetic capsule endoscope carrying, releasing, and retrieving untethered microgrippers. *IEEE Transactions on Biomedical Engineering*, 61(2): 513–521, 2014.

[39] Martin Heil and Jurriaan Ton. Long-distance signalling in plant defence. *Trends in Plant Science*, 13(6):264–272, 2008.

[40] Corne M.J. Pieterse and Marcel Dicke. Plant interactions with microbes and insects: from molecular mechanisms to ecology. *Trends in Plant Science*, 12(12):564–569, 2007.

[41] Jongyoon Han, Jianping Fu, and Reto B. Schoch. Molecular sieving using nanofilters: past, present and future. *Lab on a Chip*, 8(1):23–33, 2008.

[42] Alex M. Andrew. Nanomedicine, volume 1: Basic capabilities. *Kybernetes*, 29(9/10):1333–1340, 2000.

[43] Suk Jin Lee, Changyong Andrew Jung, Kyusun Choi, and Sungun Kim. Design of wireless nanosensor networks for intrabody application. *International*

Journal of Distributed Sensor Networks, Volume 2015 (2015), Article ID 176761, 12 pages, 2015.

[44]   M. Rosenau da Costa, O.V. Kibis, and M.E. Portnoi. Carbon nanotubes as a basis for terahertz emitters and detectors. *Microelectronics Journal*, 40(4):776–778, 2009.

[45]   Y.-M. Lin, Christos Dimitrakopoulos, Keith A. Jenkins, *et al*. 100-GHz transistors from wafer-scale epitaxial graphene. *Science*, 327(5966):662–662, 2010.

[46]   Massimiliano Pierobon, Josep Miquel Jornet, Nadine Akkari, Suleiman Almasri, and Ian F. Akyildiz. A routing framework for energy harvesting wireless nanosensor networks in the terahertz band. *Wireless Networks*, 20(5):1169–1183, 2014.

[47]   Armita Afsharinejad, Alan Davy, and Brendan Jennings. Dynamic channel allocation in electromagnetic nanonetworks for high resolution monitoring of plants. *Nano Communication Networks*, pp. 2–16, 2015.

[48]   Andrew W. Eckford. Molecular communication: physically realistic models and achievable information rates. *arXiv preprint arXiv:0812.1554*, 2008.

[49]   K.V. Srinivas, Andrew W. Eckford, and Raviraj S. Adve. Molecular communication in fluid media: the additive inverse Gaussian noise channel. *IEEE Transactions on Information Theory*, 58(7):4678–4692, 2012.

[50]   Massimiliano Pierobon and Ian F. Akyildiz. A physical end-to-end model for molecular communication in nanonetworks. *IEEE Journal on Selected Areas in Communications*, 28(4):602–611, 2010.

[51]   A. Ozan Bicen and I.F. Akyildiz. System-theoretic analysis and least-squares design of microfluidic channels for flow-induced molecular communication. *IEEE Transactions on Signal Processing*, 61(20):5000–5013, 2013.

[52]   Amit Singhal, Ranjan K. Mallik, and Brejesh Lall. Performance of amplitude modulation schemes for molecular communication over a fluid medium. In *2014 IEEE 25th Annual International Symposium on Personal, Indoor, and Mobile Radio Communication (PIMRC)*, pages 785–789. IEEE, 2014.

[53]   M. Pierobon and I.F. Akyildiz. A physical end-to-end model for molecular communication in nanonetworks. *IEEE Journal on Selected Areas in Communications*, 28(4):602–611, May 2010.

[54]   Mehmet Şükrü Kuran, H. Birkan Yilmaz, Tuna Tugcu, and Bilge Özerman. Energy model for communication via diffusion in nanonetworks. *Nano Communication Networks Journal*, 1(2):86–95, 2010.

[55]   M. Pierobon and I.F. Akyildiz. Diffusion-based noise analysis for molecular communication in nanonetworks. *IEEE Transactions on Signal Processing*, 59(6):2532–2547, June 2011.

[56]   Po-Jen Shih, Chia-Han Lee, Ping-Cheng Yeh, and Kwang-Cheng Chen. Channel codes for reliability enhancement in molecular communication. *IEEE Journal on Selected Areas in Communications*, 31(12):857–867, December 2013.

[57]   A. Einolghozati, M. Sardari, and F. Fekri. Relaying in diffusion-based molecular communication. In *IEEE International Symposium on Information Theory (ISIT)*, pages 1844–1848, July 2013.

[58]   T. Nakano and Jian-Qin Liu. Design and analysis of molecular relay channels: an information theoretic approach. *IEEE Transactions on NanoBioscience,* 9(3):213–221, September 2010.

[59]   Yuanfeng Chen, Panagiotis Kosmas, Putri Anwar, and Liwen Huang. A touch-communication framework for drug delivery based on a transient microbot system. *IEEE Transactions on Nanobioscience,* 14(4):397–408, 2015.

[60]   I.S.M. Khalil, V. Magdanz, S. Sanchez, *et al.* Magnetic control of potential microrobotic drug delivery systems: nanoparticles, magnetotactic bacteria and self-propelled microjets. In *Engineering in Medicine and Biology Society (EMBC), 2013 35th Annual International Conference of the IEEE,* pages 5299–5302, July 2013.

[61]   Hoda ShahMohammadian, Geoffrey G. Messier, and Sebastian Magierowski. Nano-machine molecular communication over a moving propagation medium. *Nano Communication Networks,* 4(3):142–153, 2013.

[62]   Tad Hogg and Robert A. Freitas Jr. Acoustic communication for medical nanorobots. *Nano Communication Networks,* 3(2):83–102, 2012.

[63]   G. Enrico Santagati and Tommaso Melodia. Opto-ultrasonic communications in wireless body area nanonetworks. In *2013 Asilomar Conference on Signals, Systems and Computers,* pages 1066–1070. IEEE, 2013.

[64]   C. Emre Koksal and Eylem Ekici. A nanoradio architecture for interacting nanonetworking tasks. *Nano Communication Networks,* 1(1):63–75, 2010.

[65]   Josep Miquel Jornet and Ian F. Akyildiz. Channel modeling and capacity analysis for electromagnetic wireless nanonetworks in the terahertz band. *IEEE Transactions on Wireless Communications,* 10(10):3211–3221, 2011.

[66]   Cecil S. Joseph, Anna N. Yaroslavsky, Victor A. Neel, Thomas M. Goyette, and Robert H. Giles. Continuous wave terahertz transmission imaging of nonmelanoma skin cancers. *Lasers in Surgery and Medicine,* 43(6):457–462, 2011.

[67]   Euna Jung, Hongkyu Park, Kiwon Moon, *et al.* Thz time-domain spectroscopic imaging of human articular cartilage. *Journal of Infrared, Millimeter, and Terahertz Waves,* 33(6):593–598, 2012.

[68]   Gerald J. Wilmink and Jessica E. Grundt. Invited review article: current state of research on biological effects of terahertz radiation. *Journal of Infrared, Millimeter, and Terahertz Waves,* 32(10):1074–1122, 2011.

[69]   C. Han, A. Bicen, and I. Akyildiz. Multi-ray channel modeling and wideband characterization for wireless communications in the terahertz band. *IEEE Transactions on Wireless Communications,* vo. 14, pp. 2402–2412, 2015.

[70]   J. Kokkoniemi, J. Lehtomaki, K. Umebayashi, and M. Juntti. Frequency and time domain channel models for nanonetworks in terahertz band. *IEEE Transactions on Antennas and Propagation,* 63(2):678–691, February 2015.

[71]   K. Kantelis, S.A. Amanatiadis, C.K. Liaskos, *et al.* On the use of FDTD and ray-tracing schemes in the nanonetwork environment. *Communications Letters, IEEE,* 18(10):1823–1826, October 2014.

[72]   J.J. Lehtomaki, A.O. Bicen, and I.F. Akyildiz. On the nanoscale electrome-chanical wireless communication in the VHF band. *IEEE Transactions on Communications*, 63(1):311–323, January 2015.

[73]   A.J. Fitzgerald, E. Berry, N.N. Zinov'ev, *et al.* Catalogue of human tissue optical properties at terahertz frequencies. *Journal of Biological Physics*, 29(2–3):123–128, 2003.

[74]   Elizabeth Berry, Anthony J. Fitzgerald, Nickolay N. Zinov'ev, *et al.* Optical properties of tissue measured using terahertz-pulsed imaging. *Proceedings of SPIE*, 5030:459–470, 2003.

[75]   S. Bush, J. Paluh, G. Piro, V. Rao, V. Prasad, and A. Eckford. Defining communication at the bottom. *IEEE Transactions on Molecular, Biological and Multi-Scale Communications*, PP(99):1–1, 2015.

[76]   Frank Box. Utilization of atmospheric transmission losses for interference-resistant communications. *IEEE Transactions on Communications*, 34(10): 1009–1015, 1986.

[77]   Atindra Nath Pal and Arindam Ghosh. Ultralow noise field-effect transistor from multilayer graphene. *Applied Physics Letters*, 95(8):082105, 2009.

[78]   A. Goldsmith. *Wireless Communication*. UK: Cambridge University Press, 2005.

[79]   William Montagna. *The Structure and Function of Skin 3E*. Elsevier, Academic Press, Inc., New York and London, 2012.

[80]   George F. Odland. Structure of the skin. *Physiology, Biochemistry, and Molecular Biology of the Skin*, 1:3–62, 1991.

[81]   Robert F. Rushmer, Konrad J.K. Buettner, John M. Short, and George F. Odland. The skin. *Science*, 154(3747):343–348, 1966.

[82]   A.N. Bashkatov, E.A. Genina, V.I. Kochubey, and V.V. Tuchin. Optical proper-ties of human skin, subcutaneous and mucous tissues in the wavelength range from 400 to 2000 nm. *Journal of Physics D: Applied Physics*, 38(15):2543, 2005.

[83]   Qammer Hussain Abbasi, Andrea Sani, Akram Alomainy, and Yang Hao. Numerical characterization and modeling of subject-specific ultrawideband body-centric radio channels and systems for healthcare applications. *IEEE Transactions on Information Technology in Biomedicine*, 16(2):221–227, 2012.

[84]   A. Knu, S. Bonev, W. Knaak, *et al.* New method for evaluation of in vivo scattering and refractive index properties obtained with optical coherence tomography. *Journal of Biomedical Optics*, 9(2):265–273, 2004.

[85]   Itai Hayut, Alexander Puzenko, Paul Ben Ishai, *et al.* The helical structure of sweat ducts: their influence on the electromagnetic reflection spectrum of the skin. *IEEE Transactions on Terahertz Science and Technology*, 3(2):207–215, 2013.

[86]   Ke Yang, Qammer H. Abbasi, Nishtha Chopra, Max Munoz, Yang Hao and Akram Alomainy. Effects of non-flat interfaces in human skin tissues on the in-vivo tera-hertz communication channel. *Nano Communication Networks*, Available online 3 November 2015.

[87]   K.M. Yaws, D.G. Mixon, and W.P. Roach. Electromagnetic properties of tissue in the optical region. In *Biomedical Optics (BiOS) 2007*, vol. 6435, pp. 643507–643513. International Society for Optics and Photonics, 2007.

[88]   Gal Shafirstein and Eduardo G. Moros. Modelling millimetre wave propagation and absorption in a high resolution skin model: the effect of sweat glands. *Physics in Medicine and Biology*, 56(5):pp. 1329–1339, 2011.

Chapter 13

The road ahead for body-centric wireless communication and networks

Masood Ur-Rehman*, Qammer Hussain Abbasi†, and Akram Alomainy‡

Wireless interaction of the human user with the computing devices has seen a profound growth in the past decade. Wearable technology has successfully moved past the adoption stage and now stands at the brink of massive diversification with an explosion in popularity and applicability. The estimated market value of the wearable technology is expected to hit \$32 billion mark by 2020 [1, 2]. It would cause the global wearable devices market it to grow from 20 million device shipments in 2015 to 187.2 million units annually by 2020 [3].

13.1 Market prospects for body-centric wireless networks

The main objective of research and development in body-centric wireless communications is to provide users with a wide range of personal electronic support systems, anywhere and at any time. The wearable technology is dominated by personal healthcare and fitness applications at the moment; however, current and future scope of this technology includes a variety of market sectors:

- *Sports and fitness*: It is the most high-profile sector for Body-Centric Wireless Networks (BCWNs) with number of applications already making their way in the market. Trackers, head bands, wrist bands, smart shoes and smart sports clothing are examples of these applications. These applications typically measure parameters such as distance travelled, calories burned, heart rate, etc. based on the observation of walking, jogging and muscle activities. Some applications also include Global Positioning System (GPS) monitoring and tracking to enhance accurate measurement of distance.
- *Infotainment*: Smart gaming consoles with motion detection and wearable hearing devices like Bluetooth headsets and smart earbuds are examples of this type of BCWN applications.

*Centre for Wireless Research, University of Bedfordshire, Luton (UK)
†Center for Remote Healthcare Technology & Systems ext. at Qatar, Department of Electrical and Computer Engineering, Texas A&M University at Qatar
‡School of Electronic Engineering and Computer Science, Queen Mary University of London

- *Personal medical, assisted living and tracking*: This sector is attracting much of the attention today and is expected to grow hugely in future. Principal reason for huge interest in this application area is due to increase in aging population that is predicted to have 101 people 60 years or older for every 100 children 0–14 years by 2050 [4]. It will bring the health sector under immense pressure in terms of economy, workforce and availability due to requirements of continuous person/ patient monitoring. The on-body and in-body BCWN devices used for this sector encompasses a wide range of physiological measurements and will incorporate enhanced functionalities of remotely accomplished testing and examination, prescription and drug delivery, and cure of chronicle diseases by monitoring and analysis of blood pressure, brain signals, sugar level, electrocardiogram (ECG), eye tracking, oxygen level, radiation exposure, respiration, sleep pattern, temperature, pulse and posture. A fully functional telemedicine system where physicians can use these wearable technologies to monitor and treat their patients remotely will bring great advantages in terms of comfort, flexibility and health improvements at personal level and economic gains at state level.

  Placing sensors on kids to track their activities, location and health by monitoring vital signs such as skin temperature, respiration and movement, as well as allowing an audio link to the parent's smartphone is another future aspect of this area. A commercial product in the form of a smart sleeping baby suit is already in the market [5]. It can also be extended for the traction of pets.

- *Fashion*: Smart clothing is one of the key application areas of BCWNs that can serve monitoring and communication purposes efficiently. Wearable textile antennas have therefore, attracted interest of researchers worldwide [6]. Although, smart clothing including shirts, trousers and smart jewellery is primarily considered for personal medical at the moment, wearable technology makes the experimentation possible with some more immersive forms of sensing. Smart social clothing can be an interesting area for future extension of the BCWNs where clothing provides social interaction such as changing the colour of the clothing by sensing and communicating the user's mood, emotion, physical sensations or location to an individual or a sport watching crowd through embedded sensors. A first step in this direction is taken in the form of the Hug Shirt [7] that contains sensors measuring the strength of touch, skin temperature and heartbeat along with actuators which can recreate them. The physical sensations can be shared between the two people wearing such shirts through a Bluetooth link.

  Smart clothing faces many technical challenges. It needs to be comfortable, flexible, washable and chargeable. There is still much work to be done to develop smart fabrics which look and feel like normal fabrics. Also, it should be available to be cut and made into garments and in different sizes without major changes in garment manufacturing techniques. Hence, new textile materials, miniaturisation and battery improvements are issues that will dictate their development in the future BCWNs.

- *Augmented reality*: Wearable devices that gather and give a user information anywhere and everywhere, overlaying it on everything he/she sees and enhancing his/her senses are another promising venue for future BCWN applications.

Although not very successful, introduction of smart eyewear like Google Glass and wearable cameras like Autographer have already set the foundation for them in the market [8, 9]. Issues of privacy, handling of huge amount of data and social awkwardness would, however, need to be looked into.

Recognition of head and eye-movement gestures has found its way in the wearable augmented reality applications for waking, turning off and making notifications. It has great potential to be used in many daily life activities with added features of security and authentication.

- *Smart watches*: This market sector is already booming with a number of devices introduced by Apple, Samsung and Motorola. It is also expected to dominate the wearable market in the near future [10]. Observation of vital physical parameters, exercise activities and location along with communication with the smart phones are key actions performed by these wrist-worn watches. The basic problem with these watches is that the features they offer are already provided by the smart phones.

13.2   Challenges and future perspective of BCWNs

Success of current and future BCWN applications depends on efficient solutions to the technological and societal factors it faces today. These factors will inhibit growth and drive innovation in the wearable technology.

13.2.1   Complex environment

BCWNs work in highly volatile environments. Path losses are high due to body absorption and cluttered operational conditions as discussed in Chapter 9. Inherent limitations of the sensor design, on/in-body sensor position, human body postures, body motions, multipath, sensor breakdown and interference could lead to erroneous and incomplete data reception. Strict power emission and specific absorption rate (SAR) regulations, design limitations due to size and shape restrictions made the antenna design very challenging. For example, a urethral valve that needs to be replaced at regular intervals without surgical operation restricts the choice of implanted antenna to a helix as the available diameter is 4–6 mm [11]. Moreover, implanted antennas need to be made up of bio-compatible and non-corrosive material like titanium and platinum instead of their more efficient counterpart copper antennas. Channel modelling is also very complex due to multipath and mobility. Techniques need to be devised to improve performance of the BCWNs in such environments such as heterogeneous and multi-hop links, efficient sensor design, use of various types of sensors and optimal positioning of the sensors.

13.2.2   Spectrum shortage

Flooding of variety of wireless technologies is leading to shrink the available spectrum. It along with benefits of large data rates for real-time communications, lower transmission time, reduction in interference and unlicensed frequencies are attracting the researchers to explore the millimetre wave (mmWave) communication in the range from 30 GHz to 300 GHz as an alternative venue for the BCWNs. It, however, brings a number of challenges including radio frequency (RF) impairments of fast fading and delay spread that makes demodulation and equalisation difficult; mmWave transceivers require data converters able to process giga-samples per second with considerable resolutions that leads to high power consumption and losses due to high penetration depth [12].

Spectrum sensing and cognition is also emerging as a viable solution to address the spectrum shortage. Cognitive radio provides a platform to control the primary (licensed) section of the electromagnetic spectrum by secondary (unlicensed) users opportunistically and cooperatively through sensing of the occupation of the spectrum [13]. The cognition-based BCWNs would lead to considerable improvement in spectrum management and would provide enhanced efficiency in terms of resources, networking and energy. Open challenges in this technology include design of intelligent sensors that can sense the spectrum with correct goal evaluation, real-time processing and energy efficient monitoring.

Efficient power and frequency allocation schemes are required to minimise the interaction of BCWNs with other networks including Wireless Local Area Network (WLAN), Ultra Wideband (UWB), Bluetooth, Global System for Mobile Communications (GSM), etc. as well as to minimise self-interference between different body-worn sensors. High overhead necessitates an intelligent routing protocol aware of spectrum changes. Energy efficiency along with quality-of-service (QoS) issues requires efficient Media Access Control (MAC) protocol to provide optimum spectrum access scheme enabling avoidance of collisions and repeated data transmissions while maintaining maximum throughput, communication reliability and minimum latency [14].

13.2.3   Body-to-body communications

Concept of internet-of-things (IoT) and device-to-device (D2D) communication also brings in the idea of body-to-body (B2B) communications. A B2B network consists of several BCWNs each communicating with its neighbour. The B2B network is envisioned as a mesh network using the human subjects to transmit data within a limited geographic region [15]. It will use BCWN devices like smart phones and smart watches to send a signal from one person to the next nearest B2B network user until it reaches the destination.

B2B communications have several challenging tasks before its efficient realisation. It includes energy efficiency and QoS considerations to support different data types, i.e. data, audio and video along with mobility of the participating BCWN devices. Coexistence and cooperation of various BCWNs operating on different technologies, i.e. Bluetooth, ZigBee, Wi-Fi, etc. sharing the same spectrum demands measures to mitigate inter-body interference resulting from co- and cross-technology interference. Transmission of sensitive data travelling from person to person in a B2B communication scenario brings in reliability, privacy and security issues. Moreover, B2B networks should be able to handle a variety of BCWN devices operating at different data rates. The human body itself acts like shadowing object for such communication scenarios. Hence, the inter-body propagation channel would also need detailed characterisation for reliable communications considering all these factors and surrounding environment conditions.

13.2.4 5G

Evolution of 5G is another venue that would require extensive research and development of wireless body area networks. 5G technology considers massive multiple input multiple output (MIMO) antenna arrays (having more than ten elements) with beamforming at the mobile receivers operating at mmWave frequencies as one of the solutions to support high data rate and interference mitigation [16]. It will require investigation of switchable and diversity antennas both in correlation and in isolation from each other to overcome fading. Chapter 8 of this book provides insight into the benefits of using diversity and MIMO antennas at relatively small scale for BCWNs.

5G is also envisioned to support advanced D2D communication that enhances spectral efficiency and reduces end-to-end latency. Not entirely depending on the cellular network, D2D devices can communicate directly with one another when they are in close proximity. Hence, D2D communication will be used for offloading data from network so that the cost of processing those data and related signalling is minimised. In D2D communication, a single radio resource can be reused among multiple groups which want to communicate with each other if the interference incurred between the groups is tolerable. Hence, the spectral efficiency and the number of simultaneous connections can be increased. B2B communication scenario (shown in Figure 13.1)

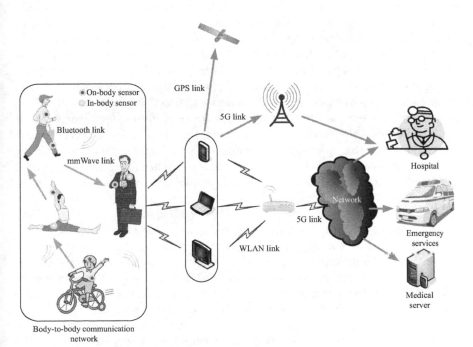

Figure 13.1 Futuristic concept view of BCWNs with B2B communication

will be the most widely used type of D2D communication as majority of the future communication devices would be body worn [17]. It will require to address the challenges of complex circuit design, high sensitivity to blocking, more spectrum needs, efficient resource allocation along with novel antenna solutions, mitigation of the human body effects (which would be high due to increased penetration depth) and channel characterisation [18].

13.2.5   Antenna design and channel modelling

In order to achieve unobtrusive communication in the BCWNs, efficient implementation of wireless modules is vital as discussed in Chapter 6. Despite extensive research and development in this area, design of BCWN modules is still very challenging due to limitations of low power requirements, reliable data transmission, compatibility with the sensor and conformal antenna design. Ever-changing requirements of body-worn sensors thanks to the emergence of new technologies such as mmWave communications, 5G, B2B communications, cognition in BCWNs and high demand of non-invasive methods for healthcare procedures (e.g. sugar detection and blood test) also necessitate new and improved solutions. Moreover, the sensors should not be designed for the sake of design but to observe and relay the vital physiological information required by the physicians to treat certain medical condition. Hence, a very close cooperation between the field of medicine and antenna design is necessary.

Antennas and associated electronics are required to be conformal to the body shape for on-body or in-body environments. They also need to show immunity from detuning in frequency and polarisation. Antennas for future BCWNs will also be made either for specialist occupations such as firefighters and paramedics, or for smart social clothing where fashion will dictate the form and design as highlighted in Chapter 10. It will bring further challenges of harsh and variable environments (such as extreme temperatures and multipath) and multi-frequency operation on top of conventional complexities of physical dimensions, directivity, efficiency and SAR. Special considerations are required to address these problems, for example, in an on-body communication scenario between multiple body-mounted antennas, the one having a monopole-like radiation pattern performs better for the surface wave coupling while the one with a patch-like radiation pattern is better for surface/space wave coupling [19]. Techniques like reconfigurability, multi-band operation, beamforming and phased-controlled arrays would also need extensive study in BCWNs scenario to support high data rate applications like mmWave communications and 5G. mmWave communications also brings the challenge of heating whose effects on the human body tissues need to be investigated as highlighted in Chapter 12.

Advent of the nanotechnology and bio-sensors has brought promising aspects in antenna miniaturisation, especially for medical implants. It can realise extremely small BCWN devices operating inside the human tissues and blood vessels to take samples, diagnose, deliver drugs and possibly avoid surgical procedures.

New techniques for the characterisation of BCWN antenna performance at mmWave and nanoscale would also be required with changing specifications. Numerical models, voxel models and phantoms play an important role in these studies [20].

A detailed study for the determination of the electric properties of the human body tissues at mmWave frequency range is therefore pertinent.

A clear understanding of wireless links between various BCWN devices, either implanted, body-worn or operating in the vicinity of the human body, is essential for the efficient and reliable deployment of the BCWNs. It requires in-depth analysis of wave attenuation in lossy human body tissues and propagation in, on and around the body along with surrounding environment effects through channel characterisation and modelling. Huge effort has been made in this area at microwave frequencies [21] while some initial steps are taken to understand the channel condition at terahertz range [22]. Chapters 3, 7, 9 and 12 of this book also offer insight into this topic considering UWB, *in vivo*, GPS multipath and terahertz transmission. However, much more needs to be done to characterise the channels for nano-implants and mmWave propagation. Moreover, measurement of implanted antennas and channels on live human subjects is very difficult. Measurements using numerical and solid phantoms are an alternative but require time and computational efficient tools with careful calibration.

13.2.6   Power consumption and battery life

One of the biggest technical challenges for the wireless applications is power. It is more vital for the BCWNs as the implantable and wearable devices are small and mobile by nature requiring miniaturised, efficient and long-lasting batteries for the continuity of operation for at least one typical usage cycle without replacement or recharging. This typical usage cycle can vary from device to device. An implanted sensor can have a usage cycle of a year while a mobile phone's usage cycle can be considered as 24 h. To realise small device sizes with reasonable battery lifetimes, typical wireless sensor nodes are designed using low-power components with modest resources. It affects the signal quality and link budget.

To overcome this problem, ambient energy harvesting from solar, vibration (kinetic), thermal and electromagnetic/RF energy sources is considered to be a viable solution. Some proposed solutions for the BCWNs include using human body warmth as a source of thermal energy [23], use of vibrations due to body movements as the source of kinetic energy employing piezoelectric devices [24, 25] and using rectennas to harvest ambient RF energy [26]. New types of clothing with built-in solar panels and smart textiles having non-invasive piezoelectric fibres and nanowire super capacitors are also proposed [27]. However, these proposals are still far from generating a realistic alternative to the conventional batteries for the BCWNs. Dissipation of heat in such devices and its effects on the human subject also needs to be mitigated. Smart textiles could provide an answer to it by employing cooling fibres based on melt/resolidify technologies. This topic would therefore need utmost attention of the researchers for the success of the BCWNs and avoid death by discharge.

13.2.7   Security and privacy

Security is considered of highest priority in other wireless networks but BCWNs have done little on it. Wearable devices can collect and transmit data ranging from the user's physical location (as presented in Chapter 11) to personal fitness and health details.

The stringent constraints of memory, power, data rate and computational capability make the solutions proposed for other wireless networks inapplicable to BCWNs. Security requirements for BCWNs include availability of the user information to the concerned people like physicians at all times, authentication of the user's data, integrity of the transmitted data in the communication channel and prevention of inter-device interference.

The IEEE 802.15.6 standard has proposed a security paradigm for BCWNs that should be implemented on MAC layer [28]. Secure key generation and distribution is considered as one of the most effective solutions for security and authentication. Biometrics technique using body parameters such as timing information of heartbeat allows the body itself to encrypt and decrypt the symmetric key for secure distribution [29, 30]. Use of random channel measurements is also proposed for the generation and distribution of cryptographic keys for secure BCWNs [31]. These efforts are, however, not enough for a full-scale implementation of BCWNs. Moreover, emergence of cloud computing and IoT have given rise to privacy of the information shared is a major concern for the users. Very private nature of the data observed by the BCWN sensors increases the concern users' multifoods and make them reluctant to share it. Assurance of secure communication is therefore key to the success of future BCWN and would require an in-depth exploration and implementation of novel security solutions.

In summary, the BCWNs will change the human life style bringing revolutionary enhancements, it has to address a number of technical challenges discussed in this chapter before being widely deployed. Moreover, numerous non-technical factors are also trivial to success of BCWNs such as pricing, aesthetics, acceptance, comfort, user friendliness, accessibility and regulatory issues dealing with social, ethical and legal problems.

References

[1] "Global home healthcare market: Industry analysis, size, share, growth, trends and forecast, 2014–2020," *Transparency Market Research*, September 2014.

[2] N. Hunn, "The market for smart wearables 2015–2020: A consumer centric approach," *WiFore Consulting*, March 2015.

[3] "Wearable device market forecasts," *Tactica*, March 2015.

[4] "World population ageing: 1950–2050," Department of Economic and Social Affairs, Population Division, United Nations, 2002.

[5] "Mimo baby monitor." Available from http://mimobaby.com. Accessed on December 22, 2015.

[6] M. Ur-Rehman, Q. Abbasi, M. Akram, and C. Parini, "Design of band-notched ultra wideband antenna for indoor and wearable wireless communications," *IET Microwaves, Antennas Propagation*, vol. 9, no. 3, pp. 243–251, 2015.

[7] "The hug shirt." Available from http://cutecircuit.com/collections/the?hug?shirt. Accessed on December 22, 2015.

[8] "Google glass." Available from https://www.wareable.com/google-glass. Accessed on December 24, 2015.

[9] "Autographer." Available from http://www.autographer.com. Accessed on December 24, 2015.

[10] T. Davona, "The wearables report: Growth trends, consumer attitudes and why smart watches will dominate," *Business Insider*, February 2015.

[11] G.-Z. E. Yang, *Body Sensor Networks*, Springer-Verlag, London, 2014.

[12] L. E. Huitema, "Progress in compact antennas," *InTech*, 2014.

[13] H. Kpojime and G. Safdar, "Interference mitigation in cognitive-radio-based femtocells," *IEEE Communications Surveys & Tutorials*, vol. 17, no. 3, pp. 1511–1534, 2015.

[14] D. Rathee, S. Rangi, S. K. Chakarvarti, and V. R. Singh, "Recent trends in wireless body area network (WBAN) research and cognition based adaptive WBAN architecture for healthcare," *Health Technology*, vol. 4, pp. 239–244, May 2014.

[15] A. Meharouech, J. Elias, and A. Mehaoua, "Future body-to-body networks for ubiquitous healthcare: A survey, taxonomy and challenges,"in *Second International Symposium on Future Information and Communication Technologies for Ubiquitous Healthcare (Ubi-HealthTech)*, Beijing, China, May 2015.

[16] S. Chen and J. Zhao, "The requirements, challenges, and technologies for 5G of terrestrial mobile telecommunication," *IEEE Communications Magazine*, vol. 52, no. 5, pp. 36–43, May 2014.

[17] "5G vision white paper," DMC R&D Center at Samsung Electronics Co., Ltd., February 2015.

[18] T. Bai, A. Alkhateeb, and R. Heath, "Coverage and capacity of millimeter-wave cellular networks," *IEEE Communications Magazine*, vol. 52, no. 9, pp. 70–77, September 2014.

[19] M. Ur-Rehman, Y. Gao, Z. Wang, *et al.* "Investigation of on-body Bluetooth transmission," *IET Microwaves, Antennas and Propagation*, vol. 4, no. 7, pp. 871–880, July 2010.

[20] M. Ur-Rehman, Q. Abbasi, X. Chen, and Z. Ying, "Numerical modelling of human body for Bluetooth body-worn applications," *Progress in Electromagnetics Research*, vol. 143, pp. 623–639, 2013.

[21] Q. Abbasi, A. Sani, A. Alomainy, and Y. Hao, "Numerical characterization and modeling of subject-specific ultrawideband body-centric radio channels and systems for healthcare applications," *IEEE Transactions on Information Technology in Biomedicine*, vol. 16, no. 2, pp. 221–227, March 2012.

[22] G. Piro, K. Yang, G. Boggia, N. Chopra, L. Grieco, and A. Alomainy, "Terahertz communications in human tissues at the nanoscale for healthcare applications," *IEEE Transactions on Nanotechnology*, vol. 14, no. 3, pp. 404–406, May 2015.

[23] D. T. H. Lai, M. Palaniswami, and R. E. Begg, *Healthcare Sensor Networks: Challenges toward Practical Implementation*, CRC Press Inc., Boca Raton, FL, 2011.

[24] S. Roundy and P. K. Wright, "A piezoelectric vibration based generator for wireless electronics," *Smart Materials and Structures*, vol. 13, pp. 1131–1142, 2004.

[25]   B. W. Ha, J. A. Park, H. J. Jin, and C. S. Cho, "Energy transfer and harvesting for RF-bio applications," *IEEE MTT-S International Microwave Workshop Series on RF and Wireless Technologies for Biomedical and Healthcare Applications (IMWS-BIO)*, London, UK, pp. 54–55, September 2015.

[26]   X. Lu, P. Wang, D. Niyato, D. I. Kim, and Z. Han, "Wireless networks with RF energy harvesting: A contemporary survey," *IEEE Communications Surveys Tutorials*, vol. 17, no. 2, pp. 757–789, 2015.

[27]   J. Bae, M. K. Song, Y. J. Park, J. M. Kim, M. Liu, and Z. L. Wang, "Fiber supercapacitors made of nanowire-fiber hybrid structures for wearable/flexible energy storage," *Wiley's Angewandte Chemie International Edition*, vol. 50, no. 7, pp. 1683–1687, 2011.

[28]   "IEEE p802.15 working group for wireless personal area networks (WPANs): MAC and security proposal documentation," *IEEE 802.15.6 Technical Contribution*, 2009.

[29]   C. Poon, Y.-T. Zhang, and S.-D. Bao, "A novel biometrics method to secure wireless body area sensor networks for telemedicine and m-health," *IEEE Communications Magazine*, vol. 44, no. 4, pp. 73–81, April 2006.

[30]   S. E. Mukhopadhyay, *Wearable Electronics Sensors*, Springer International Publishing, Berlin, 2015.

[31]   L. W. Hanlen, D. Smith, J. A. Zhang, and D. Lewis, "Key-sharing via channel randomness in narrowband body area networks: Is everyday movement sufficient?" in *Fourth International Conference on Body Area Networks*, Los Angeles, CA, USA, 2009.

Index

Printed in the USA
CPSIA information can be obtained
at www.ICGtesting.com
JSHW011508221024
72173JS00005B/1239